ANALYSIS AND DESIGN OF INFORMATION SYSTEMS

THIRD EDITION

V. RAJARAMAN
Emeritus Professor
Supercomputer Education and Research Centre
Indian Institute of Science, Bangalore

PHI Learning Private Limited
Delhi - 110092
2023

₹ 495.00

ANALYSIS AND DESIGN OF INFORMATION SYSTEMS, Third Edition
V. Rajaraman

ISBN-978-81-203-4384-9 (Print Book)
ISBN-978-93-5443-585-0 (e-book)

The export rights of this book are vested solely with the publisher.

Published by Asoke K. Ghosh, PHI Learning Private Limited, Rimjhim House, 111, Patparganj Industrial Estate, Delhi-110092 and Printed by Glorious Printer, Delhi-110092.

In memory of

Dr. M.S. Muthana

Director

Indian Institute of Technology Kanpur

Contents

Preface

One of the most important uses of computers is to provide up-to-date information to managers to efficiently run their organizations. Of the total number of computers installed in the world today, over eighty percent are used in organizations for management information systems. It is thus very important for all students of Computer Science to know how to design computer-based information systems to aid management. This introductory text gives a lucid, self-contained presentation to students on how to analyze and design information systems for use by managers.

The subject, Information Systems Analysis and Design, also known as Systems Analysis and Design, is a compulsory subject for MCA, BCA, B.Com and BE students of Computer Science and Information Technology. This book covers the syllabi of these courses and the syllabus of the DOEACC (Level A) conducted by the DOEACC Society.

I developed and taught a course on Information Systems Design to Computer Science students for twenty years, first at the Indian Institute of Technology Kanpur, and later at the Indian Institute of Science, Bangalore. This book has evolved from the class notes used in this course and has been thoroughly classroom tested. Real examples are very important to explain ideas used in designing information systems and this text uses many examples taken from the Indian context. The book also includes a large number of exercises.

I used this book to prepare learning material consisting of PPTs, solutions to all the exercises and over 500 multiple choice questions with answers. This material was prepared for a course titled "Systems Analysis and Design" and was funded by the National Programme for Technology Enhanced Learning (NPTEL) of the Ministry of Human Resource Development, Government of India. I also gave 40 lectures based on this book for the NPTEL project phase 1. Both these materials are available on the NPTEL website. The lectures are also available on YouTube.

Chapter 1 shows the distinction between data and information and classifies information as strategic, tactical and operational. It also shows how to divide management functionally and hierarchically and what type of information is required by each of these functions and levels of management. The chapter concludes by pointing out the desirable attributes of information. Chapter 2 describes various functions of organizations, their computer infrastructure needs and the variety of information systems they require. Chapter 3 describes what is meant by systems life cycle and the significance of the life cycle, and brings out the importance of user participation in arriving at requirements specification. It also analyzes the tasks of a systems analyst, the desirable attributes of an analyst, and the tools of the trade. Chapter 4 demonstrates how to gather information, the need for interviewing, and the method of consensus for formulating information requirements of an organization.

Chapter 5 discusses the use of document and data flow diagram in specifying and documenting users' information requirements. How users' requirements can be divided into operational, tactical and strategic needs is also described. Chapter 6 shows how the users' specifications are quantified and analyzes the feasibility of each alternate solution—the need for cost-benefit analysis and how it is carried out are clearly explained. Finally, the chapter shows how to document a feasibility report.

Starting from Chapter 7 the various tools used in designing information systems are dealt with. This chapter describes how data flow diagrams are developed using good style conventions. It also discusses the idea of levelling of data flow diagrams. The methods used to specify processing of information using structured English and decision tables are introduced in Chapter 8. Chapter 9 gives a detailed presentation on decision tables. I feel strongly that the use of decision tables is an excellent specification technique as it is non-procedural. Further, precise methods of detecting incompleteness, ambiguity, contradictions and redundancies in decision tables exist, which are described in this chapter.

I have introduced in this third edition of the book a new chapter (Chapter 10) on **Use Case Method**. This method is very useful in specifying users' requirements and has gained popularity in industry. Use Case method is also part of object modeling.

Chapters 11 and 12 deal with data and their organization. Chapter 11 presents the entity-relationship modelling technique to develop a conceptual model of data. It also describes the use of relations and their normalization to develop relational database for applications. Chapter 12 brings out the need for an integrated database in organizations. It also brings out the need for database management systems, their objectives and organization.

Chapter 13 discusses Object-Oriented System Modelling, as object orientation has become very important. In this chapter both the method to identify objects and the way they are used to model information systems are described. The objective here is not to discuss at length object-oriented analysis and design but to give a broad overview of this topic and compare it with structured analysis and design.

Chapter 14 shows how to input data. The topics include the design of forms and screens for data entry, data encoding, error detection and input data controls to ensure correct data entry.

Chapter 15 describes how to present processed data as information to the users of a system and presents the design of output reports, screens and graphics. It is very important to ensure that the results obtained from an information system are correct and reliable. To ensure this, systematic control and audit procedures are required. Chapter 16 discusses how to introduce controls in an information system, how to audit the system, how to test the system, and how to ensure security of the system.

Chapter 17 is on electronic commerce (or e-Commerce) which is becoming extremely important. This chapter gives a broad overview of e-Commerce so that the student can design information systems keeping this development in view.

Various tools and techniques learnt by the students are used to design a small information system in the last chapter (Chapter 18). At the end of this chapter I have given ten realistic cases which are described in reasonable detail so that a student can use the data to design a system for each of these cases.

A book of this type naturally gained a lot of ideas from the books written by many authors. I thank all of them. Mr. S. Thirumalai assisted me in teaching a course on information systems design. He read all the chapters critically and gave numerous suggestions which considerably improved the book. He also evolved the case study problems given at the end of Chapter 18. I am extremely thankful to him for his assistance. Chapters 11, 12 and 13 were reviewed by Prof. Jayant Haritsa and he gave me several valuable suggestions which helped me to update these chapters. I thank NPTEL project and its coordinator at IISc, Bangalore, Prof. N.J. Rao, for funding the preparation of the web material and video graphing the lectures based on this book. I thank Prof. R. Govindarajan, Chairman, Supercomputer Education and Research Centre, and Prof. P. Balaram, Director Indian Institute of Science, Bangalore, for the facilities provided which enabled me to write this edition of the book. The manuscript of this edition was entered on a word processor by Ms. T. Mallika. I thank her for an excellent job.

Finally, I thank my wife Dharma for proofreading the manuscript and giving numerous suggestions. Without her cheerful cooperation and assistance this book would not have been written.

V. RAJARAMAN

Bangalore
January, 2011

1

Information and Management

LEARNING GOALS

After reading this chapter a student should be able to:

1. Distinguish between data and information.
2. Identify information needs of an organization and distinguish between operational, tactical and strategic information.
3. Divide an organization based on its functions and determine information needs of each of the functional areas.
4. Identify the desirable attributes of information.

Everyone in his or her day-to-day work gathers and processes data. For example, when a housewife buys milk every morning, she writes in a notebook the number of litres she bought. At the end of the month she adds the data (litres of milk bought per day) in the notebook and multiplies it by the price per litre. The result is the information she uses to pay the milkman. The data on milk purchased each day may be processed in other ways too to obtain different information. For example, if the total milk bought in a month is divided by the number of members in the family, it gives information on average milk consumption per head. If the total monthly expense on milk is divided by the monthly income of the family, it gives information on proportion of income spent on milk. The main point is that *data and information are not the same. Data* is the raw material with which we start. *Information* is processed data which is used to trigger certain actions or gain better understanding of what the data implies. In this example the knowledge of percentage of family income spent on milk may be used by the family to manage their family budget in a more efficient manner.

As is seen above, data is the raw material with which we begin. Information is the finished product. Collecting data costs money. For example, if we want to survey the preferences of consumers regarding some products, then someone has to go out, interview prospective customers and collect data. As the number of people interviewed increases, the cost of data collected increases. This data by itself is useless unless it is processed to obtain information which can be used to arrive at marketing decisions. This information is extremely valuable. As data collection costs money in proportion to its volume, it is essential to gather data carefully. One must collect only necessary data and not redundant data.

Figure 1.1 shows that information from process 1 becomes data for process 2. Thus we must not assume that there is always a clear distinction between data and information. The main point is that mere facts and figures about activities do not enable decisions to be made or actions to be initiated. For example, the marks obtained by students in a class, by themselves, do not give any immediate idea about the class performance. By processing this data, a bar chart may be obtained which gives the number of students with marks between 100 and 90, 90 and 80, 80 and 70, etc. Such a chart gives a teacher at a glance the performance of the class and enables him to initiate appropriate action. It is clear from these examples that data are processed by computers giving information which is used by people.

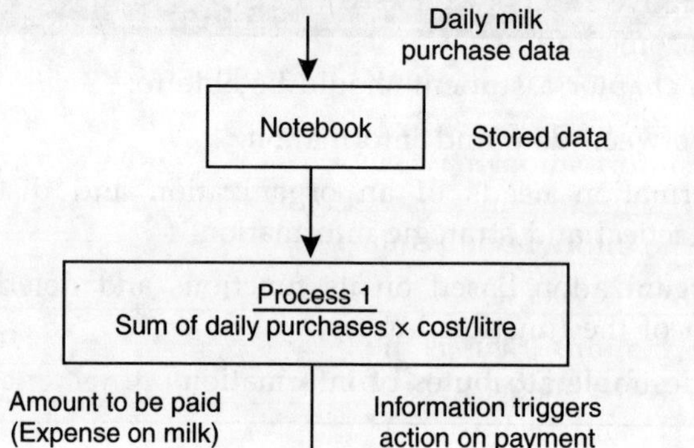

Figure 1.1 Concepts of data and information.

Arithmetic operations are not the only operations used to process data. Many of the data processing operations are non-numeric. For example, arranging students' names in alphabetic order, with last name first, uses sorting and character manipulation operations.

1.1 TYPES OF INFORMATION

We saw that data is processed to obtain information. Further, the same data may be processed in different ways to obtain different types of information. A fair amount of data is processed manually, particularly when the organization is small and the amount of data to be processed is small. Take, for example, a small retail grocery store owner.

In order to run his business efficiently he would need different types of information. He would like to know:

1. Which items in his store are about to be exhausted and need to be re-ordered.
2. Which items are sold fast so that he can order more and keep them well stocked.
3. Which items are not sold so that he can dispose them off at discounted price and perhaps not stock them in future.
4. Which bills sent to customers remain unpaid beyond due date.
5. Which wholesale merchants are reliable suppliers at competitive prices.
6. How much is the daily sale and trends in sales.
7. Profit and loss account monthly and at the end of the year.

The above list is indicative and not exhaustive. A store owner is required by law to send information on his income and profit to the income tax office; on his sales, and details of items sold and tax collected to sales tax office, etc., in addition to the information he needs to run his own business. As long as his store is small and his volume of sales is small, all the above information may be obtained by manually processing available data. When the size of the store increases, it may still be done manually by employing one or more clerks. The information required by the store owner may be classified into the following categories:

(i) Strategic information. This is the information needed for long range planning and directing the course the business should take. The store owner may like, for example, to decide whether to expand his business by stocking new varieties of items in his store, whether to work on a small profit margin to expand sales, whether to open a new branch etc. Such information cannot be obtained by mere clerical processing of data. He has to obtain information on trends, effect of reduction in price of some items, impact of advertising etc. This type of information is less structured. Further, the volume of important information in strategic planning is small and is difficult to obtain. It is often obtained by processing huge amount of archival data stored over a long period.

(ii) Tactical information. This type of information is needed to take short range decisions to run the business efficiently. For instance, the information on fast and slow moving items may be used to take the tactical decision to stock more of the former and give discount on the latter. Information on the payments of bills by customers may be used to decide appropriate credit limits to be given to customers. Tactical information requires specifically designed processing of data. Most of it is obtainable easily from day-to-day collection of routine data unlike strategic information. Further, the volume of tactical data is more than strategic data.

(iii) Operational information. This type of information is needed for day-to-day operations of a business organization. For instance, the list of items out of stock on a particular day would be used to trigger the action of ordering them. List of customers' unpaid bills beyond due date would be used to send reminders. Operational information is usually easy to obtain by straightforward clerical processing of data. Further, the volume of such information is much more than tactical information.

(iv) Statutory information. Information and reports which are required by law to be sent to government authorities are normally clearly specified and require straightforward processing of data.

In Table 1.1 we summarize the points made so far. The main point to note is that strategic information which is most important for long range growth of an organization is difficult to quantify and process from available data and consequently its volume is small. Operational information, on the other hand, is easy to process from available data and also more voluminous. Tactical and operational information are essential for efficient day-to-day operations of an organization whereas strategic information is vital for its long range well-being and growth.

Table 1.1 Information Needs in Managing a Retail Store

Type	*Examples*	*Manager*
Strategic Unstructured	Should new branches be opened? Should the business be diversified?	Chief Executive
Tactical	How to rate vendors? Should credit limits of customers be changed? How much of each item should be stocked? How much should be re-ordered and from whom? Should new discount policies be introduced?	Middle Level Managers
Operational	List of items to be reordered. List of defaulting customers. Excess or deficient supply by vendors. List of late supplies by vendors. List of outstanding bills to be paid. Daily ledger accounts.	Line Managers
Statutory	Profit and loss account Account of sales tax collected.	Compliance officer

1.2 WHY DO WE NEED A COMPUTER-BASED INFORMATION SYSTEM?

As long as organizations are small and have limited operational goals manual information systems are satisfactory. Many trends in the development of industry and commerce have made computer-based information systems essential to efficiently run organizations. These are:

- The size of organizations is becoming larger. This is particularly true in India due to increase in population and rapid rate of industrial development.
- Computer-based processing enables the same data to be processed in many ways, based on needs, thereby allowing managers to look at the performance of an organization from different angles.
- As the volume of data has increased and the variety of information and their timeliness is now of great importance, computer-based information processing has now become essential for efficiently managing organizations.
- Organizations are now distributed with many branches spread all over the country and abroad.
- Markets are becoming competitive. To maintain favourable balance of payments in a country, organizations have to be internationally competitive.

- The general socio-economic environment demands more up to date and accurate information. Human society is changing faster than ever before. Governmental regulations have become complex. Organizations have to interact with many other interested parties such as consumer groups, environmental protection groups, financial institutions, etc., which did not exist before.

All the above developments demand decision making based on up to date, well analyzed and presented information rather than thumb rules and hunches of an earlier era. This can be done only by using computer-based information systems.

1.3 MANAGEMENT STRUCTURE

The store owner of our example, besides using different types of information, also performs a set of distinct *functions*. He works as a purchaser, storekeeper, accountant, and salesman. As long as the business is small it may be possible for one person to do all the above functions. As the business grows it becomes essential to delegate responsibility to specialists in each functional area and make them accountable for their efficient functioning.

In Figure 1.2 we give the typical functions of managers of each category. The manager of each function is known as a *middle level manager*. Middle level managers

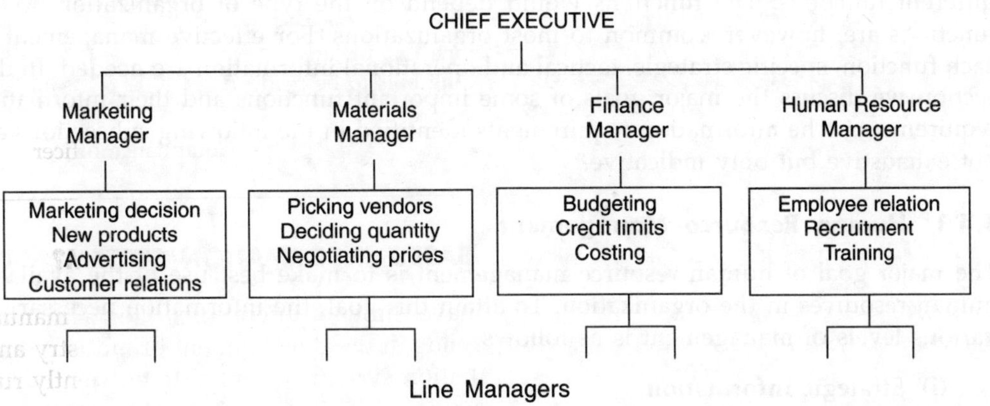

Figure 1.2 Functional responsibilities in a management system.

report to the Chief Executive who is overall in charge of the organization. The middle level managers will, in turn, have many assistants who are responsible for specific day-to-day operations. They are known as *line managers*. The management structure is thus a pyramid as shown in Figure 1.3. In this pyramid the Chief Executive, being overall in charge of policy, will require strategic information. The middle level managers require tactical information to perform their function and the line managers being responsible for day-to-day operations would require operational information. Thus we see that as a manager rises higher in the hierarchy he receives summarized information which is also less structured. More complex analysis of basic data is required as we go up in the hierarchy.

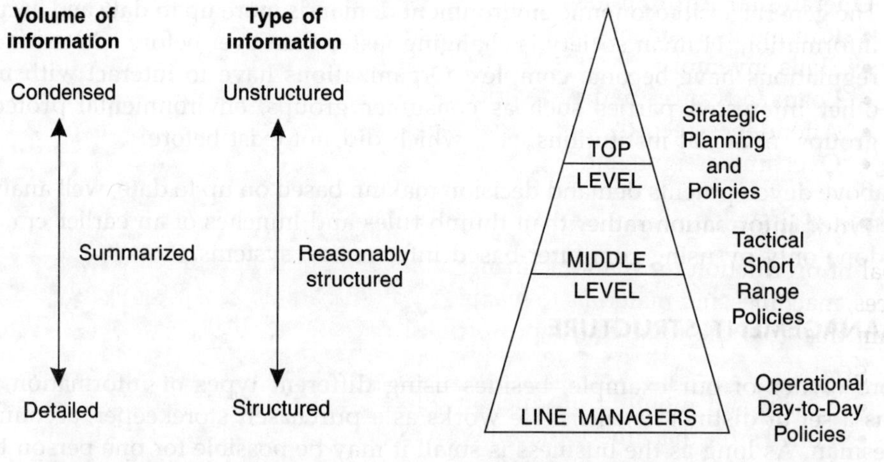

Figure 1.3 Types of information needed to manage an organization.

1.4 MANAGEMENT AND INFORMATION REQUIREMENTS

We saw in Section 1.3 that overall management of an organization would involve many different functions. The functions would depend on the type of organization. Many functions are, however, common to most organizations. For effective management of each function, specific strategic, tactical and operational information are needed. In this section we discuss the major goals of some important functions and their information requirements. The information requirements identified in the following subsections are not exhaustive but only indicative.

1.4.1 Human Resource Management

The major goal of human resource management is to make best use of the available human resources in the organization. To attain this goal, the information necessary at various levels of management is as follows:

(i) Strategic information

- Long range human resource requirements at different levels
- Policies on wages and incentives such as stock options and bonus
- Policies on human resource development and training
- Policies on personnel welfare and facilities
- Policies on outsourcing non core functions.

(ii) Tactical information

- Performance appraisal
- Demographic make-up of personnel and its impact on retirement
- Production incentives and relation to productivity
- Morale of personnel
- Absentee reduction
- Leave and overtime policies
- Personnel deployment policies.

(iii) Operational information
- Routine assessment
- Skills inventory
- Loans/advances and recoveries
- Attendance record
- Overtime assignment.

1.4.2 Production Management

The goal of production (or manufacturing) management is to optimally deploy human resources, machines and materials to maximize production of goods by the organization. To attain this goal the following types of information would be needed:

(i) Strategic
- Yearly and monthly production quotas and alternate schedules
- Policies on machine replacement, augmentation, and modernization
- Information on the introduction of new production technologies, acquistion and mergers
- Identifying best product mix
- Policy on quality assurance.

(ii) Tactical
- Identifying and controlling areas of high cost such as work in progress inventories
- Identifying critical bottlenecks in production
- Identifying alternate production schedules based on tools, machines, material, and personnel availability
- Performance measures of machines (breakdown histories and cost of repair and failure) to decide on replacement
- Quality control measures.

(iii) Operational
- Monitoring up to date production information by examining assemblies, detecting likely shortages and giving early warning
- Scheduling better production dynamically
- Preventive maintenance schedules of machines
- Monitoring tools, machines, and human resource availability.
- Ensuring quality at each step in production.

1.4.3 Materials Management

The main goal of materials management (often called supply chain management) is to have readily available materials needed by the organization and to keep optimal levels of stock of items. To achieve this goal the following types of information would be needed:

(i) Strategic
- Developing vendors for critical items
- Determining optimal levels of inventory
- Determining proportion of material to be ordered from different vendors
- Reducing varieties of inventory.

(ii) Tactical
- Developing vendor performance measures
- Determining the impact on material cost, availability and procurement with design changes and new product introduction
- Determining optimal reorder levels and reorder quantity
- Controlling high value inventory.

(iii) Operational
- Goods received, rejected and issued
- List of excess and deficient items received
- List of items rejected
- Critical items received
- Backlog of supplies
- Stores in transit and in inspection
- Value of inventory in hand and periodic stocktaking
- Order processing
- Enquire stock position.

1.4.4 Finance Management

The main goal of this function is to ensure financial viability of the organization, enforce financial discipline and plan and monitor the budget. The various levels of information required to meet these goals are as follows:

(i) Strategic
- Methods of financing
- Pricing policies
- Tax planning.

(ii) Tactical
- Variance between budget and expenses
- Large outstanding payments/receipts
- Credit and payment status
- Cost increases and pricing
- Impact of taxation on pricing.

(iii) Operational
- Periodic financial reports
- Budget status to all functional managers
- Tax returns
- Share registration and transfers
- Profit and loss account
- Accounts payable
- Accounts receivable
- Payroll, provident fund accounts
- Calculation of direct costs-overheads-standard cost.
- Cash management
- General ledger.

1.4.5 Marketing Management

The goal of this function is to maximize sales and ensure customer satisfaction. To attain this goal the various types of information needed are:

(i) Strategic
- Search for new markets and marketing strategies
- Analysis of competitors' strategy
- Technology and demographic forecasts and product changes.

(ii) Tactical
- Advertising techniques and analysis of their impact
- Customer preference and satisfaction surveys
- Correlation of prices and sales
- Sales force deployment and targets
- Exploring alternate marketing channels. Timing of special sales campaigns
- Customer relationship management.

(iii) Operational
- Sales analysis by regions, customer class, sales persons
- Sales target versus achievement
- Market share and trends
- Seasonal variations
- Effect of model changes
- Performance of sales outlets
- Costs of sales campaigns and benefit
- Call centre support
- Maintenance of home page on the world wide web
- Order entry.

1.4.6 Research, Design and Development Management

The goal of this function is to continuously improve existing products and develop new products and processes. To attain this goal following types of information are needed:

(i) Strategic
- Which products are to be developed?
- What type of improvements are required?
- What long range research is most promising?
- What technical collaboration would be appropriate?
- What areas would require new human resource inputs?

(ii) Tactical
- Setting intermediate goals for projects and assess progress
- Checking availability of equipment and appropriate selection
- Determining proportion of resources to be allocated to various parts of a project
- Deployment of personnel to projects based on talent and performance
- Information on similar and related research projects undertaken by other groups.

(iii) Operational

- Progress against goals
- Budgeted expenses versus actuals
- Status of outstanding orders for equipment and components.

1.5 QUALITIES OF INFORMATION

In the previous section we saw the nature of information managers need at various functional levels. We will now discuss the qualities of the information to be provided to managers.

- The information must be *accurate*. In other words, the correctness of the input data and that of the processing rules should be ensured so that the resulting information is accurate. Incorrect information is worse than no information.
- The information should be *complete*. That is, it should include all data and not exclude some.
- It should also be *trustworthy*. The processing should not hide some vital information which may, for example, point out the inefficiency of some individuals.
- The information should be *timely*. It should be given to the manager when he needs it. Delayed information may sometimes be of no value. For example, if a daily newspaper is delivered a day later, it becomes useless.
- Information should also be *up to date*. It should include all data available at the time of processing. A newspaper delivered in time early in the morning but reporting a week old news is timely but not up to date.
- Information should be tailored to the needs of the user and be *relevant* to him. Massive volumes of irrelevant information would waste a lot of manager's time and there is a danger of his missing important relevant information.
- It is essential to give brief *summarized* information to ensure quick action.
- The information should be presented when he needs it and where he needs it in such a way that he may immediately perceive its *significance*. For example, presentation of information in a graphical form such as bar charts, pie charts, etc., ensures quick recognition of the significance of the information. It is also essential to present the information in an attractive format which a user can immediately understand. Considerable time is spent by information system designers with the users to design the format of management reports besides their contents. Table 1.2 summarizes these points.

Table 1.2 Qualities of Information

Quality	*How to Ensure Quality*
Accurate	Ensure correct input and processing rules
Complete	Include all data
Trustworthy	Do not hide unpleasant information
Timely	Give at right time
Up to date	Include all data up to present time
Relevant	Understand user needs
Brief	Summarize relevant information
Significance understandable	Use attractive format and graphical charts

In this chapter we have discussed the need for information for managing organizations. Information is obtained by appropriate processing of available data. In fact, in practice, it is necessary to work backwards from a knowledge of required information to the data which, when processed, would give the information. Data collection and organization should be based on an extensive analysis of information requirements. We will return to this topic in later chapters. Processing of data may be done manually to obtain necessary information. That is, in fact, done in many organizations and is quite adequate if the volume of data is small or if the processing to be done is not very complex. Over the years various mechanical aids were developed and used for processing data. We will be concerned in this book on how computers may be used to process data when appropriate.

SUMMARY

1. Data and information are not synonymous. Information is processed data. Data may be processed by performing arithmetic operations on them. Data may also be processed by reorganizing them by applying appropriate rules such as sorting, merging, selection, matching, filtering, etc.
2. Information requirements for managing organizations may be classified as strategic, tactical and operational.
3. Strategic information is needed for long range planning. It is less structured and difficult to obtain by processing raw data.
4. Tactical information is used to take short range decisions and for better control of the functioning of the organization. It requires complex and ingenious processing of data.
5. Operational information is used for day-to-day management of organizations. It is obtained by simple processing of data, it is well structured, and more voluminous.
6. Statutory information consists of reports to be sent to government by law.
7. Management of organizations is divided functionally. Depending on the size of the organization, each function maybe delegated to different managers.
8. Large organizations would have a hierarchical management structure with top level managers, middle level managers and line managers.
9. Top level managers are expected to make policies and need strategic information. Middle level managers direct and control the functioning of organization to achieve optimal performance and need tactical information. Line managers supervise day-to-day operations and steer operations to meet targets set by middle level managers. They need operational information.
10. The primary functional areas of many organizations are Human Resource, Production, Materials, Finance, Marketing and Research, Design and Development.
11. The information given to managers must be accurate, complete and trustworthy. It must also be up to date and timely. It must be relevant, brief, attractively presented and its significance understandable.

EXERCISES

1.1 Distinguish between data and information. Give two examples of data and information obtained by processing data.

1.2 Give two examples of non-arithmetic data processing.

1.3 What is the main difference between strategic and tactical information? If an information system is to be designed for a hospital, what would be the strategic and tactical information?

1.4 What is operational information? In what way is it different from strategic information? For a hospital information system what would be the operational information required?

1.5 What is statutory information? What statutory information is expected from a hospital?

1.6 What type of statutory information is to be provided by a small scale industry?

1.7 What in your opinion are the strategic, tactical and operational information needed by an Electric Power Generation Corporation?

1.8 What would be the strategic, tactical and operational information needs of a State Road Transport Corporation?

1.9 What do you think are the functional management areas in a large University?

1.10 What in your opinion are the functional management areas in a large student hostel?

1.11 What would be the strategic, tactical and operational management responsibility assignments in a large hostel? What type of information would be needed for managing the kitchen of the hostel?

1.12 Give an example of incomplete information.

1.13 What is the difference between accuracy and trustworthiness of information?

1.14 Is up to date information always timely? If not, give an example of up to date but not timely information.

1.15 Does an attractive format of a report necessarily imply that it is relevant?

1.16 Give examples of graphical information presentation.

2

Examples of Information Systems

LEARNING GOALS

After reading this chapter a student should be able to:

1. Classify methods of processing data as batch processing or on-line transaction processing and apply them in appropriate contexts.
2. Classify information systems as transaction processing systems, management information systems or decision support systems.
3. Determine the type of information each of the above provide and appropriate levels of management where they are needed.

In Chapter 1 we saw that information is required for managing organizations. Further, we observed that different managers, depending upon their functions in the organization and their level in the hierarchy, require different types of information. We also specified the nature of information required for different functions such as Human Resource, Production, Materials, Finance, Marketing and Research, Design and Development. In this chapter we will broadly examine how data is processed for obtaining different types of information.

2.1 VARIOUS FUNCTIONS IN ORGANIZATIONS

All organizations are divided into many departments or sections, with each department having an assigned functional responsibility. Consider, for example, an educational institute such as a large University. It will typically have, besides academic departments, library, computer centre, and a central administrative office. The administrative office will be divided into many sections, each with an assigned function. Typically the functions will be student section (which will normally deal with student records, student admission, etc.), an accounts section, purchase section, stores section, human resource section, medical section and student hostel office. A hierarchical chart of the sections is shown in Figure. 2.1 and their functions in Table 2.1. A manufacturing organization, for

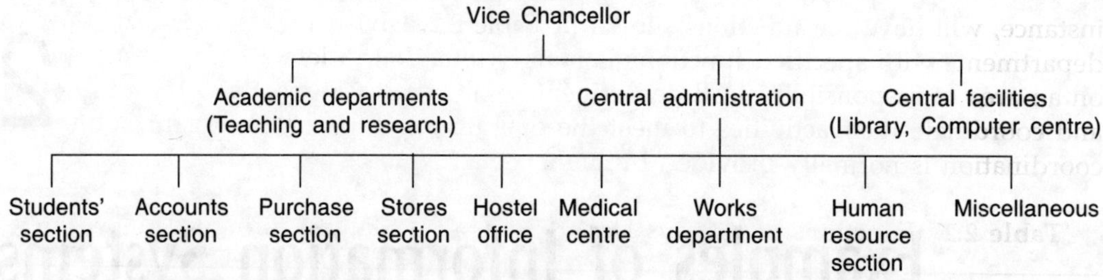

Figure 2.1 A hierarchical chart of offices in a university.

Table 2.1 Functions of Central Administration of a University

Administrative Offices	*Functions*
Students' section	Students' admission records Administering admission tests Students' academic records Students' registration information Placement
Accounts section	University budget Payroll General ledger of receipts/payments Scholarships Collection of tuition and other fees
Purchase section	Order processing Vendor selection Stock register maintenance Issues to stores Receipts from vendors
Stores section	Inventory management Processing requests for items from departments
Hostel office	Mess management Hostel purchases/stores Room assignment Residents' data
Medical centre	Medical records Medicine purchase/stores Scheduling doctors and appointments
Works department	Building construction Building maintenance Maintenance of electrical installation and water supply Maintenance of roads, gardens
Human resource	Personal records (leave, tenure) Assessment of employees Recruitment Staff discipline
Miscellaneous	Mailing Telephones Transport

instance, will have the functions shown in Table 2.2. Division of an organization into departments with specified functions is mainly intended to let each department focus on an area of responsibility. All departments will have to cooperate with one another and coordinate their activities to meet the overall objectives of the organization. This coordination is normally provided by higher level management in the organization.

Table 2.2 Functions of Various Departments of a Manufacturing Organization

Departments	*Functions*
Production	Production planning and control Maintenance management Bill of materials processing
Marketing	Advertising Customer records/follow up Sales analysis
Finance	Billing, payments Payroll Costing Share accounting Budget and finance planning Tax planning Resource mobilization
Human Resource	Recruitment Records Training Deployment of labour Assessment/promotions
Stores (Materials Management)	Stock ledger keeping, inventory control Issues/reorder Receipts Enquiry processing
Information systems	System software Applications software Hardware infrastructure Database administration Home page maintenance
Maintenance	Physical facilities Communication facilities Electricity and water supply
Research and Development	Production improvement Product development Product testing Product design

Observe that among the departments of a manufacturing organization we have also included Information Systems. Nowadays Information Systems (IS) department has become essential in all organizations as computers are used by everyone starting from a clerk receiving goods in a store to the managing director who uses it to examine progress of projects and take strategic decisions. The primary task of an IS department is to identify areas of operation of an organization which could be improved by using computer based information systems. This is done by wide consultation with end users and the management. Having identified these areas, the IS department has to convince the top management to allocate funds to develop these systems. Once the funds are allotted, the IS department should either design the application software in-house or buy off the shelf software if available or out source the job to a software company. Once the application software is implemented, the users of the system have to be trained to use it as part of their routine work. This will be readily accepted by users if it eases their work. In addition to implementing application software, the functions of IS department includes:

- Maintenance and improvement of application software.
- Design of databases for various applications, ensuring their security and continuous updating.
- Determining the configuration of hardware and system software infrastructure. Addition/replacement of these with changes in technology.
- Liaison with the hardware/software infrastructure provider to ensure reliable operation.
- Designing appropriate audit and control schemes for the application software to ensure dependability, reliability and security.
- Ensuring continuity of the operations of the organization's information systems in case there are disasters such as floods, earthquakes etc.
- Designing and maintaining the organization's website.

We looked at two diverse organizations in this section—a University and a manufacturing organization. We see that there are some common functions such as human resource, purchase, stores and accounts, and there are organization specific functions such as students' section in a university and production section in a manufacturing organization. This is a general observation. Information processing methods, however, have general features regardless of the organization for which they are designed.

2.2 COMPUTER INFRASTRUCTURE IN ORGANIZATIONS

Nowadays each department of an organization has several desktop computers connected to a Local Area Networks (LAN) which is in turn connected to several larger powerful computers called *Servers* (See Figure 2.2a). Each clerk, case worker and manager will normally have a desktop computer, called a client, connected to the LAN. Each of the servers connected to the LAN will have a specialized function such as compute server,

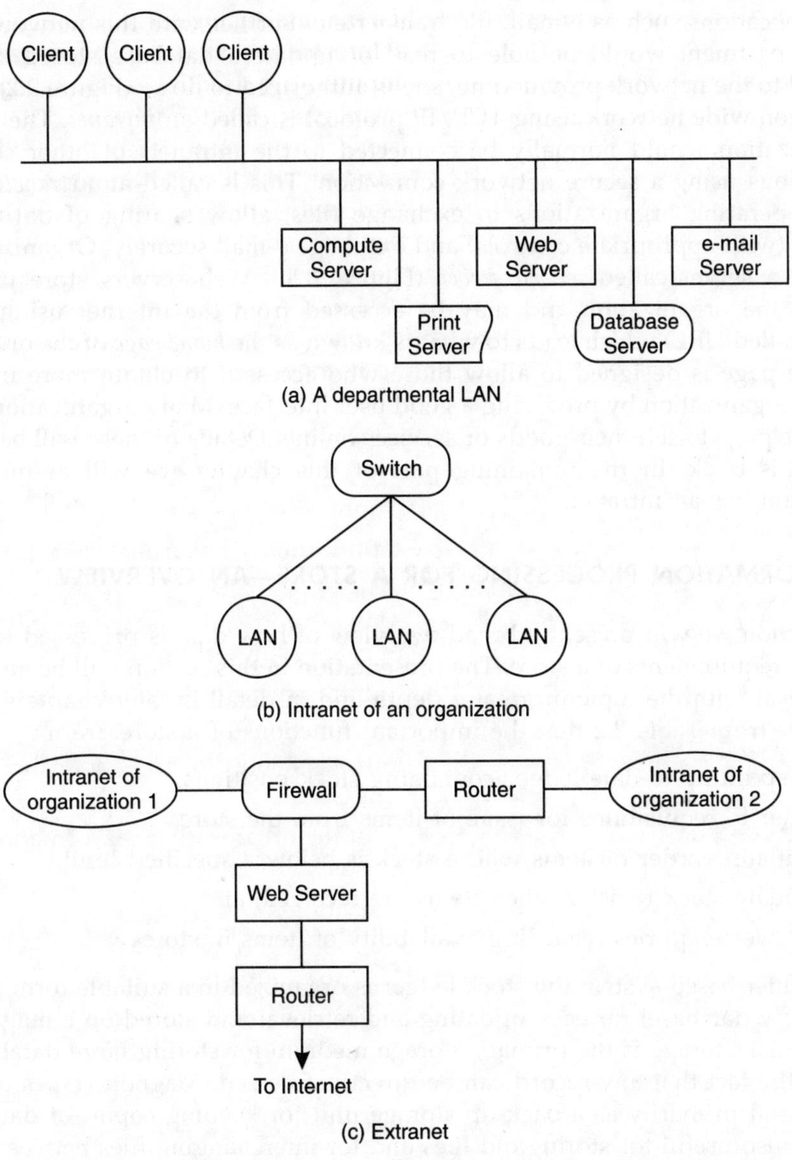

Figure 2.2 Computer infrastructure of an organization.

database server, e-mail server, print server, etc. A database server would have a large disk for storing data relevant to the function of the department. A print server, on the other hand, will have a high speed, high capacity printer attached to it to print reports. The LANs of various departments are in turn connected to one another to create an organization wide network. Usually the protocol used by this network is TCP/IP as this standard allows connection of computers of different manufacturers and also provides many applications such as e-mail, file transfer etc. Further with this network a person in any department would be able to read or update a database stored in a server connected to the network provided he/she is authorized to do so (Figure 2.2b). Such an organization wide network using TCP/IP protocol is called an *intranet*. The intranet of an organization would normally be connected to the intranets of other cooperating organizations using a secure network connection. This is called an *extranet*. Extranets allow cooperating organizations to exchange files, allow sharing of data stored in databases (with appropriate controls) and exchange e-mail securely. Organizations will also have a server called a *web server* (Figure 2.2c). Web servers store information regarding the organization and may be accessed from the Internet using a unique address (called URL) which leads to what is known as the *home page* of the organization. The home page is designed to allow those who access it to obtain more information about the organization by providing a good user interface. Many organizations also use their home page to sell their goods or services online. Details of these will be described later in this book. In the remaining part of this chapter we will assume that an organization has an intranet.

2.3 INFORMATION PROCESSING FOR A STORE—AN OVERVIEW

In this section we will present a broad overview of how data is processed to meet the functional requirements of a store. The presentation in this section will be an overview. We will deal with the topic in greater depth and in detail in later chapters.

We see from Table 2.2 that the important functions of a store are to:

- keep an up-to-date ledger containing stock positions,
- cater to requisitions for issue of items from the store,
- initiate reorder of items whose stock is below a specified limit,
- update stock register when items are received, and
- answer enquiries regarding availability of items in stores.

In a computer based system the stock ledger is organized in a suitable form (known as an inventory database) for easy updating and retrieval and stored on a magnetic disk. Magnetic disk storage is the primary storage medium for storing large databases. This is due to the fact that any record can be directly accessed. Magnetic tapes or writable CDs are used primarily as a back-up storage unit for keeping copies of data on disk. Tapes are also useful for storing old files and for interchanging files between different computers.

In order to create an inventory database for a computer-based system it is necessary to first assign unique codes for each item in the store. The unique code assigned to each item is known as the *key* of the item record and identifies the record. After that it is necessary to find out what data fields are needed for each item in the inventory. The fields are determined by working backwards, i.e., first asking what outputs are needed and based on that determining the data needed. (The detailed procedure will be discussed later in this book.) The fields are organized as a record and stored in a database. This database is the primary or *master file* for the store. (It is the computer readable version of a stock ledger used in a store.) Once the record format for each item in the store is determined, one record is created for each item in the store. These records are entered manually by a data entry operator who enters the records using a keyboard of client computer connected to the server. This is called *online* data entry. Data may also be entered on a separate computer such as Personal Computer (PC) and stored on a CD ROM. This is called *offline* data entry.

In offline data entry the data entry machine is a low cost machine. If the volume of data to be entered is very large, then a number of machines can be used and data is prepared, checked, and corrected. As opposed to this, an online data entry method uses the clients or PCs connected to a server. In such a case the server should be timeshared. Online data entry is appropriate for inserting, deleting or correcting some records in the database.

In offline data entry the data on CD ROMs are transferred to the disk connected to the main server. Offline entry is used when the database is very large and the computer used is a server or a mainframe computer. If the database is small, the desktop PC itself may be used for data entry and for data processing.

Before data is stored in a database it must be ensured that any errors made during data entry is detected and corrected. This is done by a program called an *edit program* and a control total checking program. Such programs are essential to ensure the validity of data in a master file.

A procedure similar to the one used to create the master file is also used to keep the data in the master file up to date when new items are received or new stock of items already in the inventory are received. The procedure used is summarized in Figure 2.3.

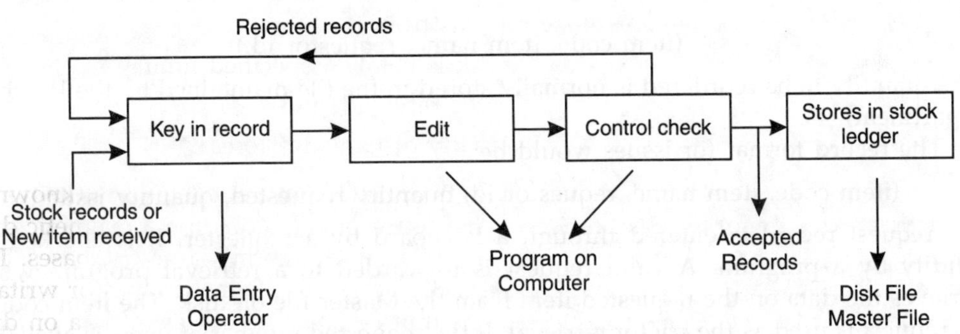

Figure 2.3 Master file creation/update.

Table 2.3 summarizes the operations performed for other functions. The format of a record for entering requests uses the same item codes assigned in creating the master file. Other fields are determined based on what outputs are needed. In this case, a reasonable format for requests is:

(item code, item name, quantity requested)

Table 2.3 Operations Performed in Stores Information Processing

1. Create inventory database
   - Codify items
   - Determine data fields needed for each item
   - Create a record for each item
   - Organize the records as a database
2. Issues/Reorder
   - Codify items
   - Determine data fields required for each request
   - Determine data fields required for each issue
   - Create record format for requests and issues
   - Create record format for reorder
3. Receipts
   - Codify items
   - Determine data fields required in each receipt
   - Create record format for receipts
4. Enquiry
   - Codify items
   - Record format for enquiry
   - Record format for response

The record format for a reorder request forwarded to the purchase department would be:

(item code, item name, requestor id.)

(The quantity to be reordered is normally stored in the file maintained by the Purchase department)

The record format for issues would be:

(item code, item name, requestor id, quantity requested, quantity issued)

The request record is entered through a keyboard by a requester. It is checked for validity by a program. A valid request is forwarded to a retrieval program which retrieves the data on the requested item from the Master file on disk. The item code in the request is used as the *key* for retrieval. If the requested number of items is available, then an issue record is sent to the requestor; otherwise a reorder request is sent to purchase department. Care must, however, be taken to ensure that once a reorder is requested it is not requested again till the item reordered is taken into stock. Figure 2.4 summarizes this operation. Each request record processed by the computer is called a *transaction*. If each request is processed as and when it arrives and the terminal on which the request is entered is connected to the computer, then the processing method is called *OnLine Transaction Processing* (OLTP).

As seen in this example a transaction consists of a number of elementary operation. In the case described above, namely, processing a request for issuing an item from the store, the operations are:

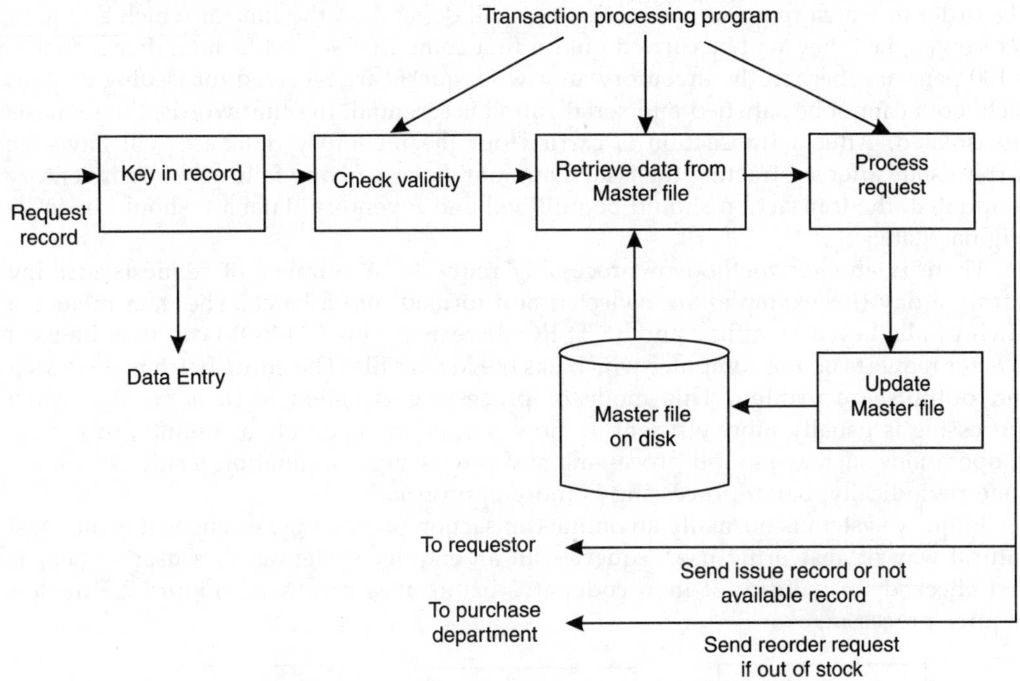

Figure 2.4 Processing of requests to stores.

1. Check the validity of request
2. Compare key of the request with keys of data stored in the master file (inventory database)
3. Retrieve record with matching key
4. Process request (e.g. Allow issue of item if in stock)
5. Update master file (database)

A transaction is complete only if all the operations are carried out successfully. A transaction has four important properties. They are: *atomic, consistent, isolated* and *durable* (commonly known as ACID properties). By atomic we mean that a transaction is successful only if *all the* elementary operations succeed. If any of the operations fail the rest should be undone, i.e., they should not persist. If the database is consistent before a transaction begins, it will be consistent after it ends. A transaction is isolated in the sense that if more than one transaction is in progress, the effect of each transaction is hidden from all other transactions. By durable we mean that when a transaction is completed its result is persistent and will survive any malfunctions of software or hardware.

In the example of issuing an item from a store it is clear that the transaction of issuing an item is not complete unless all the five steps are complete. The inventory database is changed only if the transaction is completed. Thus the consistency property is satisfied. If two different requests for issue of the same item are received then it is necessary to do them one after the other. In other words, the requests should be isolated.

The order in which requests are carried out will depend on the time at which a request is received, i.e., they will be carried out on first come first served fashion. For instance, if 100 pens are there in the inventory and two requests are received for issuing 75 pens each, both cannot be satisfied and serialization is essential. In other words, the requests are isolated. After a transaction is carried out the inventory database will have the current state after subtracting the issued amount. In case of any failure before inventory is updated, the transaction should be nullified and inventory database should be in its original state.

There is another method of processing requests. A number of requests arriving during a day (for example) are collected and formed into a batch. The data in such a batch can be keyed-in offline and a CD ROM created. This CD ROM can then be used to enter requests on the computer which has the Master file. The entire batch is processed and outputs are printed. This mode of processing is called *batch processing*. Batch processing is usually more efficient. It, however, is not as timely as online processing. In operations such as payroll processing and processing examination results which are done periodically, batch processing is more appropriate.

Enquiry system is normally an online transaction processing system as it is the most natural way of answering users' queries. In an enquiry system also, a user's query is first checked for validity of item code, etc., before it is processed. Figure 2.5 depicts enquiry processing.

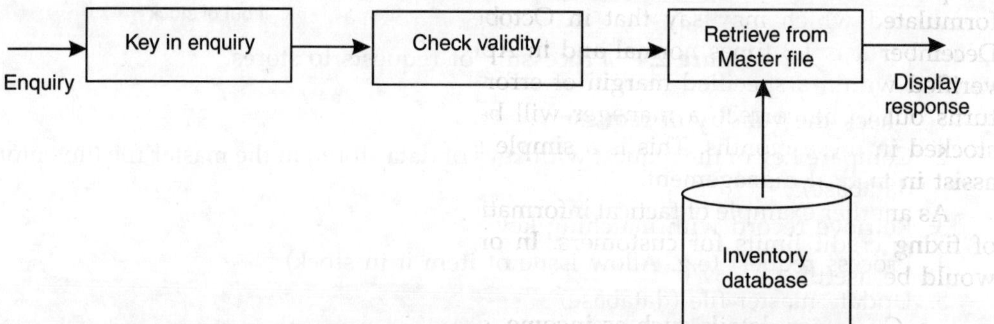

Figure 2.5 Processing steps in an enquiry system.

A variety of information systems used in practice are primarily online transaction processing systems. Common examples are airlines and railway ticket reservation systems. Designing such systems require special care to ensure that response to enquiries are fast and that the system has a "hot standby" if there is a failure. High reliability is required as failures can be catastrophic (imagine many persons getting the same berth reserved on a train). Similarly, reliability and availability is essential in online banking systems.

2.4 VARIETIES OF INFORMATION SYSTEMS

The processing methods presented in the last section are called *transaction processing systems* and are used for *routine data processing*. The information they provide is operational information. The system required to obtain tactical information require further processing. Such systems are known as *Management Information Systems* (MIS). In the stores processing example, some tactical decisions would be: at what stock level should reorder be initiated? How much should be reordered? These are determined based on data such as rate of issue of each item, time needed for delivery by vendor from date of order, transport cost, storage cost, shelf life, and loss incurred if an item is out-of-stock. These data have to be collected separately over a period of time, often as a byproduct of a transaction processing system. In the stores example daily issues of some critical items can be abstracted and the average issue can be computed. Data on delivery times, transport cost, etc., can be separately collected. Well known methods of operations research can then be used to compute stock level for initiating reorder and the optimal quantity to be reordered.

Operational data collected over a period of time from several routine applications is called *data archives* and the process of collecting it is called *data archiving*. Data archives are important resource of an organization as they can be analyzed in various ways to obtain information which can be effectively used by managers. With the availability of massive disks in which terabytes (10^{12} bytes) of data can be stored, it has become feasible to analyze the archived data. Analyzing archived data to observe patterns assist in management decision making. A stores manager may, based on his experience, think that in the months of October, December and April, the sale of sugar is very high compared to other months. This conjecture may be verified by data analysis. A rule is formulated which may say that in October sugar sale is 1.5 times the average, in December it is 1.3 times normal and in April it is 1.4 times normal. This rule may be verified within a specified margin of error by examining the data archive. If the rule turns out to be correct, a manager will be able to decide how much sugar is to be stocked in these months. This is a simple example of the analysis of archival data to assist in tactical management.

As another example of tactical information requirement, let us consider the question of fixing credit limits for customers. In order to arrive at this, the following inputs would be useful:

- Customer details such as income, occupation, etc.
- Customer payment history
- Volume of purchase by customer
- Outstanding dues (if any) from the customer.

Using these one may formulate some rules to arrive at the credit limits and also predict their possible effect such as:

- anticipated effect on sales caused by varying credit limits
- anticipated loss/profit due to credit limits.

By analyzing the impact of credit limits as specified above, a decision may be arrived at to fix credit limits.

The primary point to note is that one has to formulate a model and sometimes simulate a system to obtain tactical information. The operational information from routine processing becomes an input to obtain tactical information.

Strategic information is obtained through what are known as *Decision Support Systems (DSS).* In the stores example a strategic decision would be to reduce variety in inventory by discontinuing some items in store, deciding what new items to introduce in the store, and when and where to open a new branch. Decisions such as these require provisions for a variety of data transformations and representations. Nowadays professionals also use the term *Business Intelligence* (BI) for Decision Support Systems. BI may be defined as methods to aid informed decision making based on analysis of functional data obtained during operation of a system.

Strategic information is often unstructured. Strategic decisions are made after trying to answer questions such as "What will be the profit if I take a decision and what will be the long range loss if I don't take it?". In a complex decision many parameters will be involved. Identifying these and predicting their impact on a decision needs judgements coupled with analysis. For example, taking a strategic decision of whether to open a new branch or not would require the following information:

- Projected demands in the new branch
- Impact on current branch
- Pricing in new branch
- Competition in new location
- Profitability of new branch

These have to be provided using analytical and simulation model known as decision support models. Some books use the term Business Analysis for this. These models are more difficult to evolve than those needed in tactical information development. Decision support systems should also provide aids to the manager for conceptualization such as charts, graphs, etc. They should also provide facilities to ask a variety of queries on the database. A variety of summary reports should be made available on request. The overall purpose of decision support systems is to aid in strategic, unstructured decision making. Developing such systems is much more difficult than developing operational systems. They, however, are the ones required by the top management of organizations.

SUMMARY

1. Organizations are divided into many departments, each with a specific set of functions.
2. Even though an organization may have some specialized functions, many functions such as Accounts, Human Resource, Stores, Purchase are common among organizations.
3. Each function in an organization needs operational, tactical and strategic information.
4. Operational data is codified and stored in a Master file (or a central database).
5. A request for retrieving a record from a Master file is called a *transaction.*
6. The method of processing data in which transactions are grouped together and processed as a batch is called *batch processing.*

7. The method of processing data in which transactions are processed by a computer as and when they arise is called *OnLine Transaction Processing* (OLTP).
8. Transaction processing systems (TPS) are used to obtain operational information. These are normally repetitive.
9. Data abstracted from routine data processing are further processed using well thought out rules to obtain tactical information. Such systems are called *Management Information Systems* (MIS).
10. Information from multiple sources in an organization are normally presented in summarized forms such as graphs and charts to aid top management to take strategic decisions. Such systems are called *Decision Support Systems* (DSS).
11. Data used in operational data processing is normally stored in many organizations as back up or for legal reasons. These are called *data archives*.
12. The archived data may be analyzed to obtain tactical and strategic information. The term Business Analytics is also used nowadays for such analysis.

EXERCISES

2.1 Why are organizations divided into departments and sections?

2.2 What are the primary functions of a marketing department of an organization?

2.3 In what ways do different departments in an organization interact? Give an example.

2.4 Enumerate some functions which are common among many organizations.

2.5 Enumerate some functions which are common between a hostel and a hospital.

2.6 Explain how data will be processed to meet the functional requirements of a purchase department.

2.7 Explain how data will be processed in a bank when a cheque is presented by a customer and payment made to him (assume that the bank uses computers).

2.8 Why is editing important in data entry?

2.9 What is the difference between on-line transaction processing and batch processing?

2.10 When is batch processing preferred?

2.11 What is the basic difference between transaction processing systems and MIS?

2.12 What type of information is provided by MIS?

2.13 What is the difference between MIS and DSS?

2.14 What will MIS provide in a marketing function?

2.15 What will a DSS provide in a marketing function?

2.16 What do you understand by archival data? How can archival data be used?

2.17 What is data analysis? Give an example of data analysis.

2.18 What is business intelligence?

2.19 What is business analytics?

2.20 What will MIS provide in production function?

3

Information Systems Analysis Overview

LEARNING GOALS

After reading this chapter a student should be able to:

1. Explain what is meant by a system life cycle.
2. Exemplify the need for user participation in designing information systems.
3. Describe the nature of tasks performed by information systems analysts.
4. Identify the attributes of a good information systems analyst.
5. Evaluate different approaches to system development.

In Chapter 1 we saw the need for an information system to aid the management of organizations. In this chapter we will discuss how information systems are designed. There are a number of steps in the design and we will first give an overview using an analogy.

3.1 OVERVIEW OF DESIGN OF AN INFORMATION SYSTEM

3.1.1 Requirements Determination

An information system is intended to meet the needs of an organization. Thus the first step in the design is to specify these needs or requirements. We will take as an analogy, an example familiar to all, namely building a house for a family.

Before beginning to build a house the head of the household has to specify the needs of his family. These needs are arrived at by consulting the family members. Each member of the family would state his or her requirements. Children would like to have a large playing area, mother a big kitchen, grandfather a big pooja room, etc. The available plot of land and money are, however, limited and it would not be normally possible to meet all the requirements. Thus the head of the family considers all the requests, sets priorities amongst them and arrives at a consensus by mutual consultation. These modified requirements are expected to be fulfilled with the available plot of land and money. Determining the requirements to be met by a computer-based information system in an organization is similar. The top manager of the organization takes the basic decision to use a computer-based information system for managing the organization. Having done this, the next step is to determine the requirements to be met by the system. This is arrived at by a consensus reached among managers in the organization. Meetings of prospective user departments are held and, through discussions, priorities among various applications are determined, subject to the constraints of available computer system, human resource, time and budget.

3.1.2 Requirements Specification

Coming back to our analogy, the head of the household, after reaching a consensus on his family's requirements, will consult an architect. The architect studies the requirements, talks to the head of the family, and some of the family members. Knowing the area of the plot for the house, available money, the requirements and preferences of the family members, he draws a set of sketches as plans of the house. These plans give not only the floor plan but also the side view, landscape, etc. These sketches can be easily understood by any intelligent person. They are the initial specifications of how the proposed house will look like. The prospective house owner need not have any specialized training to understand the architect's sketches. The house owner looks at the sketches, consults his family and gives his suggestions to the architect for changes. After discussions the architect revises the sketches and gives them back to the house owner for further comments, if any. Once the sketches are approved, the architect draws up the final building plans more precisely, determines the materials and quantities required to build the house and works out the cost estimates.

A similar procedure is followed in designing an information system for an organization. The top management of an organization first decides that a computer-based information system would be desirable to improve the operations of the organization. Once this basic decision is taken, a systems analyst is consulted. His role is similar to that of the architect of our example. The first job of the systems analyst is to understand the existing system which could be manual or computer-based. During this stage he understands the organization by means of extensive discussions with all levels in the organization hierarchy, namely, top managers to clerks. Based on this he identifies what aspects of the operations of the organization need computer-based information systems. The analyst discusses with managers and users their functions and determines the areas where a computer-based information system can be effective. The applications where a computer can be employed are listed. By consensus among managers and users priorities are assigned to these applications. It is extremely important to get the users involved right from the initial stages of the development of an information system and seriously consider their suggestions. After discussions the analyst finally arrives at the specifications of users' requirements by consensus. This ensures that all potential users feel involved in the project. This is essential to obtain cooperation in the long run in the development and operation of the information system.

3.1.3 Feasibility Analysis

Having drawn up the rough specification, the next step is to check whether it is feasible to implement the system. A feasibility study takes into account various constraints within which the system should be implemented and operated. The resources needed for implementation such as computing equipment, human resource and cost are estimated, based on the specifications of users' requirements. These estimates are compared with the available resources . A comparison of the cost of the system and the benefits which will accrue is also made. This document, known as the feasibility report, is given to the management of the organization.

3.1.4 Final Specifications

The management of the organization studies this feasibility report and suggests modifications in the requirements, if any. Knowing the constraints on available resources, and the modified requirements specified by the organization, the final specifications of the information system to be developed are drawn up by the systems analyst. These specifications should be in a form which can be easily understood by the users. Just as a sketch and building plans are in a form understood by anyone, the language used to specify an information system must be understood by any person with very little knowledge about computers. The specifications state what the system would achieve. It does not describe how the system would do it. These specifications are given back to the users who study them, consult their colleagues and offer suggestions to the systems analyst for appropriate changes. These changes are incorporated by the systems analyst and a new set of specifications are given back to the users. After discussions between the systems analyst and the users the final specifications are drawn up which are approved for implementation. This is known as System Requirements Specification (SRS). Along with this, criteria for system approval are specified, which will normally include a system test plan.

3.1.5 Hardware and Software Study

Based on the finalized specifications it is necessary to determine the configuration of hardware and support software essential to execute the specified application. This is based on volumes of data to be processed, frequency of reports, whether online enquiry is required, response time for online jobs, etc. Whether off the shelf software can be used or whether it is to be developed from start, whether to outsource it or to develop it in-house are also parts of this study.

3.1.6 System Design

The next step is to develop the logical design of the system. The inputs to the system design phase are functional specifications of the system and details about the computer configuration. During this phase the logic of the programs is designed, files or database are designed, and program test plans and an implementation plan are drawn up.

An important point to be kept in mind is that the entire systems analysis and design steps we described so far are not unidirectional. At every step we may have to go back and review and often modify an earlier step. This is called *spiral development* as opposed to a unidirectional step-by-step development which is called *waterfall model*.

3.1.7 System Implementation

The next phase is implementation of the system. In this phase all the programs are written, database is created, user operational document is written, user interface is designed, users are trained, and the system tested with operational data.

3.1.8 System Evaluation

After the system has been in operation for a reasonable period, it is evaluated and a plan for its improvement is drawn up. Table 3.1 and Figure 3.1 summarize the information system analysis and design procedure. This is called *system life cycle*. The shortcomings of a system—namely, what a user expected from the system and what he actually got—are realized only after a system is used for a reasonable time. Going back to our analogy of a house, the house owner finds out all deficiencies in a house only after living in it for a few months. In fact he can advise others on how not to plan a house from his experience! Similarly, the shortcomings in an information system are realized only after it is implemented and used for some time.

Table 3.1 System Life Cycle

Phase	*What is Done During this Phase*	*End Product of this Phase*
1. Requirements determination	Determine requirements to be met by system being contemplated.	Set of requirements and their priorities.
2. Requirements specification	Draw up understandable plan of what the system will provide as outputs. Determine needs and priority by consensus among end users.	Detailed specifications of information to be provided. (revised user requirement)
3. Feasibility analysis	Taking into account available resources such as human, computer, time and money find whether specified requirements can be met.	Feasibility document specifying resource needs and availability. Expected cost vs. benefits of system.
4. System specification	Obtain functional specification based on revised user requirements and feasibility analysis.	Functional specification. Budget, time schedule. Physical requirements such as storage and processor.
5. Hardware study	Determine hardware requirements for system.	Hardware configuration—disk space, CPU power, Computer network design.
6. System design	Logical design of programs, design of databases, test and implementation plan.	Logical design of programs, data bases and test plan.
7. System implementation	Writing programs, creating databases, developing graphical user interfaces, report generation testing programs and operation plans. Documenting system. Training users. Data conversion if needed. Installing system. Trial of system and parallel runs with existing system. Tests to accept system.	Programs, databases, graphical user interfaces, output report formats, user manual and operational manual.
8. System evaluation	Find out from users if system meets their needs.	Evaluation report with suggestion for improvement.
9. System modification/ maintenance	Change system, adding or deleting features to satisfy users (modified) needs.	Improved system containing modifications and improvements.

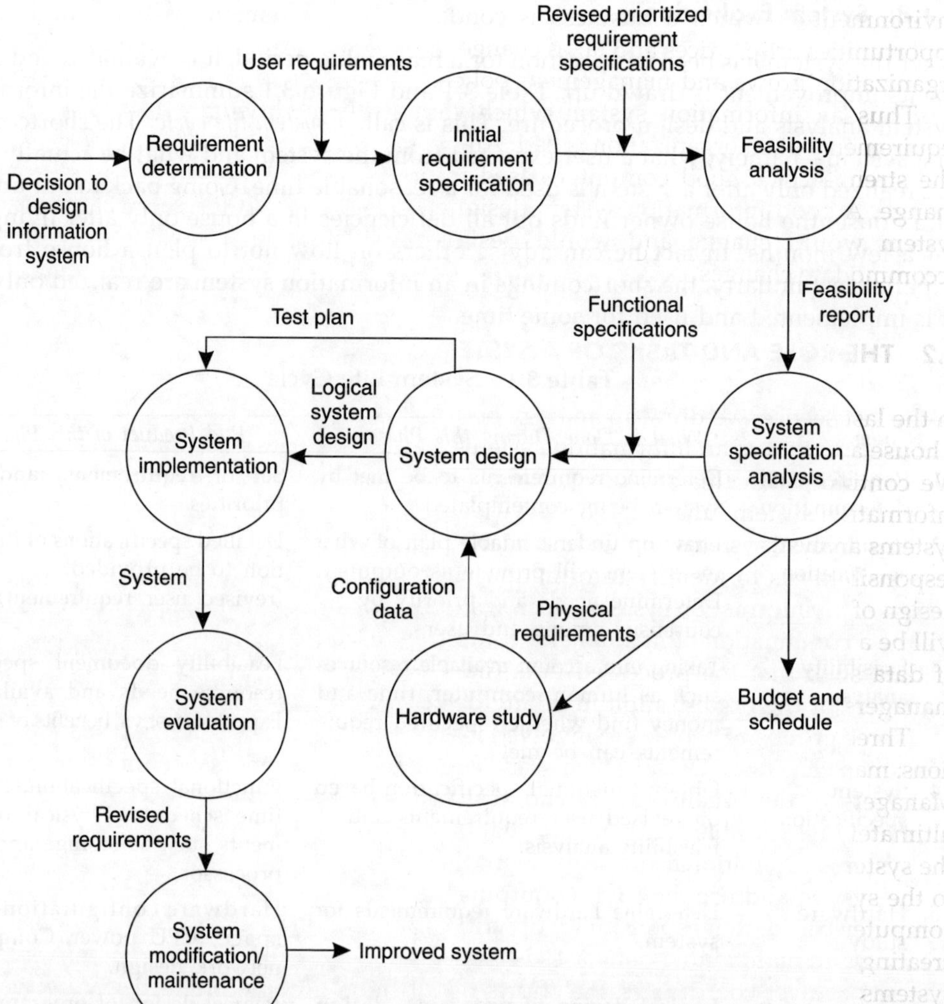

Figure 3.1 Stages in a system's life cycle.

3.1.9 System Modification

In the case of a house, the house owner, even though he finds out the shortcomings of the house and can give suggestions on how to modify and improve the house, does not demand these modifications as these would cost a lot of money and often it may be impossible to make any changes. A house is built with bricks and cement and is a hard system. It cannot be easily modified. On the other hand, a computer-based information system is a piece of software. It can be modified. Modifications will definitely cost time and money, but users expect modifications to be made as the name 'software' itself implies it is soft and hence changeable.

Further, information systems designed for use by organizations cannot be static. These systems are intended for the management to effectively carry out their business. The environment in which a business is conducted never remains static. New business opportunities arise, prices and taxes change. New competitive organizations emerge. The organization grows and management policies change.

Thus an information system which cannot be modified to fulfill the changing requirements of an organization is bad. An *information system should be designed for change.* The strength of a good computer-based information system is that it is amenable to change. A good information system designer is one who can foresee what aspects of a system would change and would design the system in a flexible way to easily accommodate changes.

3.2 THE ROLE AND TASKS OF A SYSTEMS ANALYST

In the last section we drew an analogy between the functions of an architect in building a house and that of an information systems analyst who designs an information system. We continue with this analogy in this section and describe the responsibilities of an information systems analyst. (For brevity we use, in what follows, the terms information systems analyst, systems analyst and analyst as synonymous.) A system analyst's primary responsibility is to identify information needs of an organization and obtain a logical design of an information system which will meet these needs. Such an information system will be a combination of manual and computer-based procedures to process a collection of data such that the processed data (namely, reports/information) are useful to the managers of the organization in taking decisions.

Three groups of people are involved in developing information systems for organizations: managers, users of the systems, and computer programmers who implement systems. Managers of the organizations who are responsible for running the organization are ultimately responsible for their information systems. They are also the primary users of the systems. Operational staff work with the systems daily. They provide the data input to the system and use the routine outputs. As most information systems nowadays are computer-based, the responsibility of writing programs to implement the systems and of creating and maintaining databases is that of computer programming professionals. The systems analyst coordinates the efforts of all these groups to effectively develop and operate computer-based information systems.

3.2.1 Defining Requirements

The most important and difficult task of an analyst is to understand the users' requirements. It involves interviewing users and finding out what information they use in the current systems and how they use it. They are then asked what information they lack, which if provided, will allow them to meet a set of quantified goals. The analyst has to then determine how such information can be generated.

3.2.2 Prioritizing Requirements by Consensus

In any organization there are many users. Each user has some special information needs. It may not be possible to satisfy needs of every one due to limited computing capacity, time constraints, availability of programmers, etc. Even if computer capacity is available, it would be necessary to take up some of the tasks first and others later. Thus there is a need to set priorities among the requirements of various users. This is best achieved by having a common meeting with all the users and arriving at a consensus. This responsibility of the system analyst requires good interpersonal relations and diplomacy. He must be able to convince all the users about the soundness of the group decision and obtain their cooperation.

3.2.3 Gathering Data, Facts and Opinions of Users

Having determined the information needs and their priority, the systems analyst must develop the system with the active and willing cooperation of all the users. The users must be made aware of what information they will get, how it will be derived, and how they can make use of it. The analyst designing the system must make use of the users' long experience and expertise. With the users' help he must gather the necessary data. While developing a system the analyst must continually consult the users and get their views. He must be the users' student in learning various facts about the organization, and their teacher in telling them how best to use the computer-based system.

3.2.4 Analysis and Evaluation

The system analyst analyzes the working of the current information system in the organization and finds out to what extent they meet users' needs. He then sifts through the facts and opinions gathered by him and finds the best characteristics of a new or modified system which will meet the users' stated information needs. The key to good analysis is to reject redundant data and focus on important data. Various graphical means of data analysis are useful in this task.

3.2.5 Solving Problems

Very often a problem such as the following is presented to a systems analyst: "We find that we are not able to quickly meet the demands for spare parts to repair our trucks sold to customers all over the country. We do not know how much spare parts we must stock at various branch workshops so that we have the parts when and where required without incurring high cost of storing many unused parts. What is the best systems which will solve this problem?"

An analyst must study the problem in depth and suggest alternate solutions to the management. The relative difficulties in implementing each of the alternatives and benefits of each must be determined so that a manager can pick what he considers as the best solution.

3.2.6 Drawing up Specifications

A key job of an analyst is to obtain the functional specification of the system to be designed in a form which can be understood by users. The specification must be non-technical so that users and managers understand it. The analyst must win the acceptance of all levels of users of the system such as clerks, middle level managers and top managers. The specification must be precise and detailed so that it can be used by system implementers. It must also take into account expansion envisaged in the near future.

3.2.7 Designing Systems

Once the specifications are accepted, the analyst designs the system. The design must be understandable to the system implementer. The design must be modular to accommodate changes easily. The analyst must know the latest design tools to assist him in his task. As part of the design he must also create a system test plan.

3.2.8 Evaluating Systems

An analyst must critically evaluate a system after it has been in use for a reasonable period of time. The time at which evaluation is to be done, how it is to be done, and how users' comments are to be gathered and used must be decided by the analyst. He must have an open mind to accept valid criticism and he must be a good critic himself to enable him to carry out necessary improvements.

3.3 ATTRIBUTES OF A SYSTEMS ANALYST

To be effective, a systems analyst must have several attributes. These are now discussed.

3.3.1 Knowledge of Organizations

A systems analyst must understand the way in which various organizations function. He must understand the management structure and the relationship between the departments in organizations. He must also find out how day-to-day operations are performed in the organization for which the information system is being developed. As many systems are built for marketing, accounting, and materials management, he must have a working knowledge of these areas. As a generalist he must try to familiarize himself with a variety of organizations.

3.3.2 Knowledge of Computer Systems and Software

The analyst must also know about recent developments in computer systems and software. He need not be an expert programmer or computer manager, but know enough technical details to interact effectively with program designers and computer managers. He acts as the intermediary in understanding users' needs and advising an applications programmer on how the needs can be realized on a computer system. He must also advise on the information presentation techniques, appropriate use of packaged software, newer computer languages and environments.

An analyst's knowledge of computer systems must be deep enough to determine the feasibility of developing the required systems on a given computer network configuration. Conversely, an analyst must be able to advise on the computer network configuration needed to develop the required applications. Sometimes, the analyst may also advise on augmentation of existing computer network to implement new systems.

3.3.3 Good Inter-personal Relations

An analyst must be able to interpret and sharpen fuzzily stated user needs, must be a good listener, a good diplomat and win users over as friends. He must be able to resolve conflicting requirements and arrive at a consensus. He must understand people and be able to influence them to change their minds and attitudes. He must understand their needs and motivate them to work under stressful conditions such as meeting deadlines.

3.3.4 Ability to Communicate

An analyst is also required to orally present his design to groups of users. Such oral presentations are often made to non-technical management personnel. He must thus be able to organize his thoughts and present them in a language easily understood by users. Good oral presentations and satisfactory replies to questions are essential to convince management about the usefulness of computer-based information systems.

3.3.5 An Analytical Mind

Analysts are required to find solutions to problems. A good analyst must be able to perceive the core of a problem and discard redundant data and information in the problem statement. Any practical solution is normally non-ideal and requires appropriate trade-offs. A good analyst would use appropriate analytical tools as necessary and use commonsense.

3.3.6 Breadth of Knowledge

A systems analyst has to work with persons performing various jobs in an organization. He may have to interact with accountants, sales persons, clerical staff, production supervisors, stores officers, purchase officers, directors, etc. A system analyst must understand how they do their jobs and design systems to enable them to do their job better. During his career he will be working with a variety of organizations such as hospitals, hotels, departmental stores, transport companies, educational institutions, etc. Thus a general broad-base education is very useful.

3.4 TOOLS USED BY SYSTEMS ANALYST

As pointed out in the previous sections, one of the most important tasks of a systems analyst is to understand users' requirements. There are two aspects in this understanding:

1. Understanding the flow of documents in an organization. (Documents incorporate the basic data available in the organization and hence it is essential to know where they originate and how they flow.)
2. Understanding the rules used to process data. (Very often the rules are complex and orally stated, leading to misunderstanding.)

Two important tools are used to assist in the above tasks:

1. Data flow diagrams which specify the origin of data and how they flow through a system and where they are processed.
2. Decision tables which are used to specify complex processing rules known as business rules in a concise and easily understood form.

Besides these tools used in specifying users' requirements, there are also other tools used in a system design. These are database management systems and associated languages which allow rapid prototyping of systems. Quickly obtaining a prototype is very useful to get good user feedback. Tools such as spreadsheets are very useful in answering questions such as, "If the cost of raw materials x, y, z change by p, q, r per cent how will it affect the cost of the final product?" Further, report generators and graphics systems are used in visualizing information. A system analyst must be conversant with all these tools and know how to effectively use them.

In succeeding chapters we will describe in detail these tools and their use.

3.5 APPROACHES TO SYSTEMS DEVELOPMENT

Over the past three decades several approaches to development of computer-based systems have been proposed and tried. A particular approach is not necessarily better than another. Organizations select an approach which best suits their environment. We describe two major approaches both of which depend on use of what are known as Computer Aided Software Engineering (CASE) tools. CASE tools are a set of integrated tools (i.e., computer programs) which assist a systems analyst to carry out his task faster. Typical tools are used to:

- Draw diagrams used in analysis such as hierarchy charts, data flow diagrams (Chapter 7), entity-relationship diagrams in database design (Chapter 11), flow charts, etc.
- Produce standard system documents.
- Assist in developing graphical user interfaces for system users such as clerks, managers, etc.
- Design input data forms (Chapter 14) and output reports (Chapter 15).
- Analyse process specifications (Chapter 8) and decision tables (Chapter 9) for ambiguity, incompleteness , contradictions and other logical errors to the extent feasible.
- Generate programs automatically from the specifications in several cases.

3.5.1 Joint Application Design (JAD)

In this method proposed by IBM in the late 80s a JAD session analyst, who is usually a senior systems analyst, has the primary task of finalizing systems requirement specifications (SRS) and based on that work out design-details such as user interfaces, system development times, organizational commitment of resources including human resources. In order to do this a selected group of users, managers and system developers are grouped together as a team. The team meets regularly as determined by the session coordinator. The team finalises SRS and all other important details of development. The session leader may develop small prototypes and demonstrate them during the meetings so that a consensus is reached on the look and feel of the final system, time lines, milestones to be achieved, and intermediates reviews of the project. This methodology is used by several financial institutions such as banks, insurance companies, mutual funds, etc.

3.5.2 Rapid Application Development

In Rapid Application Development (RAD) the primary goal is to create prototypes of sub systems quickly and show it to the concerned personnel in an organization to get their feedback. This is one way of eliciting requirements of the system while it is being development rather than develop SRS first and then design a system to meet SRS. The disadvantage of the latter method is that very often users do not understand what a computer-based systems can do and what they really want. After the system is developed users request new features which are difficult to implement. With the advent of CASE tools and languages such as Visual Basic which allow easy development of good graphical user interfaces it is possible to develop almost in parallel various parts of a system to illustrate their "look and feel" to system users. This allows users to refine their requirements statements. Further when the full system is implemented they know what to expect.

SUMMARY

1. There are nine distinct phases in the development of an information system. These phases constitute what is known as the system life cycle.
2. A summary of what is done in each phase and the outputs obtained at the end of each phase is given in Table 3.1.
3. It should be remembered that in a design one may have to go back to an earlier phase in the design based on results obtained in a later phase. The phases are primarily intended as milestones to assess progress in design.
4. A systems analyst should interact with managers, users, and application programmers in designing a system.
5. A systems analyst must, through discussions with users, determine their information requirements, interact with them during the design phase and explain to them what the system will provide. He must assign priorities among different requirements, analyze and evaluate existing systems and improve them. An analyst must be able to identify and solve management problems in organizations, draw up specifications and oversee implementation. An analyst should evaluate the designed system and modify them if needed.
6. A good system analyst must know the operation and management structure of diverse organizations, must understand both hardware and software features of computers, must exhibit good interpersonal relations, be able to express his thoughts well, and capable of life-long learning.
7. A system analyst should know the use of tools such as data flow diagrams, decision tables, prototyping systems, spreadsheets, database systems, report generators and graphics systems.
8. There are several approaches to systems development among which two are widely used. In Joint Application Development, team of users work closely with a system analyst (project leader) to evolve Systems Requirements Specifications before a system is developed. In Rapid Application Development prototypes of

sub systems are developed with graphical user interfaces to demonstrate to users the look and feel of the system to be delivered. It allows user to clarify their requirements while various subsystems are developed leading to ready acceptance when the final system is completed and delivered.

EXERCISES

3.1 How are information requirements arrived at in an organization?

3.2 Who do you think should participate in determining the information requirements of a student hostel?

3.3 As a student what information would you consider as important in a students hostel?

3.4 What is the difference between information requirement determination and specification?

3.5 Specify the information requirement you stated in answer to Exercise 3.3.

3.6 What do you understand by the term *consensus*?

3.7 Why is feasibility analysis necessary before designing a system?

3.8 In designing an information system for a hostel what constraints should be taken into account during feasibility analysis?

3.9 What benefits do you expect if an information system for a hostel is designed?

3.10 What factors are taken into account in arriving at the final specifications of an information system?

3.11 What should be the important characteristics of the final specifications?

3.12 What are the inputs and outputs of the system design phase?

3.13 What activities are carried out in the system implementation phase?

3.14 What activities will you carry out during implementation of a hostel information system?

3.15 When should a system be evaluated?

3.16 What is the basic difference in the design philosophy of a house and that of an information system?

3.17 What are the characteristics of a good information system?

3.18 What special skills should a systems analyst possess?

3.19 Why should a systems analyst be able to communicate well?

3.20 Are excellent programmers necessarily excellent systems analysts? Justify your answer.

3.21 List eight tasks performed by systems analysts.

3.22 List six desirable attributes of a systems analyst.

3.23 The qualifications and experience of three candidates for the post of a systems analyst are given in Table 3.2. Who would you choose as the most promising candidate? Why?

Table 3.2 Data for Problem 3.23

Name	*Age*	*Education*	*Experience*	*Aptitude Test Stores*
Mr. A. Krishnan	30	B.Sc. Diploma in Computer Applications	Operator 3 years. Programmer 4 years Senior Programmer 2 years	Verbal 60%. Quantitative 90%
Ms. K. Vasantha	32	B.Sc. Diploma in Management	Office Manager 6 years	Verbal 90%. Quantitative 70%
Mr. P. Kumar	25	Master of Computer Applications. Diploma in Management	Management Trainees 1 year	Verbal 85% Quantitative 80%

3.24 Why can't managers design their own systems? What are the advantages and disadvantages of such a decision?

3.25 What are the tools currently available to design systems? What are the advantages of using these tools?

3.26 What are the two main approaches to system development? Compare and contrast the two.

3.27 What are CASE tools? For what applications are they used?

4

Information Gathering

LEARNING GOALS

After reading this chapter a student should be able to:

1. Evolve strategy to gather information for computerization.
2. Identify sources of information.
3. Devise methods of searching for information.
4. Plan interviews to gather information from employees to top managers.

During the requirements determination phase, a system analyst has to find out how the current system works and what is expected from a new system. He has to spend considerable time in talking with users and gathering all relevant information on the project. In this chapter we will explain how this is done.

4.1 STRATEGY TO GATHER INFORMATION

Gathering information in large and complex organizations is difficult and takes time. All relevant personnel should be consulted and no information overlooked. Very often what hurts an analyst most is what is not stated by users because they assume that it is "obvious and everyone knows it". The analyst should perceive such situations. A clear strategy should be evolved by the analyst to gather information. The strategy consists of

- identifying information sources,
- evolving a method of obtaining information from the identified sources, and
- developing an information flow model of the organization.

4.2 INFORMATION SOURCES

The main sources of information are:

1. Users of the system
2. Forms and documents used in the organization
3. Procedure manuals and rule books which specify how various activities are carried out in the organization
4. Various reports used in the organization
5. Computer programs of existing systems with their documentation.

Information should be obtained from all the above sources. The consistency between them should be checked and then they should be summarized and consolidated.

4.3 METHODS OF SEARCHING FOR INFORMATION

Information gathering first starts with interviewing top level management. An overview of the organization, available information, and objectives to be met by the proposed system are normally gathered from top management. A gross system model is then worked out and verified.

The next step is to interview middle level management and operational staff. At this level day-to-day operational requirements and information required for current use are gathered. The earlier gross model is now expanded and refined.

It should be remembered that operational staff at the day-to-day working level such as clerks, storekeepers and foremen can make or break a system. It is essential to gain their confidence and explain to them what the intended system would do. They should understand their responsibilities when the system is implemented. Being deeply involved in day-to-day operation, these personnel will have a lot of information on operational problems. Many persons at this level would have been on their jobs for many years and have rich experience. Qualitative information based on their experience is often very valuable.

It is therefore clear that interviewing is the most important method of gathering information. Before starting interviewing an analyst must decide on:

- Who will be interviewed,
- In what order they will be interviewed, and
- What specific questions will be asked in each interview.

Besides individual interviews, we also saw that a consensus has to be reached by all users in assigning priorities among objectives of the system. Thus leading group discussions and encouraging frank expression of opinions are also the responsibilities of an analyst. During such group discussions many bits of information are revealed. Thus organizing and steering these discussions need to be carefully planned.

It is not normally feasible to obtain all information during the course of a single interview with a user. Usually two or three interviews are needed. The first interview is normally to get to know a person and his areas of responsibilities and the type of information he has. Subsequent interviews are used to verify facts and gather more information of relevance.

4.4 INTERVIEWING TECHNIQUE

The primary purpose of interviewing is to obtain both quantitative and qualitative data (current as well as desired data) regarding user requirements, policies, procedures and practices. It is essential to gain the confidence of the interviewee and obtain frank and complete answers to questions. No user likes an analyst to tell him how to run his business. Obviously an analyst's job is not this. His job is to let the user clearly define for themselves the goals of an effective information system.

We have given below some guidelines to ensure a successful interview.

1. Make a prior appointment with the person to be interviewed and inform the purpose of the interview, and how much time you would require. An interview should be brief and not exceed 40 minutes.
2. Read the background material and go prepared with a checklist.
3. State again the purpose of the interview at the beginning of the interview.
4. Good manners are essential. Introduce yourself. Be punctual and pay attention to what the user says. Obtain permission to take notes or record the conversation.
5. Do not use computer jargon. Explain all technical terms. If a user employs a word (jargon) not known to you, immediately ask what it means.
6. Avoid yes/no answers. Try to obtain both quantitative and qualitative information. Obtain both facts and opinions. Discriminate between essential requirements and desirable requirements.
7. State what you understood as a fact and confirm this.
8. Watch for signs of impatience of the interviewee and do not prolong the interview in such a case.
9. Summarize the information gathered by you during the interview and verify this with the user.

4.5 QUESTIONNAIRES

Detailed questionnaires are useful to gather quantitative data such as sales statistics. They are no substitutes for interviews as questionnaires do not get qualitative information. When quantitative data is to be collected from a large number of persons or organizations, questionnaires are useful. Usually questionnaires sent by mail are not promptly answered. Continuous follow-up by letters and sometimes personal calls are needed to get back the filled-up questionnaires. The shorter a questionnaire, the higher is the probability of getting it back quickly.

4.6 OTHER METHODS OF INFORMATION SEARCH

Other methods of information search are to examine:

1. systems used in other similar organizations;
2. trade journals and reports of conferences describing similar systems; and
3. trade and national statistics.

Apart from interviewing, if an analyst spends some time in the work place carefully observing the existing system, he will learn many important facts. Some of these facts may not be revealed in interviews since a person who is part of the system may think it is "obvious". Sometimes some facts are hidden and no one may be willing to tell the whole truth.

When available records are too many, an analyst may sample a smaller subset which is representative of the entire data set.

Fact recording. The results of fact finding are summarized as a report. If a large number of analysts are taking part in a project, a standard format may be used for recording facts. The report should contain the following data:

- Background information
- Sources used in obtaining information
- Method used in gathering information
- Current procedures
- Currently available data.

4.7 CASE EXAMPLE—HOSTEL INFORMATION SYSTEM

System analyst's diary: "I was called at 9 a.m. today (20 March 2009) by the Chief Warden of Raman Hostel, Prof. V. Ramakrishna. He wants to examine if computers can help in hostel administration. I made an appointment to meet him tomorrow at 10 a.m. in the hostel office. I told him I will need 40 minutes to get an overview of his requirements". I made the following checklist before meeting him.

Checklist

- Find out what aspects of hostel operation he wants to analyze and improve
- Find out why he wants this and what benefits he foresees
- Find out his priorities
- Find out how early he wants the system
- Find out about the organization of the hostel administration
- Find out who else he wants me to talk to
- Find out if any documentation on existing procedures exists.

I met Prof. V. Ramakrishna at 10 a.m. on 21 March 2009 in his office in the hostel and asked his permission to take notes. The summary of my notes which I made and sent to him on 23 March 2009 for his comments are:

- The total mess expenses are divided among members. Policy is set by the Chief Warden in consultation with a student mess committee.
- He wants correct mess bills to be prepared and despatched within 5 days of the end of the month to the accounts office for collection.
- He wishes to know which students have large unpaid bills.

- He is eager to control inventories to minimize stock.
- He intends to control issue of stores to cooks to minimize wastage.
- He wants to do some planning about purchase of vegetables and fruits to minimize cost and wastage.
- He desires to have an up to date database of students in hostel.
- The primary objective is to minimize students' mess bills, reduce outstanding payments, and have ready information about residents.
- The priority order of the Chief Warden is: control inventory, control stores issues, ensure speedy bill collection, student database.
- He would like a new system to be implemented within 4 months.
- The organization chart of the hostel is given in Figure 4.1.
- The Chief Warden wants me to talk to assistant wardens, mess supervisors and students' mess committee members.
- There is no procedure manual. Most operations are done routinely using a commonsense approach.

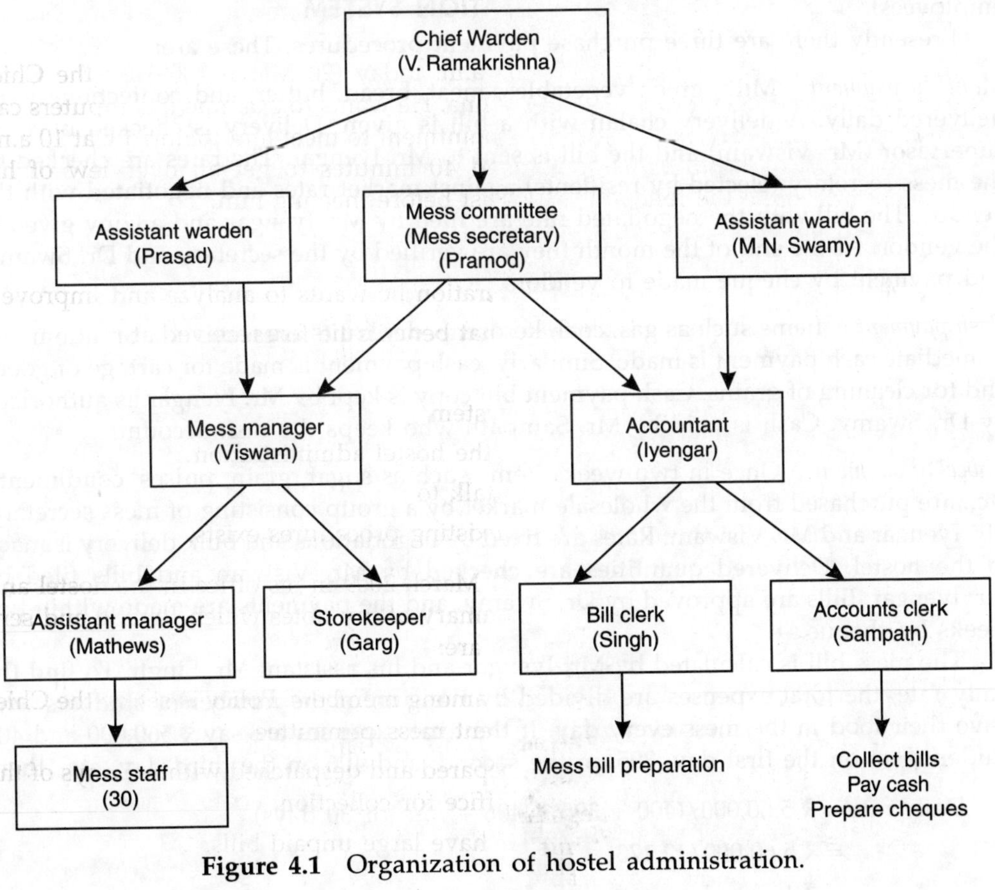

Figure 4.1 Organization of hostel administration.

Professor Ramakrishna agreed that my notes, on the whole, reflected our conversation. He added that the number of residents in the hostel was 400, all staying in single rooms. He further indicated that he could afford to install a Personal Computer if benefits can justify cost. I then made an appointment to meet Dr. M.N. Swamy, the assistant warden on 23 March 2009 at 4 p.m. in his office. I requested half an hour of his time for discussions. I made the following checklist before meeting him:

- Find out details of accounting and billing procedure currently used
- Ascertain his ideas about the current system and improvements
- Try to know his ideas on deployment of personnel if a computer is used.

The following notes were prepared by me on 23 March 2009 evening. I used the jottings in my notebook made with Dr. Swamy's permission on 23 March 2009 during the interview.

Dr. Swamy is in-charge of all hostel accounts and is assisted by Mr. R. Sampath whose responsibilities are to collect mess dues from residents and issue receipts, prepare cheques for vendors' payments and keep petty cash account. Besides Mr. Sampath, there is Mr. K.S. Iyengar, an accountant, who keeps accounts of (a) daily purchases for the mess, (b) periodical purchases for mess, and (c) employees' salaries (total around 30 employees).

Presently there are three purchase payment procedures. These are:

Monthly payment. Milk, green vegetables, meat, bread, butter, and confectionery are delivered daily. A delivery chalan with a bill is given. Delivery is checked by mess supervisor (Mr. Viswam) and the bill is sent to Mr. Iyengar. The rates are checked by the mess secretary (elected by residents) against market rates and negotiated with the vendor. The bill with the negotiated rate are filed by Mr. Iyengar and a copy given to the vendor. At the end of the month they are verified by the secretary and Dr. Swamy, and payment by cheque made to vendors.

Cash payment. Items such as gas, coal, kerosene, dry fruits are received at random and immediate cash payment is made. Similarly, cash payment is made for cartage of goods and for cleaning of grains. Cash payment bill copy is kept by Mr. Iyengar as authorized by Dr. Swamy. Cash is paid by Mr. Sampath who keeps the cash accounts.

Biweekly payment. Once in two weeks items such as sugar, grain, pulses, condiments, etc., are purchased from the wholesale market by a group consisting of mess secretary, Mr. Iyengar and Mr. Viswam. Rates are fixed by negotiations and bulk delivery is made to the hostel. Delivered quantities are checked by Mr. Viswam and bills filed by Mr. Iyengar. Bills are approved by Dr. Swamy, and the payments are made within two weeks by cheque.

The mess bill is calculated by Mr. Iyengar and his assistant Mr. Singh. To find the daily rate, the total expenses are divided by the sum of the number of students who have their food in the mess every day. If the total expenses are say ₹ 560,000 and 400 students eat on the first day, 395 on the second, and 406 on the third day, etc., then

$$\begin{aligned}\text{Daily rate} &= ₹\,5{,}60{,}000/(400 + 395 + 406 + \ldots \text{ for 30 days})\\ &= ₹\,5{,}60{,}000/12{,}173\\ &= ₹\,46\end{aligned}$$

Currently, the manual calculation of bills takes over 6 days and many errors are found. Bills do not go on time to students. This needs a better system.

Students are allowed to claim rebate if they do not eat for 2 days consecutively in the mess, provided they give minimum 3 days notice in writing in the register kept with the mess supervisor. Students can bring guests for meals and they have to enter the number of guests for meals (not more than 4 at a time) in a register. They can also order "extras" such as milk and eggs, which are entered in a register. Such entries from a daily register are transferred to another students' extras/guests/rebate register which has one page for each student, and all entries for a student are written date-wise in this by Mr. Singh. This register is used in calculating individual bills.

Dr. Swamy feels that this procedure complicates bill preparation. The student body wants this facility to be continued. Dr. Swamy feels that if a proper system is designed, Mr. Iyengar and Mr. Singh can be easily trained to use it, as both are intelligent and also know typing.

The above notes were sent to Dr. Swamy who approved them. He added that a speedier and accurate billing method with full details will be welcomed by students. Currently full details of extras/guests are not provided in the bill as manual entry is too slow. Further follow-up with students who have not paid their mess bill is currently not methodical.

I then decided to meet Dr. Prasad who is the assistant warden in charge of mess affairs. I fixed a meeting with him on 24 March 2009 at 3 p.m. Before the interview the following points were noted in the checklist:

- Find out how items are issued to cooks each day
- Find out if any inventory control is exercised
- Find out how wastage is reduced
- Find out his ideas for an improved system and its operation.

The following notes were prepared by me after the interview and sent to Dr. Prasad on 25 March 2009. (Dr. Prasad oversees the work of Mr. Viswam, the mess manager, Mr. Garg, the storekeeper, and coordinates with the student mess committee).

Dr. Prasad is very unhappy with the way the inventory is controlled. The storekeeper waits till an item is exhausted and only then he tells the mess manager about this. It leads to many trips to the city to order items and piecemeal arrival of items, leading to high travel/transport cost. ₹ 20,000 is spent in transport which can be considerably reduced. Items are issued by the mess manager to the cooks based on nominal number of students in the hostel. Proper care is not taken to reduce issues to cooks when many students are absent. This leads to wastage. Over 100 different items are stocked in the store and a systematic technique of replenishing stock should be used. Mess bill could be reduced by bulk purchase of non-perishable items when their cost is low. This is not done. Dr. Prasad feels that use of a computer in inventory control and issue control can be very effective.

The above notes were approved by Dr. Prasad who suggested that I should meet the secretary of the student mess committee Mr. A. Pramod. I found Mr. Pramod in his room in the hostel on 26th morning and he invited me to have a breakfast meeting. (Mr. Pramod was elected secretary 8 months ago and has 4 more months in his current term. He wants improvements within that period). He reiterated the points made by Dr. Prasad and Dr. Swamy on the need to systematize issues, inventory and billing. He pointed out one of the main difficulties he faced. Currently the bill is calculated at the end of the month. Till then he has no idea of what the daily rate will be. Very often the daily rate goes up sharply from say ₹45 to ₹55 which the students do not like. This leads to hasty economy measures in the mess without proper knowledge of the areas in which these measures will be effective. The student body would like to have uniform rates. This can be ensured if estimates of daily rates are available at the end of each week. If the rate tends to become high, some planned economy can be effected. Weekly adjustment can lead to uniform rates. The daily menu is decided by the mess manager in consultation with the mess secretary. Currently it is somewhat repetitive. Students joke "It is beans curry and therefore it is Wednesday". There must be some way of randomizing menus, keeping nutritional balance, variety of items and of effectively using seasonal low cost vegetables, thereby reducing the mess bill. He thinks aloud whether some operations research techniques can be used in this problem. I am somewhat intrigued by this suggestion.

I summarized the above discussion and sent it to the mess secretary and Dr. Prasad for their comments. I next met employees of the mess to get their views. I met Mr. Viswam, the mess manager on 27 March morning. He explained his responsibilities as:

- Preparing daily menu based on advise from the mess committee.
- Preparing an "issue list" of items and quantities for each meal and sending it to the hostel store for issue to the kitchen.
- Preparing a "return list" of items and quantities not used during the day by the cooks to the store.
- Stock register updating based on issue and return list.
- Keeping track of rebates to students absent from mess and extras taken by students which are to be separately charged. Relate rebates to stores issue.
- Ordering items when stock level is low. In practice, however, ordering is done only when an item is exhausted. As delivery at door may be delayed, unnecessary special trips are made to do cash purchases.

Mr. Viswam felt that some help in automatically preparing issue list based on number of persons who will eat in the mess will be very helpful. He would also appreciate assistance in error free recording of rebates and extras. An automatic warning mechanism when stocks become low would be useful. Consolidation of purchases to reduce trips would be desirable.

A summary of the above discussion was sent to Mr. Viswam for his comments.

Next I met Mr. Mathews, the assistant mess manager. His responsibilities include supervising mess staff and day-to-day maintenance of the hostel. He also assists the warden in room allotment by keeping a list of available rooms, preferences, applications for change of rooms, etc. He feels that mess bill preparation is time consuming as it takes a week. Room allotment, on the other hand, is not difficult and the current system is satisfactory.

I then met Mr. Garg, the storekeeper. He keeps the stores stock register and helps the mess manager in preparing the issue and return lists. He prepares lists of items to be bought and monitors quality of supplies.

I finally held a group meeting with the Chief Warden, assistant wardens, mess manager, assistant manager, mess secretary, accountant and storekeeper. I showed the data and document flow diagrams (Figures 4.2–4.4) prepared by me and got their approval. I then listed the critical needs which various persons identified. These are given in Table 4.1. Based on this a consensus was reached on what aspects needed computerization, and priority among them was also determined. The following priority list was prepared as system to be computerized.

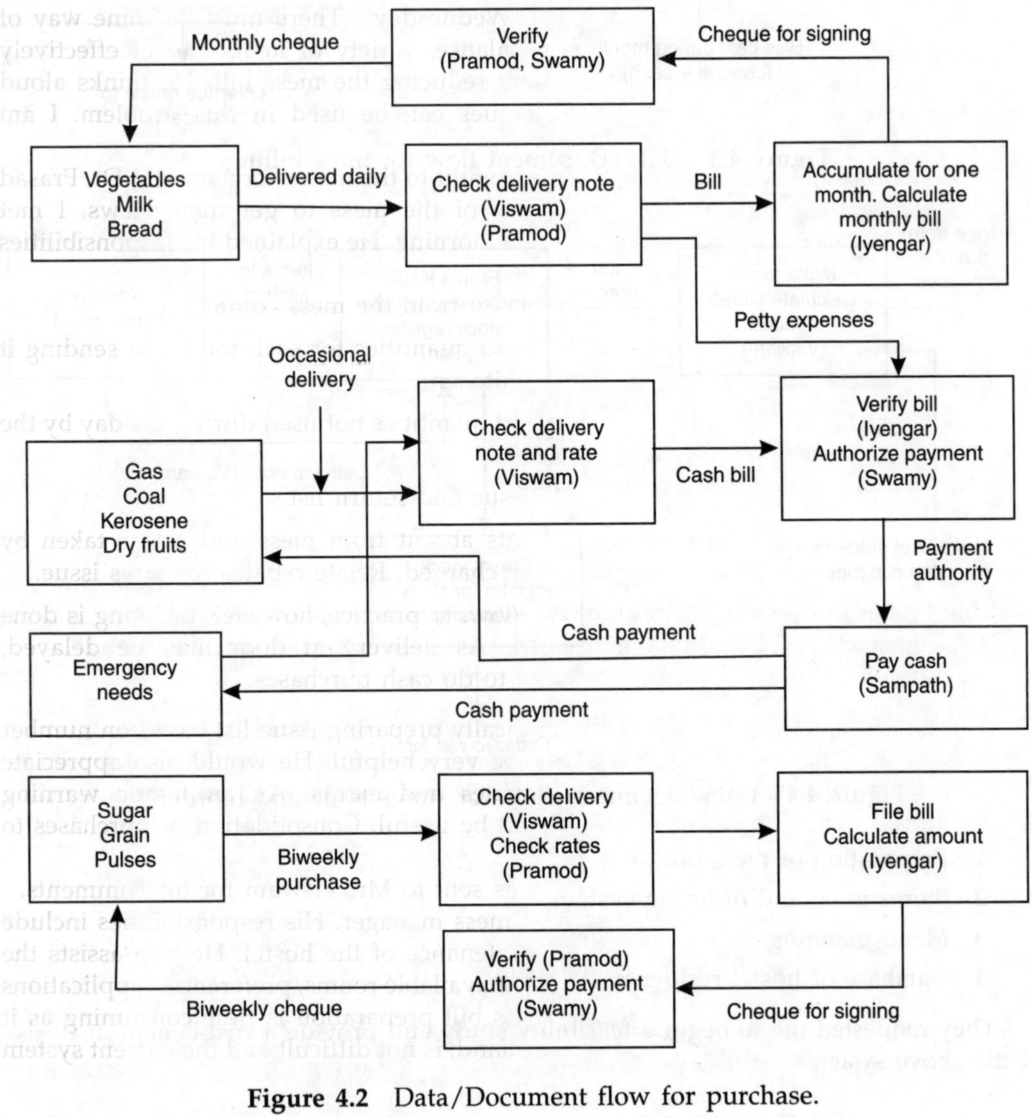

Figure 4.2 Data/Document flow for purchase.

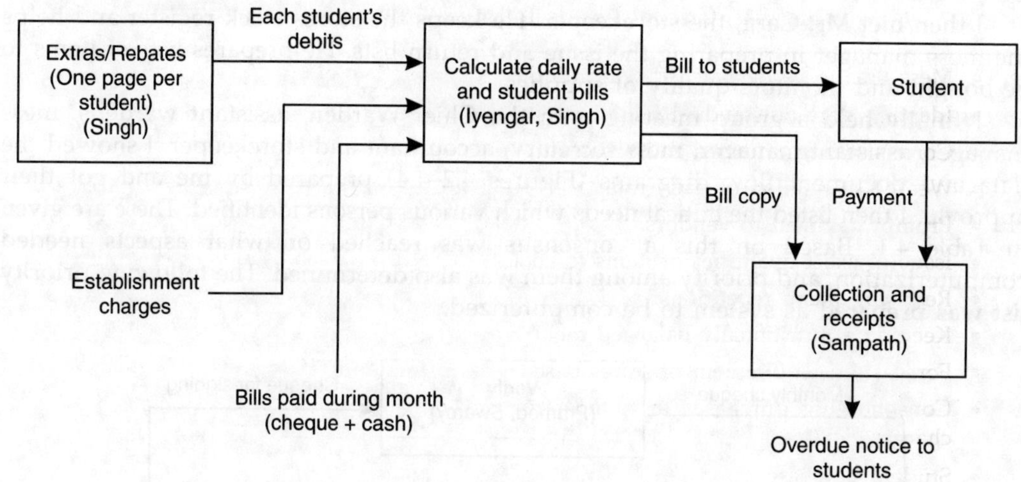

Figure 4.3 Data/Document flow for mess billing.

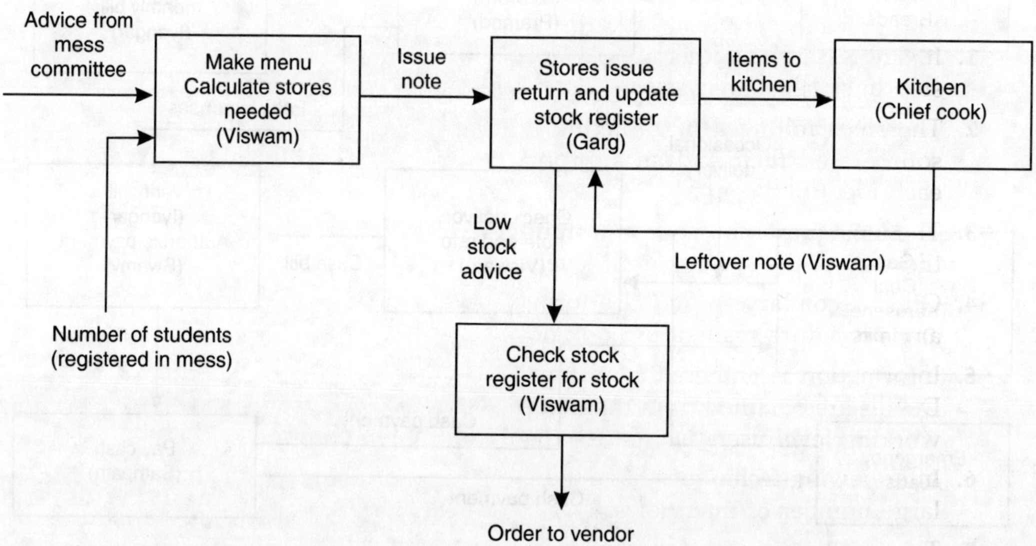

Figure 4.4 Data/Document flow for stores issue and reorder.

1. Preparation of mess bill
2. Stores issue and ordering system
3. Menu planning
4. Database of hostel residents.

They requested me to begin a feasibility study and provide a cost-benefit analysis of the above systems.

Table 4.1 Critical Needs Identified by Various Persons

- Mess bills to be despatched within 5 days to students.
- Identify large outstanding bills from students.
- Correct billing of extras and rebates to students.
- Weekly calculation of daily rate.
- Prompt payment to vendors.
- Issue of exact quantities of items to cooks based on menu and number of students.
- Reducing menu repetitions.
- Keeping a nutritionally balanced menu and minimizing cost.
- Forecasting requirement of stores based on menu and students.
- Consolidating purchases to reduce frequent trips for purchase and reducing cartage charges.
- Student database.

SUMMARY

1. It is necessary to gather all relevant information about an organization as a first step in systems analysis.
2. The steps followed in gathering information are to first identify information sources and then find an appropriate method of obtaining information from each identified source.
3. The most important source of information, both qualitative and quantitative, is the user of the system at all levels.
4. Other secondary sources are forms and reports used by the organization and any procedure manual or book of rules.
5. Information is gathered from top down. An overview is obtained at the top. Details are obtained from those at the working level. Gaining the confidence of working level users is vital for the success of the project.
6. Interviewing techniques have to be learnt by an analyst as he has to conduct a large number of interviews.
7. The key to a successful interview is to observe good manners, have an open mind, and encourage the person being interviewed to reveal all important details.
8. Distribution of questionnaires is not a substitute for interviews. Questionnaires are most useful when quantitative data are to be gathered from a large number of persons.

EXERCISES

4.1 Describe the general strategy an analyst should use to gather information.

4.2 What are the main sources of information in an organization?

4.3 What are the methods of gathering information?

4.4 Why are interviews essential in gathering information?

4.5 How should an analyst prepare before an interview?

4.6 Are group discussions useful in information gathering? Give reasons.

4.7 Are there some guidelines for good interviewing? State them.

4.8 What is the advantage of using a questionnaire in fact gathering?

4.9 When should questionnaires be used?

4.10 When should questionnaires be not used?

4.11 State as many sources of information as you can think of.

4.12 What is the last step in information gathering?

4.13 Before interviewing the Chief Warden, what checklist did the systems analyst make?

4.14 Make an organizational chart of the hostel and separately specify the responsibilities of each of the officials in the hostel.

4.15 How can actual issues of items to cooks be controlled?

4.16 Is a daily update of daily rate calculation feasible?

4.17 How many students stay in the hostel and how many employees are there?

4.18 What interesting suggestions did the mess secretary make?

4.19 Among applications noted in Table 4.1, give priorities as perceived by you.

4.20 Why is the lowest priority assigned to student database?

4.21 Go to a medical store in your locality. Interview the owner and find out his information needs and prioritize them.

5

System Requirements Specification

LEARNING GOALS

After reading this chapter a student should be able to:

1. Draw document flow diagrams to depict the flow of documents through offices in an organization.
2. Draw data flow diagrams to describe the logical flow of records and how they are processed in an organization.
3. Specify operational, tactical and strategic information requirements by refining user's statement.
4. Refine data flow diagrams to meet user's information requirements.
5. Determine data requirements of a system.

When an organization feels dissatisfied with its current information system, it would normally request an information systems analyst to design a system which will meet its requirements in a more satisfactory fashion. There are also occasions when an organization wants information systems designed from the beginning. The first case is more common.

In order to design an information system, the analyst requests the user to state his requirements. He also finds out how they are met (may be partially) in the current system. Normally user's statement of requirements is descriptive and not precise. Further, many points are not explicitly stated by the user. A systems analyst should also try to find out whether he has understood fully the user's statements. He must therefore use a method that will enable him to precisely inform the user his understanding of the requirements. Two graphical specification tools are useful in doing this job:

1. Physical document flow diagram
2. A logical data flow diagram.

In this chapter we will illustrate how system requirements specification (commonly known as SRS) is obtained from a descriptive statement of requirements.

5.1 SYSTEM REQUIREMENTS SPECIFICATION: EXAMPLE

We will illustrate the method used to obtain a specification with an example. Usually an imprecise statement of the needs of an organization is given by the management. We now give a sample of such a statement, based on which a more precise definition of the requirements is evolved by an analyst.

(i) User's narrative of requirements. "Our company receives a number of items from many vendors and they are received at the receiving office. As we receive over 1000 items a day it is now virtually impossible for the receiving office to check whether the vendor has supplied items against an order, or sent a wrong item and inform the purchase office. We are also not able to find out if there are excesses or deficiencies in delivery and whether the vendor adhered to the delivery schedule as per the order. The items received at the receiving office are sent for physical inspection. The physical inspection consists of checking whether the quantities stated in the delivery note agree with the physical count, whether the item is the correct one ordered, and a check on the quality of item. We would like to keep a record of rejections due to bad quality, incorrect items, excess/deficient supply, etc., determined during inspection. This will enable us to assess vendors' reliability to guide us in placing orders in the future, besides keeping track of supplies. Items cleared by the inspection office are taken into the inventory by the stores office which keeps a ledger of items stocked and quantity available of each item. Customers send requisitions to the stores. The stores fulfill the requests based on availability and updates the ledger. Currently we are not able to meet some of our customers' requests. We would like to incorporate automatic reordering by our purchase office if the inventory level of an item is low. We would also like to keep track of unfulfilled requests and meet them when items reach the store. Currently we are not able to pay our vendors promptly due to delays in payment order reaching our accounts office. We would like to rectify this. We would also like to bill our customers promptly and keep track of customers' payments".

(ii) Analysis of statement. Reading the statement carefully we find that the first three sentences starting with "our company receives a number of items..." and ending with "...vendor adhered to the delivery schedule as per order" relate to what happens at the receiving office of the company. This is depicted in Figure 5.1 which shows the physical entities, namely, the vendor(s), the receiving office and the purchase office, and the documents which flow between them. The diagram is called the physical *document flow diagram.* In this diagram rectangles are used to depict offices, and a straight line with an arrow is used to show flow of documents. The arrow shows the direction of flow. We give below the line, the name of the document flowing between offices. In the figure we also show a dashed line to depict flow of *physical items*. This is not relevant to the development of a computer-based information system but is useful to the user/manager to verify whether the analyst has understood him. In this diagram a line connecting purchase office and vendor depicts the actual order. This is not clearly stated by the user as it is assumed that a vendor supplies items against orders. This point should be told to the user as situations may arise when a vendor by mistake sends items not ordered at all and action is needed in such a case.

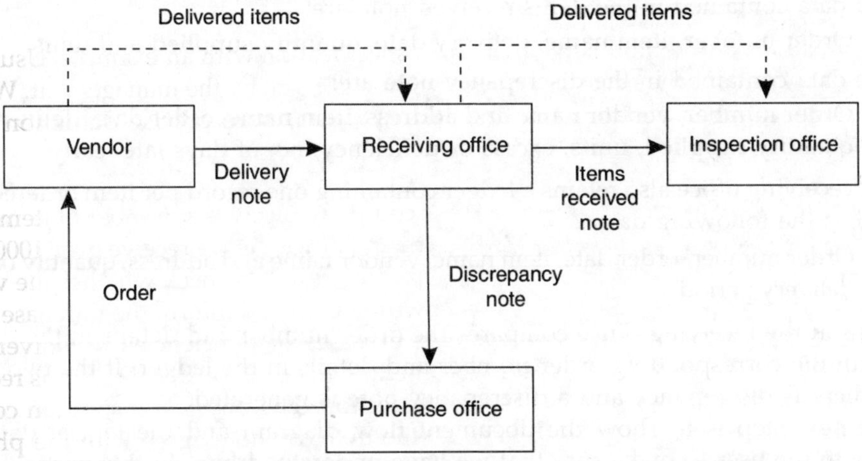

Figure 5.1 Document flow diagram in an organization.

The *input* to the receiving office is a delivery note and the *outputs* are items received note and discrepancy note. We now find out the data contained in these notes and the data processing rules followed by the receiving office. This is again depicted as a diagram called a *data flow* diagram (Figure 5.2) in which squares are used to indicate physical entities from which data originates or terminates. Labelled directed lines are used to indicate data records which flow in and flow out of sections where data processing is performed. A label is used to name the data record. The arrow indicates the direction of flow of data. A circle is used to indicate processing of data carried out on input data to produce output data. Two parallel lines depicts a data store. If data from the store is to be only read by a process, an arrow is shown from it to the process. If a data store is modified by the process, an arrow from the process points to the data store. Currently the data contained in the delivery note are:

- Order number, vendor name and address, item name, delivery date, quantity supplied and units.

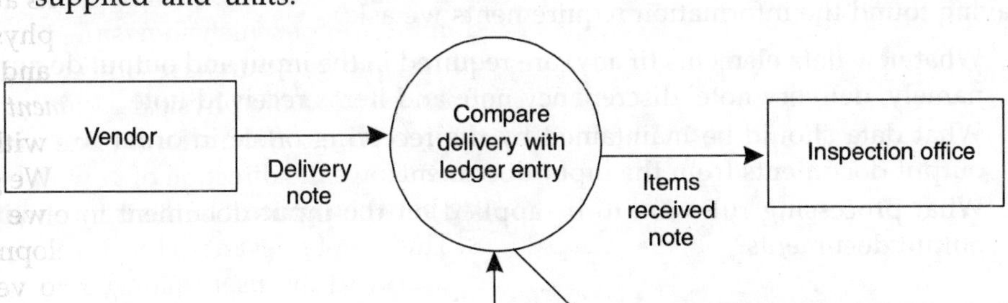

Figure 5.2 Logical data flow diagram of manual process.

The data contained in the items received note are:

- Order number, item name, delivery date, quantity supplied and units.

The data contained in the discrepancy note are:

- Order number, vendor name and address, item name, order date, delivery date, quantity supplied, units, excess or deficiency, no. of days late/early.

The receiving office also retains a ledger containing one record per item ordered. Each record has the following data:

- Order number, order date, item name, vendor name and address, quantity ordered, delivery period.

The clerk at the receiving office compares the order number and details in the delivery note with the corresponding order number and details in the ledger. If the two do not agree, there is discrepancy and a discrepancy note is generated.

The next step is to show the document flow diagram and the logical data flow diagram to the user to make sure that we have understood him. In this particular case it is very simple, but if we had taken the entire word statement given at the beginning of Section 5.1, the diagram would have been complex (we will examine this later). Having made sure that we have understood what the user said, the next step is to clarify the user's information requirements. For this example, the following information requirements for the receiving office are arrived at after discussion with users.

(i) **Operational information requirements.** Automatically check vendor's delivery notes with the orders placed with the vendor. Find whether delivery is for an order placed with vendor. Determine excess, short, late or early delivery. Store data on deviations from purchase order.

(ii) **Tactical information requirements.** Vendor performance index may be evolved using information on excess, short, early and late deliveries. The index must also use data on rejections if any.

(iii) **Strategic information requirements.** We may use vendor performance index as one criterion in deciding on proportion of items to be ordered from different vendors.

Having found the information requirements we ask:

1. What new data elements (if any) are required in the input and output documents, namely, delivery note, discrepancy note and items received note?
2. What data should be maintained by the receiving office in order to create the output documents from the input document?
3. What processing rules are to be applied on the input document to obtain the output documents?

5.2 DATA REQUIREMENTS

In order to find the data elements required in the various documents, we work backwards. In other words, we ask what data elements are required to meet the information needs and find out the data elements required in the input documents and that required to be maintained at the receiving office. The data elements in the output are also determined by the specifications of information requirements. In this example the data elements in the input documents, namely, the delivery note, are:

(i) Delivery note.

- Order number, vendor name and address, item name, delivery date, quantity supplied, units.

Examining this with operational information requirements we see that item delivered is to be checked. The item name is not enough as a spelling mistake in it will make the comparison invalid. Thus we need a unique identification code for items. This must be included in the delivery note. To meet the tactical information need we require data on vendors to be accumulated. Vendor name and address is not suitable as any spelling mistakes or minor change of address will distract a computer-based system. We need a unique vendor identification. Hence vendors must be given a unique code which we will call vendor code. Thus the modified delivery note is:

- Order number, vendor code, vendor name and address, item code, item name, delivery date, quantity supplied, units.

One may ask whether both vendor code, and name and address are necessary. They are used for different purposes. It is necessary to have the name and address of the vendor if a rejected item is to be returned.

(ii) Discrepancy note. Using the same arguments, the modified discrepancy note must contain the following data items:

- Order number, vendor code, vendor name and address, item code, item name, delivery date, quantity supplied, units, excess or deficiency, no. of days late/early.

The receiving process has to create the discrepancy note using the delivery note and a data store containing records of all orders. Individual records of this data store must have the data to create the discrepancy note given the delivery note. The data items in the data store containing the order records are:

Order records: Order number, vendor code, vendor name and address, item code, item name, order date, quantity ordered, units, specified delivery period (in days).

The specified delivery period is necessary to determine late/early delivery. The items received note generated by the receiving process must contain the following data:

(iii) Items received note. Order number, vendor code, vendor name and address, item code, item name, delivery date, quantity supplied, units.

The codes for vendor and item are needed for easy computer processing as explained earlier.

The data elements in each of the documents are listed below:

Delivery note	=	Order no. + vendor code + vendor name + vendor address + item code + item name + delivery date + quantity supplied + units.
(Discrepancy note)	=	Order no. + vendor code + vendor name + vendor address + item code + item name + delivery date + quantity supplied + units + excess/deficiency + no. of days late/early

This can also be written as:

(Discrepancy note) = Delivery note + (excess/deficiency) + (no. of days late/early)

The data element discrepancy note is written within parentheses to indicate that it may be empty, that is, not generated at all. The discrepancy note will not be generated if there is no discrepancy, that is, exact quantity is delivered as per order and on time. In the discrepancy note, one of the elements, namely, excess/deficiency or no. of days late/early, may not be present. The *items received note* has the same data items as the *delivery note.*

Item-received note = Delivery note

Order records = Order no. + vendor code + vendor name + vendor address + item code + item name + order date + quantity ordered + units + delivery period.

After individual data elements are determined, they are consolidated in a *data dictionary.* A data dictionary contains data about data which is called *metadata*. In the data dictionary the data elements are arranged in alphabetical order. Information on each data element is shown in Table 5.1. The data dictionary thus forms an essential part of the documentation of a system. The detailed data dictionary is developed during the system design phase.

Table 5.1 Information Specified in Data Dictionary

Data Name	*Where Used*	*Length*	*Type*	*Range of Value*
		Characters	Numeric, alpha, alphanumeric, bit, etc.	(If applicable)

Having obtained the data elements in the input, outputs and the order data store at the receiving office, the next step is to specify how the input will be *processed* to obtain the outputs. A logical data flow diagram (Figure 5.3) depicts this. *Processing rules* may be specified in English so that a lay person can understand it. The English used is more

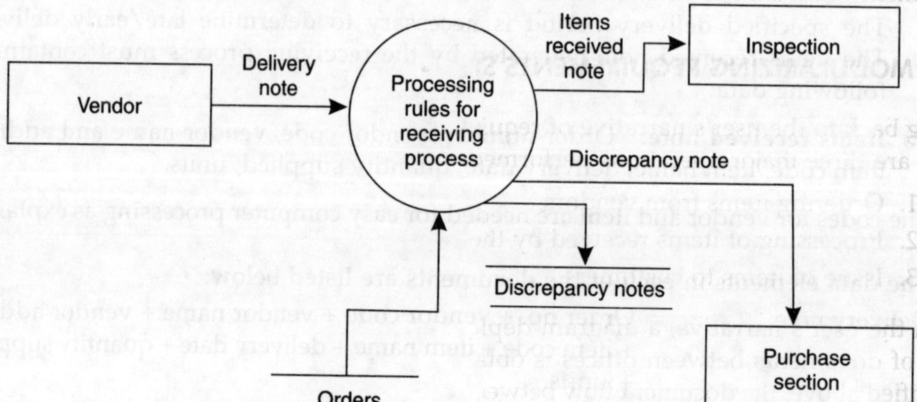

Figure 5.3 Data flow diagram for receiving process.

precise and is known as *Structured English*. Whenever many conditions have to be checked to arrive at actions, a tabular specification tool called *Decision Table* is used. We will illustrate the processing rules for this example in Structured English and as a Decision table in Chapters 8 and 9, respectively.

Thus the logical specification consists of:

1. Data elements in the input and output documents.
2. Data elements in the data store (if any) required for processing inputs to obtain specified outputs.
3. Document and data flow diagrams which graphically depict the flow of data and the places where they get transformed.
4. A statement of the processing rules used to transform the input document to output document (the processing rules are expressed in Structured English or as one or more Decision tables).
5. A data dictionary which consolidates all the data in the documents and data store.
6. A statement (mostly descriptive) of requirements met by the system.

5.3 STEPS IN SYSTEMS ANALYSIS

In Section 5.2, we have seen that given a user's stated requirements, we first study the current system and how documents flow in the system. Using this model we derive a logical equivalent of a system and represent it by a logical data flow diagram. This is refined after finding out from the user his information requirements and a new logical data flow diagram is obtained.

Along with this the processing rules (expressed in Structured English or Decision tables), the data dictionary and information provided by a system are obtained. This design process is shown in Figure 5.4. In Figure 5.4 is also shown a feasibility document as input in modelling the system. In our example we assumed that the system was feasible and we did not impose any constraints. In future studies, however, feasibility document will be an important input.

5.4 MODULARIZING REQUIREMENTS SPECIFICATIONS

Going back to the user's narrative of requirements given in Section 5.1, we observe that there are three major operations performed in the organization. They are:

1. Ordering items from vendors
2. Processing of items received by the company
3. Issue of items to customers

From the user's narrative, a diagram depicting various offices in the company and the flow of documents between offices is obtained. For each of the three major operations identified above, the document flow between offices is depicted in Figures 5.5, 5.6 and 5.7, respectively. Such a division of a large problem into smaller ones is called *modularisation*. We obtain next, after discussion with users, their operational, tactical and strategic information requirements.

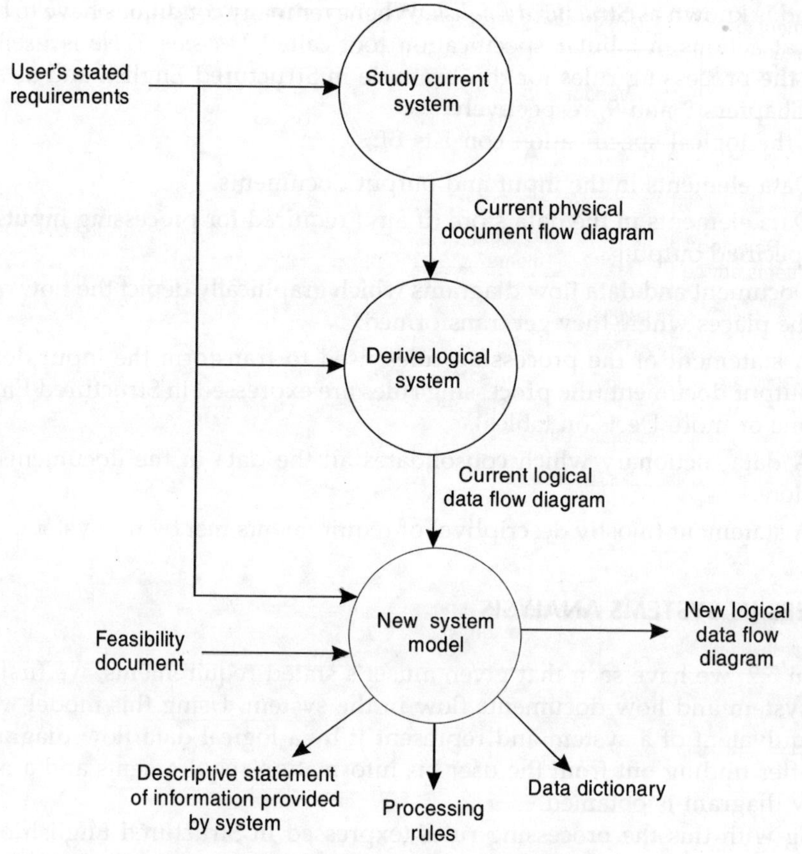

Figure 5.4 Steps in system analysis.

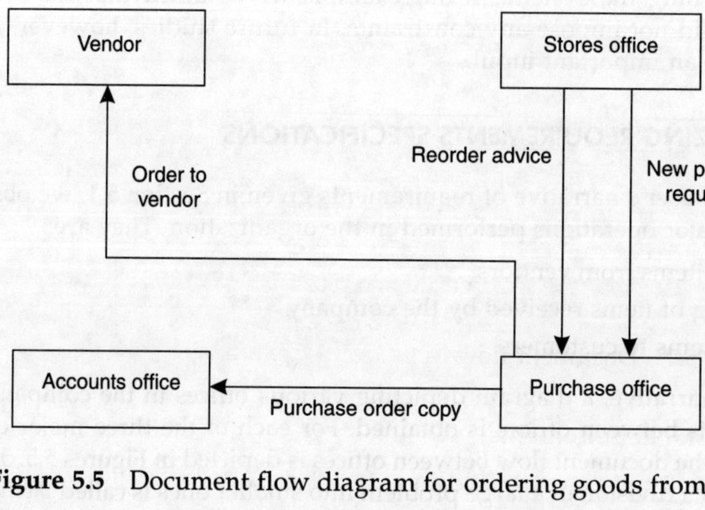

Figure 5.5 Document flow diagram for ordering goods from vendor.

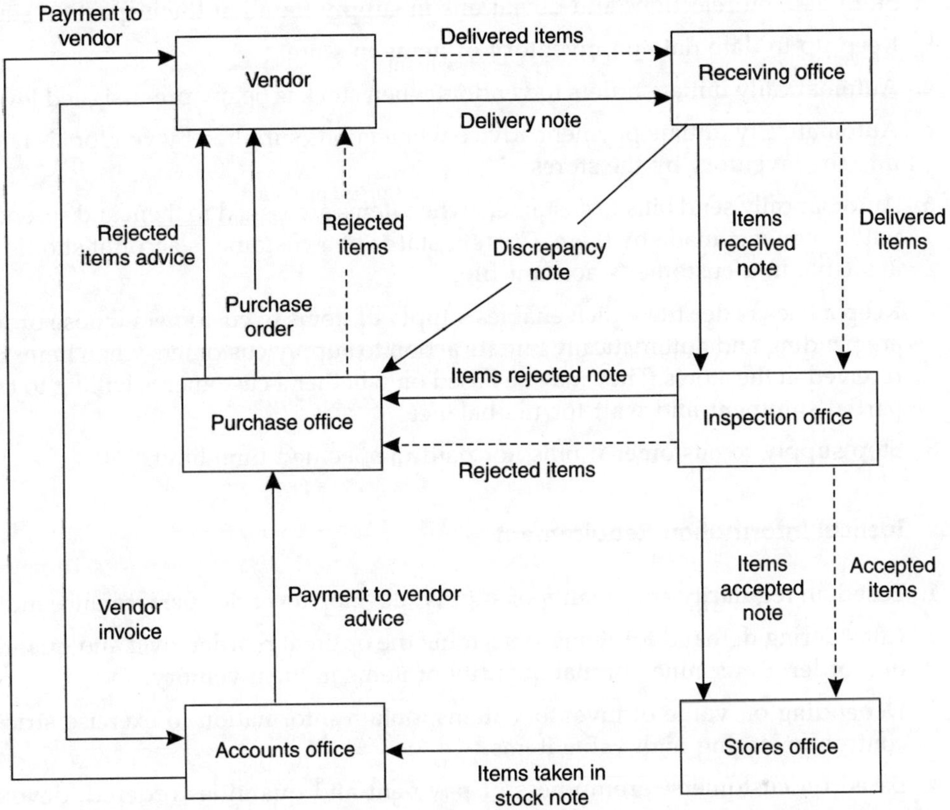

Figure 5.6 Document flow diagram for goods received.

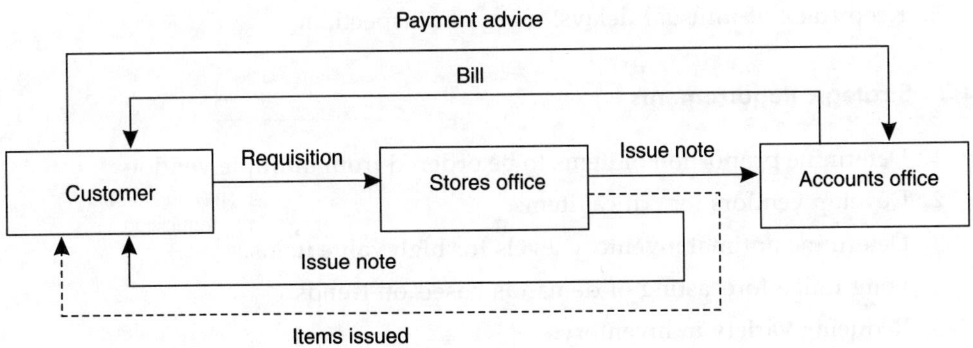

Figure 5.7 Document flow diagram for items issued to customer.

5.4.1 Operational Requirements

1. Automatically check vendors' delivery notes against orders placed with them and initiate action if there is an excess or short delivery or delay or early delivery. Simultaneously store data on the deviations of supplies.

2. Store data on rejections and deviations in supply found at the inspection stage.
3. Keep up to date data on inventory of items in store.
4. Automatically initiate orders to vendors when stock is below pre-assigned levels.
5. Automatically initiate payment advice when goods supplied by vendors is taken into the inventory by the stores.
6. Automatically send bills to customers when items are issued to them and reconcile with payments made by them. Current status of a customer's account should be available in a customer's account file.
7. Keep a back order file which enables supply of items to customers whose orders are pending and automatically initiate action to supply customers when items are received in the stores. They may be based on whether a customer is willing to take partial shipment and wait for the balance.
8. Stop supply to customer if bills not paid in specified time limit.

5.4.2 Tactical Information Requirements

1. Based on regularity and quality of supply, develop a vendor performance index.
2. Considering demand for items, determine the optimal reorder level and quantity of reorder. Determine optimal quantity of items in the inventory.
3. Depending on value of inventory items, obtain information to exercise stricter control on storing high value items.
4. Based on customers' promptness of payment and quantities ordered, develop credit ratings of customers and customer rating.
5. Keep track of large outstanding payments and receipts.
6. Keep track of unusual delays, if any, in inspection.

5.4.3 Strategic Requirements

1. Determine proportion of items to be ordered from multiple vendors.
2. Develop vendors for critical items.
3. Determine optimal inventory levels for high value items.
4. Long range forecasting of demands based on trends.
5. Reducing variety in inventory.
6. Build better customer relations.

Using the information requirements and preliminary document and data flow diagram, we obtain the revised data flow diagram. We show it for the stores process in Figure 5.8. The data elements in stores process are:

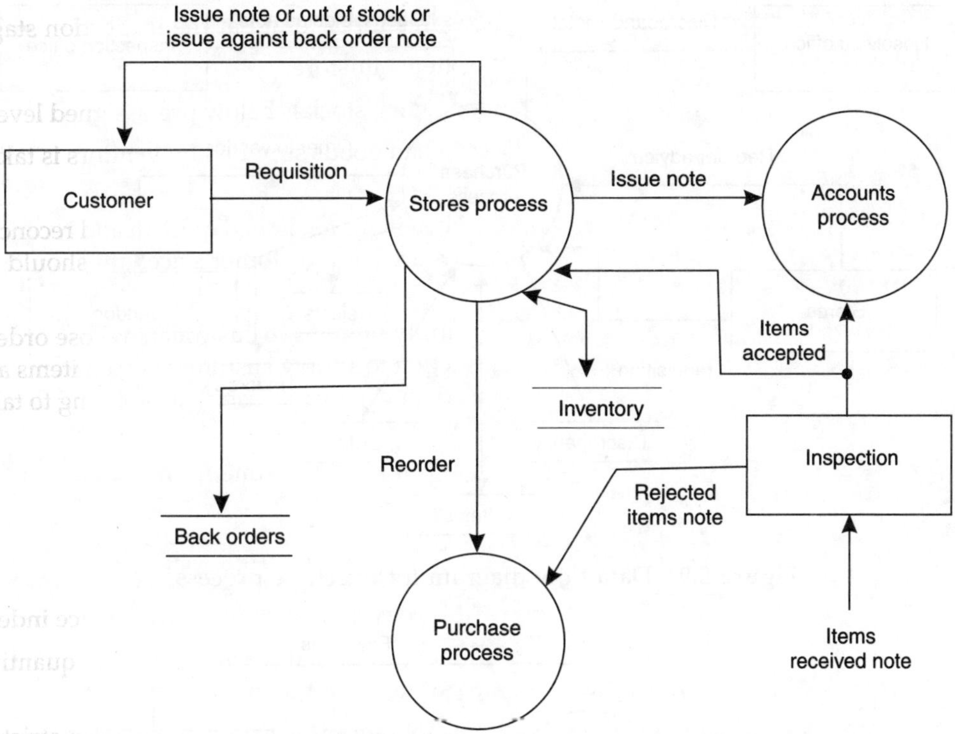

Figure 5.8 Logical data flow diagram for stores process.

Requisition	=	Customer code + customer name + customer address + item code + item name + quantity ordered + units
Items accepted	=	Order no. + vendor code + vendor name + vendor address + item code + item name + quantity accepted
Issue note to accounts	=	Customer code + customer name + customer address + item code + item name + quantity issued + units
Issue note to customer	=	Customer code + customer name + customer address + item code + item name + quantity issued or back order or out of stock + units
Reorder	=	Item code + reorder quantity + units
Inventory record	=	Item code + item name + quantity in stock + units + reorder level + reorder quantity
Back order record	=	Customer code + items code + quantity to be supplied + units

To complete this example, we give the data flow diagrams for the other two modules, namely, the purchase process and the accounts process as Figures 5.9 and 5.10, respectively.

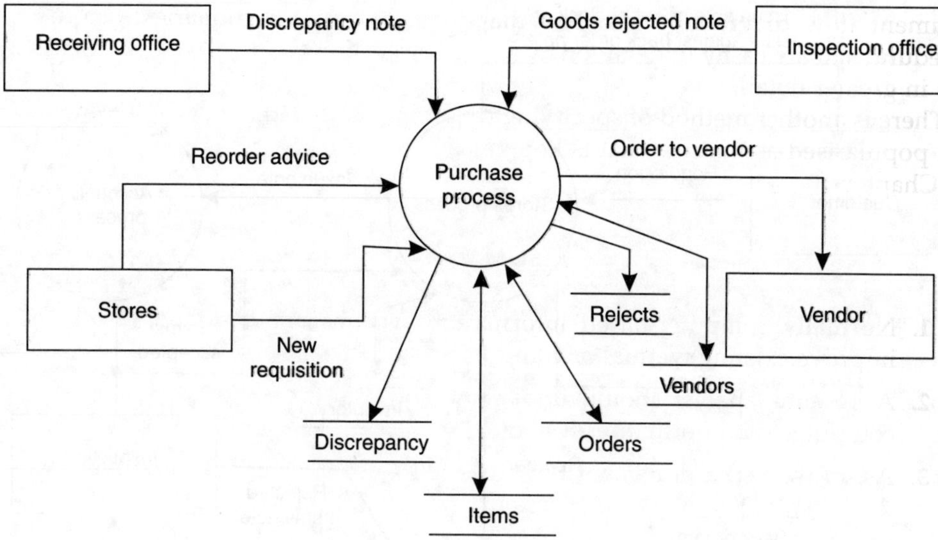

Figure 5.9 Data flow diagram for purchase process.

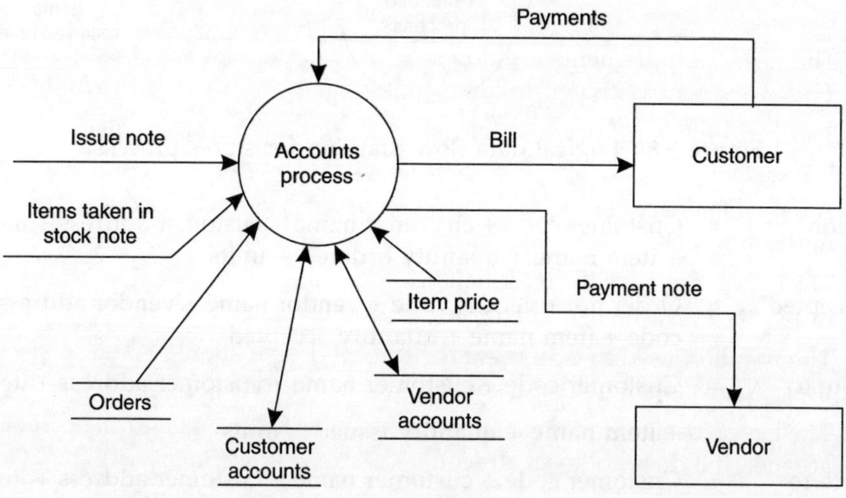

Figure 5.10 Data flow diagram for accounts process.

5.5 CONCLUSIONS

We saw in this chapter that before designing an information system, it is essential to understand user's requirements. To find out if the systems analyst understood the user's requirements correctly, it is necessary to prepare a document which specifies the user's requirements. This specification document must use a notation which any non-specialist user (i.e., a person who does not know much about computers) can understand. Data flow diagrams, structured English and decision tables constitute such a specification tool.

Document flow diagrams, data flow diagrams and data dictionaries also are non-procedural and are easily understood by a lay user. In later chapters we will discuss these tools in greater detail.

There is another method of specifying requirements called *use case method* which has been popularised as part of what is known as object oriented analysis. We will discuss it is Chapter 10.

SUMMARY

1. Normally computer-based information systems are designed to replace and improve existing manual or computer-based systems.
2. A systems analyst should understand what users expect from the proposed computerized information system.
3. As a first step a document flow diagram is used as a tool to understand what documents flow between offices in an organization.
4. This is refined and expanded into a data flow diagram. A data flow diagram depicts the logical structure of data flowing in the system and the processing rules which transform data. The refinement and expansion is carried out such that all the stated requirements of users are met.
5. The users' requirements are documented as operational, tactical, and strategic. This document is shown to appropriate users and their approval is obtained.
6. The processing rules are specified either in structured English or as decision tables.
7. A data dictionary is used to document information about all data records flowing in the system.
8. The logical design of the information system is complete when the tasks mentioned in points 3-6 in this summary are accomplished.
9. The notation used to document the logical design should be understood by any user. A user need not be a specialist in computers.
10. The logical design can be examined by the user to see whether all his requirements are met. If it does not meet all requirements, it should be modified by the analyst.

EXERCISES

5.1 What are the initial two steps followed by a systems analyst when he is requested to design an information system?

5.2 What graphical tools are useful to a systems analyst to understand user's requirements?

5.3 A hostel warden states the following requirements for a hostel information system: "Our hostel has 500 rooms and 4 messes. Currently, there are 1000 students in all in 2 seated rooms. They eat in any one of the messes but can get a rebate if they inform and do not eat for at least 4 consecutive days. Besides normal menu, extra items are also given to students when they ask for it. Such extras are entered in an *extras book*. At the end of the month a bill is prepared based on the normal daily rate and extras and given to each student. We find that bill preparation is delayed. We are also not able to keep proper track of payments and billing for extras. We need a system to streamline this".

Obtain a document flow diagram for the problem described above.

5.4 What are the operational, tactical and strategic information needed for Exercise 5.3?

5.5 Using your answers to Exercises 5.3 and 5.4, obtain a data flow diagram for meeting the information needs of the hostel. Obtain the data dictionary.

5.6 Develop the data dictionary for the stores process expressed as a data flow diagram in Figure 5.8.

5.7 Develop the data dictionary for the accounts process whose data flow diagram is given as Figure 5.10. The accounts process needs data on prices for items purchased from vendors and charged to customers. This should be included in your data dictionary at appropriate places.

5.8 Develop the data dictionary for the purchase process whose data flow diagram is given as Figure 5.9.

6

Feasibility Analysis

LEARNING GOALS

After reading this chapter a student should be able to:

1. Formulate goals of a project and quantify them.
2. Obtain alternative solutions to satisfy the goals.
3. Examine the feasibility of implementing each of the alternative solutions.
4. Evaluate cost vs benefit of each feasible solution.
5. Prepare a system proposal.

At the end of the information gathering phase, we have a reasonable idea about the data available currently and the deficiencies of the current system. We also reach a consensus on requirements and priorities among these requirements. Using these data, it is possible to define broad goals for the project and detailed sub-goals. It is necessary at this stage to quantify the goals and sub-goals. For instance, a goal "Prepare student mess bills within 5 days of the end of the month" is well quantified, whereas a goal "Ensure prompt payment to vendors" is not quantified. If we say "Ensure that payment is made to vendors within 10 days of their presenting their bill", it is quantified. Once these goals are quantified, the next step is to find out whether these goals can be met, and if yes, how they will be met and at what cost. Feasibility analysis is mainly concerned with these questions. The steps to be followed before feasibility analysis are given in Figure 6.1.

At the end of feasibility analysis, an analyst is able to find an appropriate solution and its cost-benefit. Given the cost-benefit the management of the organization would request the analyst to prepare a detailed system proposal, if they are satisfied with the feasibility analysis. At the end of this chapter we describe how a system proposal is prepared.

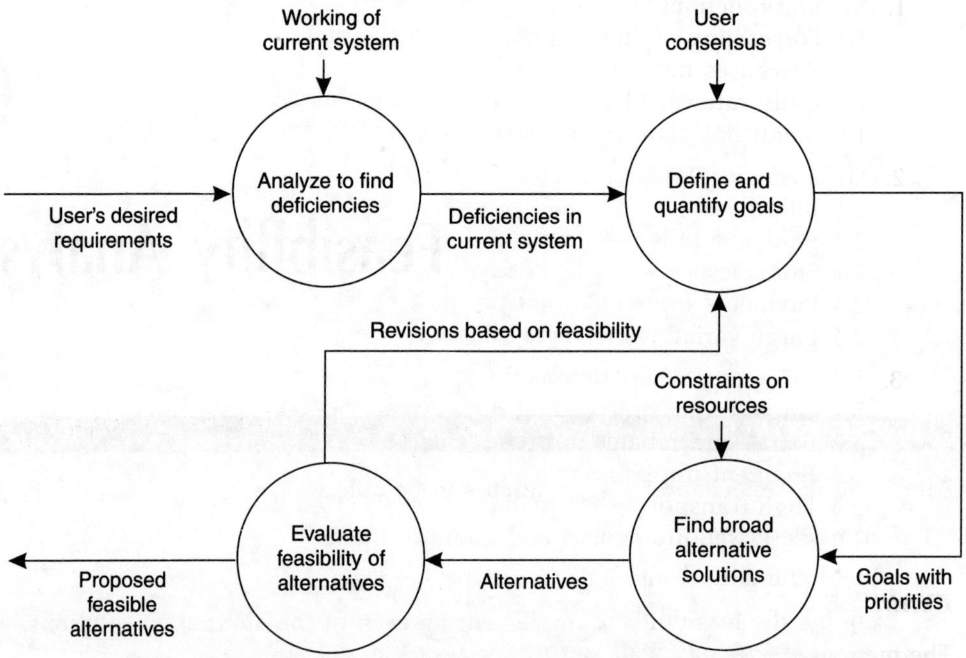

Figure 6.1 Steps in feasibility analysis.

6.1 DECIDING ON PROJECT GOALS

Deciding on the goals of a project, quantifying them and obtaining consensus of users is an important task of a system analyst. The important guidelines for arriving at goals are:

1. Identify deficiency by pinpointing
   - missing functions,
   - unsatisfactory performance, and
   - excessive cost of operations.
2. Set goals to remove these deficiencies
3. Goals must be
   - Quantified,
   - realizable within the constraints of an organization,
   - broken down into sub-goals, and
   - agreeable to all concerned.
4. Set goals not only to remove deficiencies but also to effectively meet competition. For instance, goals may be based on what competitors do.

EXAMPLE 6.1.1

In the case study given in Chapter 4 (Section 4.7 and Table 4.1), the deficiencies identified in the current manual system are as follows:

1. Missing functions.
   1.1 Forecasting of stores requirements not done.
   1.2 Purchases not consolidated.
   1.3 Daily rate calculation not updated frequently.
   1.4 Menu not planned for balanced nutrition and low cost.
2. Unsatisfactory performance.
   2.1 Billing not accurate and prompt.
   2.2 Students' bills not itemized.
   2.3 Stores issues to cooks not systematic.
   2.4 Payments to vendors not prompt.
   2.5 Large variations in mess bills every month.
3. Excessive cost of operations due to the following:
   3.1 Unpaid long outstanding dues from students.
   3.2 Extras and rebates not reflected in stores issues.
   3.3 Frequent unplanned visits to city to buy out-of-stock items.
   3.4 High transport cost (small purchases at high cost) due to not consolidating stores requirements.

EXAMPLE 6.1.2

The main goal and sub-goals to remove deficiencies 1.3, 2.1, 2.2, 2.5, 3.1 (mentioned in Example 6.1.1) are given below. (Deficiencies removed shown in parantheses)

Main goal

M1 Send the bills to students within 5 days of the end of the month. (2.1, 3.1)

Sub-goals

S1.1 Itemize bills indicating extras and rebates with dates. (2.2)

S1.2 Ensure less than 5% variation of bills from month to month. (2.5)

S1.3 Bring to the attention of the Chief Warden the bills which are not paid within 10 days of issue. (3.1)

S1.4 Update daily rate. Calculate bills every day. (1.3, 2.5)

The main goals and sub-goals to remove deficiencies 1.1, 1.2, 2.3, 2.4 and 3.2 to 3.4 (mentioned in Example 6.1.1) are given below:

Main goal

M2 Control inventory of items and issues to cooks so as to bring down mess bill by 10%. (1.1, 1.2)

Sub-goals

S2.1 Ensure payment to vendors within five days of supply of items. (2.4)

S2.2 Make sure that at most four trips per month are made to the town for purchase and consequently transport cost is less than 1% of item cost. (3.3)

S2.3 See that inventory levels are not more than 10% of monthly requirements. (3.4)

S2.4 Make sure that cooks do not get more than 5% of calculated requirements. (2.3, 3.2)

Goals to add function 1.4.

Main goals

M3.1 Balanced menu to meet nutritional requirements.
M3.2 Cost of food not to exceed current cost.

Sub-goal

S3.1 Randomize menu.

6.2 EXAMINING ALTERNATIVE SOLUTIONS

During the feasibility phase, broad alternative solutions to problems are examined. It is not necessary to get into details at this stage. For each alternative solution the cost and the benefits have to be examined before deciding one of the alternatives.

Broad solutions will consist of:

1. specification of information to be made available by the system,
2. description of what will be done manually and what will be done by the computer, and
3. specification of new computing equipment needed or specification of expansion of an existing computer.

EXAMPLE 6.2

For the case study considered, we will examine three alternative solutions:

Solution A : Improved manual system.
Solution B : A PC-based periodic update system.
Solution C : An on-line transaction oriented system with a server and clients.

We will consider each solution in turn.

Solution A

One can improve the existing manual system by carrying out the following:

- Keeping the up to date running total of extras and rebates of each student. This is done by the assistant mess manager between lunch and tea each day.
- Keeping a quick reference table of requirements of each item for a specified number of members expected to take food on a day.
- Costing each day's issues and keeping a running total of expenses. This will enable quick preparation of mess bills.
- Calculating *standard* values of stores needs by averaging issues of a few months. These standard values may be used for consolidating orders to vendors.
- Keeping duplicates of bills issued to students in a box and transferring paid bills to a "paid" box.
- The attention of the Chief Warden can be drawn to the bills which are unpaid after 10 days.

This modification would meet main goals M1, M2 and sub-goals S1.3, S2.2, S2.3 and S2.4 (see Section 6.1). It will not satisfy other goals. The extra cost will be negligible.

The solution is rejected as it does not meet the important goals M3.1, M3.2 and sub-goals S1.1, S1.2, S2.1, S3.1.

Solution B

The subsystems involved in this solution are now described.

(i) System for Student Billing

(i) *Billing subsystem.* An extras/rebates register will be maintained in the mess for students to enter the requests. One page per student is assigned. The register entries are codified for the use of computer processing. The blank register pages are printed on the computer each month and stapled together and kept in the mess for entries by students in appropriate columns.

Entries in the register are keyed into the PC each day by the mess clerk. A program will check whether sufficient notice has been given for rebates, whether number of days rebate asked is as per rules. The data entered is used by a program to create a summary statement giving expected number of meals to be prepared 48 hours from now. This is passed on to the mess manager to control issues. This program also updates each student's record with details of extras and dates of absence.

(ii) *Subsystems in billing system.* Subsystem for finding meals to be cooked.

INPUT: Extras/rebates requests of each student each day.

PROCEDURE

- Check each input record to see if rebate is admissible and extras are correct.
- Enter admissible rebates and extras in student billing file.
- Find number of meals to be cooked two days from now. Print note to mess manager.

OUTPUT: Note to mess manager giving number of members and meals to be cooked 48 hours from now.

(This subsystem runs daily on the PC)

(iii) *Subsystem for mess bill calculation.*

INPUT: Items issued each day (perishable and non-perishable)

PROCEDURE:

- From items issued each day find total expenses each day.
- Accumulate daily expense and add fixed overhead (salaries, etc.)
- Using number of members calculate up to date daily rate.
- Using student billing information file and expense file, calculate the itemized bill for each student at the end of the month.
- Enter in student billing file.

OUTPUT: Itemized bills to students.

(Updating daily rate is run each day whereas student billing system is run once a month).

(iv) Subsystem for unpaid bills

INPUT: Payments information.

PROCEDURE

- Enter all payments in student billing file.
- Five days after last date for payment, examine this file and create a list of students with unpaid bills.

OUTPUT: List of overdue bills sent to Chief Warden
(This system runs once a month)

The billing system described above satisfies main goal M1 and sub-goals S1.1 and S1.3. To satisfy sub-goal S1.2, the mess secretary who gets an updated daily rate each day must adjust menu, using a menu planning system.

(ii) System for stores issue and control

Subsystem for stores issues to cooks

INPUT

- Menu for (today's date + 2) i.e., if today's date is 20.04.09 menu is given for 22.04.09.
- Number of meal to be cooked for (today's date + 2).

PROCEDURE: Using inputs find items to be issued on (today's date + 2).

OUTPUT: Items to be issued on (today's date + 2) sent to mess manager.

Subsystem for inventory control

INPUT: Daily use of stores items, perishables and vendor supplies.

PROCEDURE

- Based on past month's total consumption of each item, estimate consumption of current month. Add 2 days consumption as buffer.
- Work out minimum level (= 4 days needs) of all critical items.
- Decide on order quantities, and place orders.
- When goods come in, check in order file and update inventory if items are accepted.
- Send the cheques to vendors on item acceptance.
- Update stores inventory after each issue.
- When critical items go below minimum level inform mess manager.

OUTPUTS

- Orders to vendors
- Cheques to vendors
- Low item stock message to mess manager

(This system is run once a day)
This system satisfies main goal M2 and sub-goals S2.1 to S2.4 (see Section 6.1).

(iii) System for menu planning

INPUT

- Number of items in menu
- Alternative for each item
- Cost of each item
- Maximum cost.

PROCEDURE

- Reduce total cost by trying alternative items in menu.
- Try an optimization package such as Linear Programming.
- Rank menus by cost and keep 14 different menus for each meal.
- Randomize menus each week.

OUTPUT: Balanced menu at reduced cost.
(This system is run once a month)

This system satisfied main goals M3.1, M3.2 and sub-goal S3.1 (see Section 6.1).

Equipment needed: File size estimate

Items sorted

Non-perishable: 500

Perishable: 100

Record size per item – 200 bytes

Storage needed = 600 * 200 = 1,20,000 bytes

Inventory file size = 120 KB

Student file

No. of students = 400

Record size for student = 200 bytes

Storage for students' file = 80 Kbytes

Vendor file

No. of vendors = 20

Record size per vendor = 400 bytes

Storage for vendor file = 8 Kbytes

Storage needed = 118 Kbytes

Processing is not extensive and database is small.

One PC with 20 Mbytes hard disk, 1 floppy drive, a CD drive and a printer (80 column dot matrix) would be more than adequate. Even though this configuration is sufficient such a computer cannot be bought today (2011). The minimum configuration now is 400 GB disk, 1 GB main memory and a 32 MB pen drive. It is advisable to have a writable CD ROM for backing up of files. We have to settle for this configuration. In future years the capacity of computers is bound to increase but the cost will remain almost same; around ₹50,000 with printer and software such as Windows 7O.S., open office and anti-virus software.

Solution C

This solution will require entering data online to a server from the mess, stores and accounts offices as and when a transaction such as stores issue, money receipt, and extras request takes place. The advantage is up to the minute information available on disk. As the number of transactions is small and online query answering is not needed, such a solution is not appropriate. It will also cost more as a server and 3 clients PCs are required. The operating system on server should handle multiple transactions simultaneously.

6.3 EVALUATING PROPOSED SOLUTION

After looking at broad alternative solutions, a short list of solutions is kept. These solutions are further evaluated to find out the following:

- Their *technical feasibility*. By this we mean whether the technology needed is available and if available whether it is useable.
- Their *operational feasibility*. To find this, we ask whether the proposed solution can fit in with existing operations and whether the right information at the right time is provided to users.
- Their *economic feasibility*. To find this we ask whether finances are available for implementing the proposed solution and whether the money spent is recovered by the savings or by better user satisfaction.

EXAMPLE 6.3

In Example 6.2, Solution A was rejected as it is not operationally feasible to meet the proposed new goals. Solution C is rejected as it is an expensive solution which is not appropriate for the problem. Thus in this case only solution B is appropriate.

6.4 COST-BENEFIT ANALYSIS

A cost-benefit analysis is necessary to determine economic feasibility. The primary objective of cost-benefit analysis is to find out whether it is economically worthwhile to invest in the project. If the returns on the investment is good, then the project is considered economically worthwhile.

Cost-benefit analysis is performed by first listing all the costs associated with the project. Costs consist of both direct costs and indirect costs. Direct costs are those incurred in buying equipment, employing people, cost of consumable items, rent for accommodation, etc. Indirect costs include those involving time spent by user in discussing problems with system analysts, gathering data about problem, etc. Details of direct costs are:

1. Cost of computer, peripherals and software. It could be either a capital cost for buying a computer or the cost of renting one.
2. Cost of space such as rent, furniture, etc. In a place like Bombay the cost of space occupied by a system analyst (5 sq. metres) in prime location could be ₹10,000 per month!

3. Cost of systems analysts and programmers (salary during the period of assignment).
4. Cost of materials such as stationery, CD ROMs, toner, ribbon, etc.
5. Cost of designing and printing new forms, user manuals, documentation, etc.
6. Cost of secretarial services, travel, telephone, etc. An estimate is sometimes made of indirect cost if it is very high and added to the direct cost.
7. Cost of training analysts and users.

Benefits can be broadly classified as *tangible benefits* and *intangible benefits.* Tangible benefits are directly measurable. These are:

1. Direct savings made due to reducing (a) inventories, (b) delays in collecting outstanding payments, (c) wastage, (d) cost of production, and increasing production, as also its speed.
2. Savings due to reduction in human resources or increasing volume of work with the same human resources.

Intangible benefits are:

1. Better service to customers
2. Superior quality of products
3. Accurate, reliable and up to date strategic, tactical and operational information which ensures better management and thereby more profits.

The sum of all costs (direct and indirect) is compared with the sum of all savings (tangible and intangible). It is not always easy to assign money value to intangible benefits. It is arrived at by discussion amongst users of the information system.

If the project is a high cost one, extending over a period of time, then it is necessary to estimate costs during various phases of development of the system so that they can be budgeted by the management.

EXAMPLE 6.4

Consider Solution B of the hostel mess management problem. The direct costs are:

1. Cost of PC, printer, uninterrupted power supply = ₹70,000.
2. Cost of space (nil). No extra space allocated.
3. Cost of Systems analysts/Programmers/Consultants for 3 months = ₹60,000.
4. Recurring costs: Stationery cost, CD ROM cost, Maintenance/Electricity = ₹2000 per month.
5. One time cost = ₹1,30,000.

Benefits (Direct savings)

1. Savings per month due to inventory reduction and wastage
   = 5% of mess bill of 400 students
   = 2.25 * 400 * 30
   = ₹27,000
   (Assume ₹45 bill per day per student. Savings per day is 2.25 per student. 30 days in a month).

2. Savings in transport cost (estimate) ₹800 per month.
3. Savings due to early payment to vendors
   = 1.2% of total billing to vendors
   = 37.5 * 400 * 30 * 0.012
   = ₹5400 per month
   (₹37.50 per day is assumed to be material cost in mess bill)
4. Savings due to better collection (40 defaulting students, 1% interest per month)
   = 40 * 1350 * 0.01
   = ₹540 per month

Intangible benefits

1. Student satisfaction due to itemized bills and less variation
2. Better menu planning

Total benefits = ₹33,740 per month
Recurring cost = ₹2000 per month
Net benefit per month = ₹31,740
Total capital cost = ₹1,30,000

6.5 PAYBACK PERIOD

Simple Payback Method

Once the cost of the project and the benefits have been quantified, the next step is to find out whether the benefits justify the cost. There are two ways of finding out this. They are known as the *payback method* and the *present value method*. The payback method is used to find out in how many years the money spent is recovered as benefits. In the example considered, we found that the cost was ₹1,30,000 and the benefits ₹31,740 per month. Thus in 4.1 months we recover ₹1,30,134 which exceeds the cost. The payback is thus 4.1 months.

Payback Method with Interest

The payback method is a simple method of computing benefits. In this method we have not taken into account the fact that if ₹1,30,000 is invested in a bank, we will get interest. Hence the benefit should be reduced by the amount of interest which would be earned. If an interest rate of 1.5% per month is used, then the interest per month on ₹ 1,30,000 will be ₹1950. Therefore, net monthly benefit in the example being considered is ₹29,790. Thus the payback period for a payback method with interest becomes nearly 4.36 months. Even this method is not entirely correct as benefits themselves would earn interest.

Present Value Method

The correct application of the present value method is to ask the question: What is the present value of earnings which may accrue after n years? Thus if $r\%$ is the interest rate per annum, the present value of earnings x accruing in the nth year is

$$\text{Present value} = x/(1 + r/100)^n$$

In Table 6.1a, we show the present value of benefits for the mess management system. The cumulative benefits are also shown. Thus we see that with 1.5% interest rate, per month, the cumulative benefit at the end of 5 months exceeds the cost. Table 6.1b is another example with net benefits varying every year.

Table 6.1a Cost-benefit (Present value method) for the Mess Management System

Month	*Cost (₹)*	*Net Benefit*	*Present Value of Benefit*	*Cumulative Present Value*
0	1,30,000	–	–	–
1		31,740	31,271	31,271
2		31,740	30,809	62,080
3		31,740	30,354	92,434
4		31,740	29,905	1,22,339
5		31,740	29,463	1,51,802

Table 6.1b Cost-benefit (Present value method) for the Mess Management System

Month	*Cost (₹)*	*Net Benefit*	*Present Value of Benefit – 8% interest*	*Cumulative Present Value*
0	1,50,000	–	–	–
1		40,000	33,898	33,898
2		60,000	43,091	76,989
3		80,000	48,691	1,25,680
4		70,000	36,105	1,61,785

6.6 FEASIBILITY REPORT

At the end of feasibility study, it is necessary to prepare a feasibility report. This report is meant for the management. The primary objectives of this report are to inform the management about the following matters:

1. What the proposed system will achieve
2. Who will be involved in operating it in the organization
3. The organizational changes needed for its successful implementation
4. The benefits the system will give
5. The estimated cost of the system.

These are given in a simple non-technical language at the beginning of the report as an "Executive Summary". It is then followed by a detailed System Proposal if the management approves the feasibility report and requests a detailed proposal. The detailed proposal will contain the following items:

1. An introduction with an outline of proposal
2. A broad data flow diagram of the existing system
3. A modified data flow diagram of the proposed system
4. Alternative solutions
5. Any new equipment to be installed
6. Expected benefits
7. Cost of the system
8. New procedures to be implemented and involvement of persons in operating the new system
9. Any anticipated problem in implementation
10. Plan for implementing the system. This will include time schedules for prototyping and final implementation.

We will give as Example 6.5 Executive Summary based on the case study considered.

EXAMPLE 6.5

Executive summary

Information system for a hostel mess. An information system for efficient operation of the mess will be designed, and will have the following features:

- It will bring down mess bills by around 10%
- It will ensure that inventory levels are within 10% of requirements
- Itemized bills will be sent to students within 5 days of the end of the month
- Monthly variation of mess bills will not be more than 5%
- It will bring to the attention of the Chief Warden bills which are overdue by more than 10 days
- It will give a more balanced variety of menus.

It is proposed to install a Personal Computer (IBM compatible PC) in the mess supervisor's office. This will be used to bill students including extras/rebates, monitor payment of bills by students, control inventory, ensure prompt payment to vendors and plan menus. The requisite data is already available in ledgers and will be entered into the computer's disk storage by the existing personnel who will be trained to do this. No major organizational changes are required to implement the system. The cost of a new computer with requisite peripherals will be ₹ 1,30,000 and a recurring cost of ₹ 2000 per month is estimated. The benefits will be (a) reduce inventory cost by ₹ 27,000 per month, (b) a saving of ₹ 540 per month due to better collection of bills, and (c) ₹ 6200 per month due to faster billing/receipts and supply consolidation.

Other unquantified benefits are: better student satisfaction and better menu planning. The direct benefits will repay the cost of installing a computer within the first 5 months of operation.

Based on this executive summary we assume that the hostel management requests a System Proposal. The System Proposal follows.

6.7 SYSTEM PROPOSAL

Introduction

On a request from the Chief Warden of the student hostel, a study was initiated to find out how the operation of the hostel could be improved by using a better information system. After lengthy discussion with various personnel in the hostel, it was concluded that the problem which had to be tackled on a high priority basis was a better information system for billing, accounting, inventory control and stores issues in the students' mess. The deficiencies identified in the current system are as given in Example 6.1.1. These deficiencies are to be removed in the proposed information system.

Goals of the proposed system. The main goals and the sub-goals of the proposed system are as given in Example 6.1.2.

Existing document flow in the hostel mess. The existing organizational structure and data/document flow in the hostel mess is as given in Figures 4.1–4.4.

Modified data flow for proposed system. The data flow diagram for the proposed system is as given in Figures 6.2 and 6.3.

Alternative solutions examined. Three alternative solutions were examined. These are given in Example 6.2.

Selected solution. Solutions B and C are technically and operationally feasible. The cost-benefit of Solution B is as in Example 6.4. Solution C costs around 3 times solution B with no commensurate increase in benefits. It is thus rejected. The final system to be implemented is Solution B.

New equipment to be installed. PC and printer with the following features:

Hardware

IBM compatible PC
1 GB main memory
1 GB pen drive
400 GB hard disk
15" colour monitor
Read/write CD ROM
132 character 80 cps dot matrix printer

Software

Windows 7 Operating System
Open office
C and Visual BASIC
Anti-virus

Infrastructure

500 VA, Uninterrupted Power Supply with 30 minutes backup battery.
Need a dust-free room to locate PC

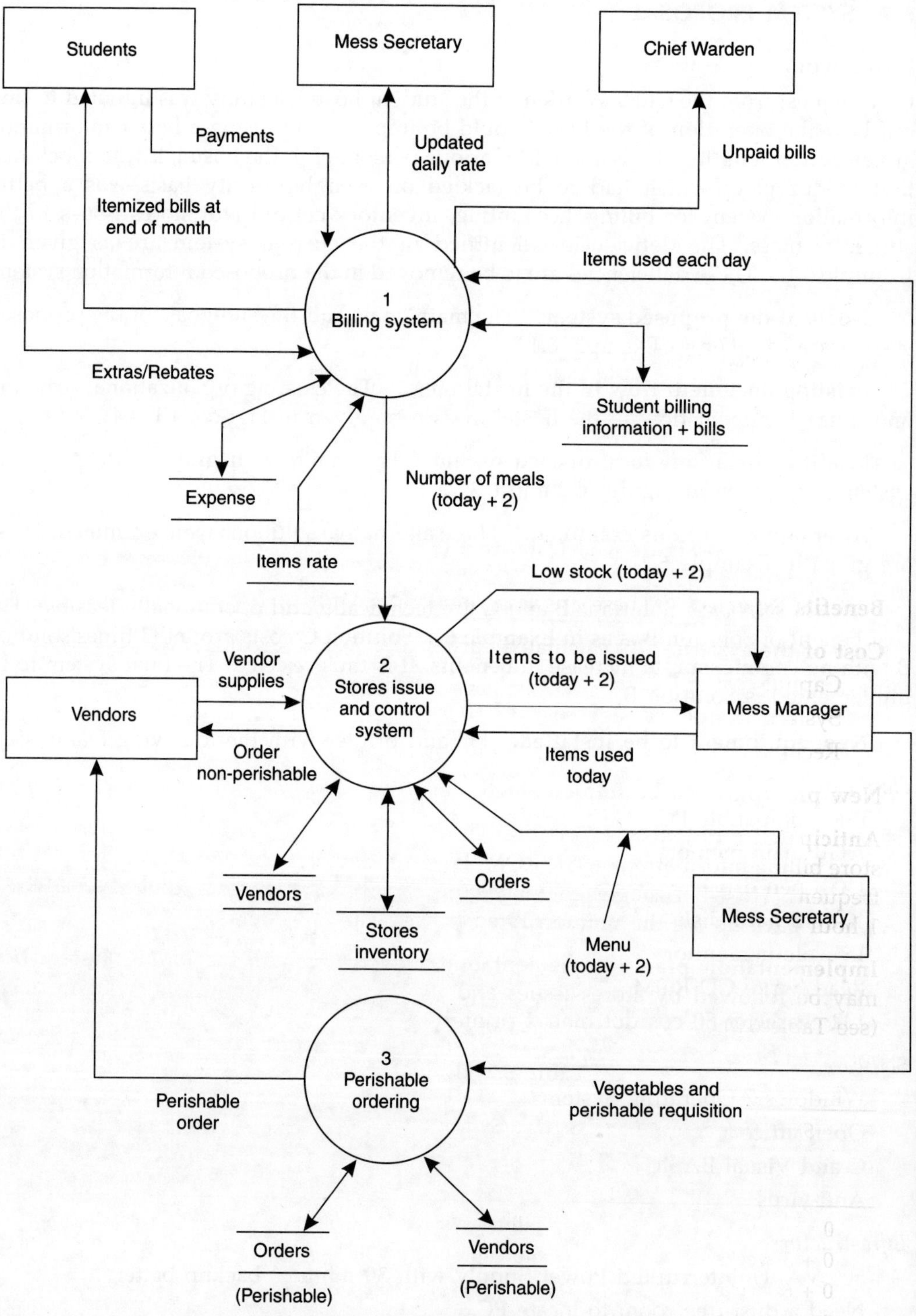

Figure 6.2 Data flow diagram for a hostel mess.

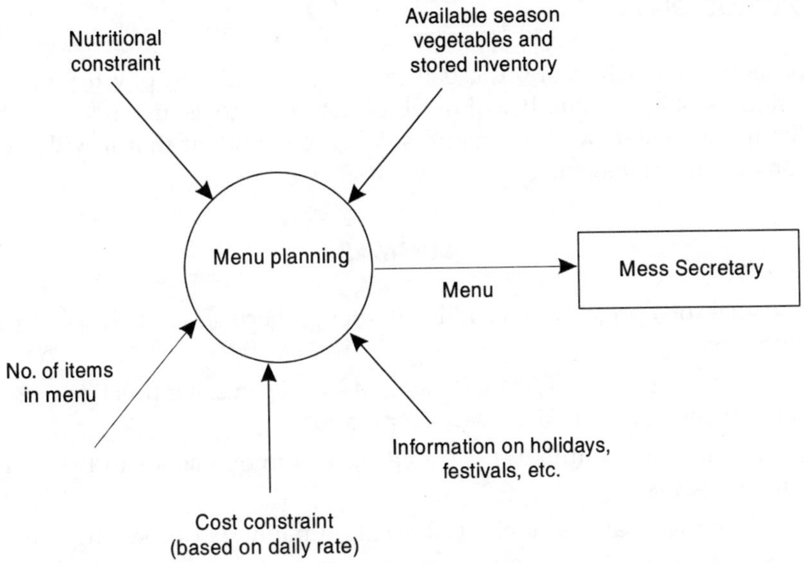

Figure 6.3 Data flow diagram for menu planning.

Benefits expected. (These are as given in Example 6.4)

Cost of the system. As calculated earlier, the cost of the new system will be

Capital cost = ₹ 70,000
System analysis/design cost = ₹ 60,000
Recurring cost = ₹ 2000 per month

New procedures to be implemented. See Solution B of Example 6.2.

Anticipated problems. Procedures should be set up to regularly back up all files and store billing information on CD ROM. Power failure may lead to difficulties, if it is very frequent. Hostel management may consider installing higher capacity batteries for 1 hour back up for the uninterrupted power supply system.

Implementation plan. Implementation may begin with the billing system first. It may be followed by stores issues and inventory control and finally menu planning (see Table 6.2).

Table 6.2 Implementation Plan

Start Time	*System*	*Time Schedule Design, Implement and Test (man weeks)*
0	Billing system	3
0 + 3 weeks	Stores issue system	3
0 + 6 weeks	Inventory system	3
0 + 9 weeks	Menu planning	3

6.8 CONCLUSIONS

The proposed system will cost ₹ 1,30,000 capital cost and will pay for itself within the first five months of operation. It will meet all the main goals approved by the wardens and can be implemented within 3 months. We are confident that it will fully meet the expectations of the management.

SUMMARY

1. The goals of a project are formulated after gathering facts and having discussions with users.
2. Goals are set by identifying deficiencies such as missing functions, unsatisfactory performance or excessive cost of operation.
3. Goals must be quantified, realizable, agreed to by users and then broken down into sub-goals.
4. After setting goals we look at alternative methods of meeting these goals.
5. Each alternative method is broadly formulated by specifying the inputs, outputs and procedures to be used. Both manual and computer-based procedures are described.
6. How each method will meet the goal is examined.
7. Next the feasibility of implementing each method is examined.
8. A method must be technically, operationally and economically feasible.
9. Technical feasibility examines whether technology is available to carry out the project.
10. Operational feasibility examines whether the proposed method can fit in with existing operations or is easy to operate with a modified system.
11. Economic feasibility examines whether the returns (in terms of benefits from the project) are commensurate with the cost of the project.
12. It is necessary to examine both tangible and intangible benefits of each alternative solution before picking a solution.
13. At the end of feasibility analysis, a document consisting of an executive summary on the feasibility of the project is submitted to the management which requisitioned the study.
14. When management approves the feasibility of the project, a detailed System Proposal is submitted.

EXERCISES

6.1 When are the goals of a project specified?

6.2 How are goals determined?

6.3 What should be the characteristics of goals?

6.4 What is the difference between main goals and sub-goals? Illustrate with an example.

6.5 Once the goals are formulated, what is the next step followed by a system analyst?

6.6 Is it essential to use computers in all information systems?

6.7 Distinguish between technical, operational and economic feasibility.

6.8 What do you understand by the term *feasibility study of a solution?*

6.9 Give an example of a solution which is technically feasible, but not operationally feasible.

6.10 Give an example of a technically feasible solution which is not economically feasible.

6.11 Give an example of an operationally feasible solution which is not economically feasible.

6.12 Is it essential that an operationally feasible solution should be technically feasible? Discuss with examples.

6.13 What is the difference between tangible and intangible benefits?

6.14 Give examples of tangible and intangible benefits.

6.15 For the problem of periodical information system (Exercise 6.23), specify tangible and intangible benefits.

6.16 What is meant by cost-benefit analysis?

6.17 A system costs ₹1 lakh to install and ₹10,000 per month as recurring expenses. The benefit per year is ₹1.5 lakhs. Assuming an interest rate of 12%, what is the payback period of the investment?

6.18 How does inflation affect payback period?

6.19 Is it essential to have tangible benefits to justify an information system? If your answer is no, justify your answer by giving an example.

6.20 A project costs ₹2 lakhs and the net benefits are ₹50,000 (1st year), ₹80,000 (2nd year), ₹90,000 (3rd year), ₹70,000 (4th year), ₹50,000 (5th year), and ₹30,000 (6th year). Assuming 10% interest rate, would you proceed with this project if your criterion is cost-benefit?

6.21 A manager states the following as the goals of a production planning system:

1. Reduce stock of semi-finished products.
2. Provide better information for the production planning.
3. Prevent overproduction.

How would you quantify the goals? How would you obtain subgoals and quantify them if appropriate?

6.22 A university administrator calls a systems analyst to improve the administration of sponsored research projects. The main problems are delay in obtaining latest financial position to project co-ordinators, reconciliation of advances given to co-ordinators, prompt demands not sent to sponsors to collect promised grants and lack of information to answer following questions:

Which areas of research get maximum grants?

Which agency aids which type of projects?

What trends can be seen in the nature of grants?

Now:

1. Classify the above problems into missing functions, unsatisfactory performance and excessive cost of operation.
2. How would you set goals to meet the deficiencies?
3. How would you quantify them?

6.23 A library receives 1300 journals of varying periodicities. The journals received have to be recorded and displayed. Action has to be taken when journals are not received in time or lost in mail. Unless request for replacement is sent quickly, it may not be possible to get the replacement. Periodicals have to be ordered at different times during the year and subscriptions renewed in time. Late payment of subscription may lead to non-availability of earlier issues or paying higher amounts for those issues. Current manual system is not able to meet these requirements.

1. Specify what should be the goals and sub-goals of an information system for ordering periodicals.
2. Quantify these goals.
3. Suggest alternative means of achieving the goals specified by you.

6.24 What operational, tactical, and strategic information should be provided by the mess billing system mentioned in the text? (case study).

6.25 When is a detailed system proposal prepared. What are the contents of a system proposal?

7

Data Flow Diagrams

LEARNING GOALS

After reading this chapter a student should be able to:

1. Model systems using data flow diagram.
2. Expand data flow diagrams giving more details.
3. Develop good style conventions to draw data flow diagrams.
4. Distinguish between physical and logical data flow diagrams.

In Chapter 6, we used document and data flow diagrams informally. In this chapter we will describe in greater detail this tool and how it is used to model systems.

Data Flow Diagram (abbreviated as DFD) was introduced by De Marco (1978) and Gane and Sarson (1979), and is an important tool used by systems analysts. A data flow diagram models a system by using external entities from which data flows to a process which transforms the data and creates output data flows which go to other processes or external entities or data stores. Stored data may also flow to processes as inputs. The main merit of DFD is that it can provide an overview of what data a system would process, what transformations of data are done, what data are stored and which stored data are used, and where the results flow. The graphical representation of the system makes it a good communication tool between a user and an analyst on the one hand and the analyst and a system designer on the other. DFDs are structured in such a way that starting from a diagram which gives a broad overview at a glance, they can be expanded to a hierarchy of diagrams giving more and more detail.

7.1 SYMBOLS USED IN DFDs

Four symbols are used in drawing data flow diagrams. A *circle* is used to depict a *process*. A process describes what a system does. Both inputs and outputs to a process are data flows. Processes are numbered and given a name (Figure 7.1). The term *bubble* is also sometimes used instead of circle.

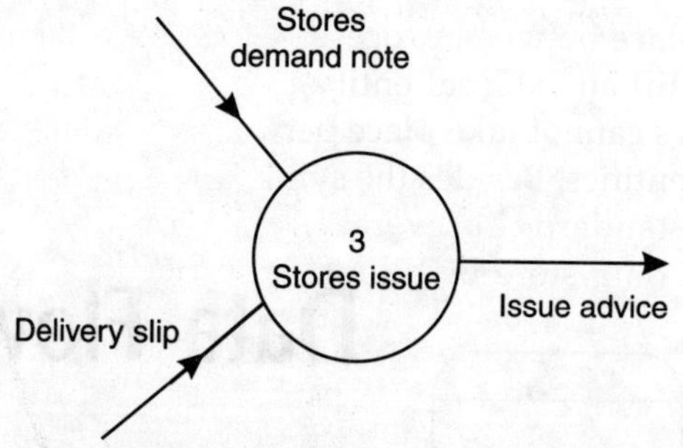

Figure 7.1 A process with data flows.

Data flows are represented by a *line with an arrow*. The arrow shows the direction of flow of data. The name of the data appears below the line. In Figure 7.1, the data flows are labelled: stores demand note, delivery slip, and issue advice. *External entities* (Figure 7.2) are represented by *rectangles,* and are outside the system, such as vendors or customers with whom the system interacts. The designer has no control over them. They either supply or consume data. Entities supplying data are known as *sources* and those that consume data are called *sinks*.

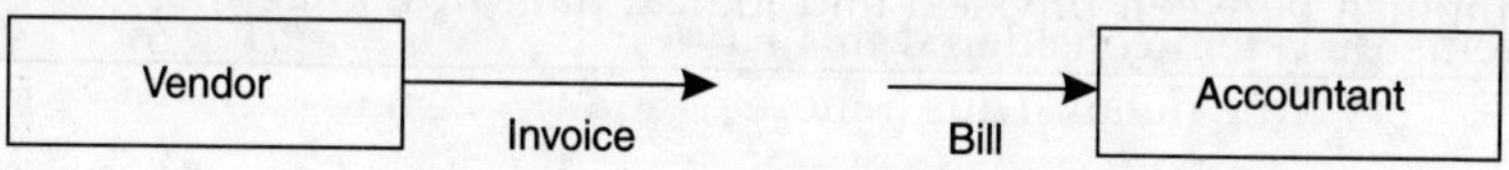

Figure 7.2 External entities.

A *data store* stores data. Processes may store or retrieve data from a data store. A data store is depicted by *two parallel lines* (see Figure 7.3). If an arrow points to the store, it indicates operation of writing in the store. If it points away from the store, it indicates operation of reading from the store. If both arrows are there, it indicates that one may write in or read from the store.

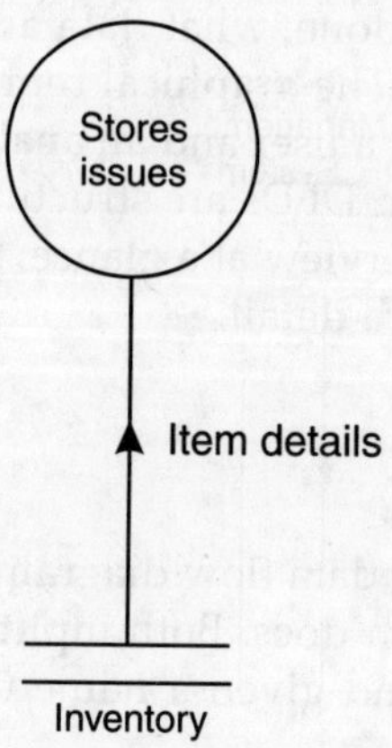

Figure 7.3 A data store.

Data flow can take place between processes, from (i) a data store to a process, (ii) a process to a data store, (iii) an external entity to a process, or (iv) from a process to an external entity. Data flows cannot take place between two data stores as they are passive or between two external entities. Besides the symbols we have used, other variations exist as there is no universal standard. In several books a process is depicted by a rectangle with round edges and a data store by a long open rectangle as shown in Figure 7.4.

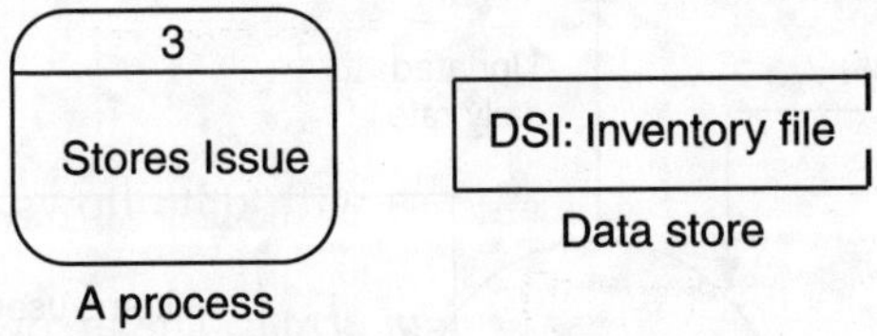

Figure 7.4 Alternate notations used in DFDs.

7.2 DESCRIBING A SYSTEM WITH A DFD

As a first step, an entire system can be modelled by a data flow diagram with one process which gives a system overview. It is called a *context diagram*. The context diagram for the mess management system considered in Chapters 4 and 6 is shown in Figure 7.5. This diagram gives very little detail. Thus a more detailed diagram with the process split up

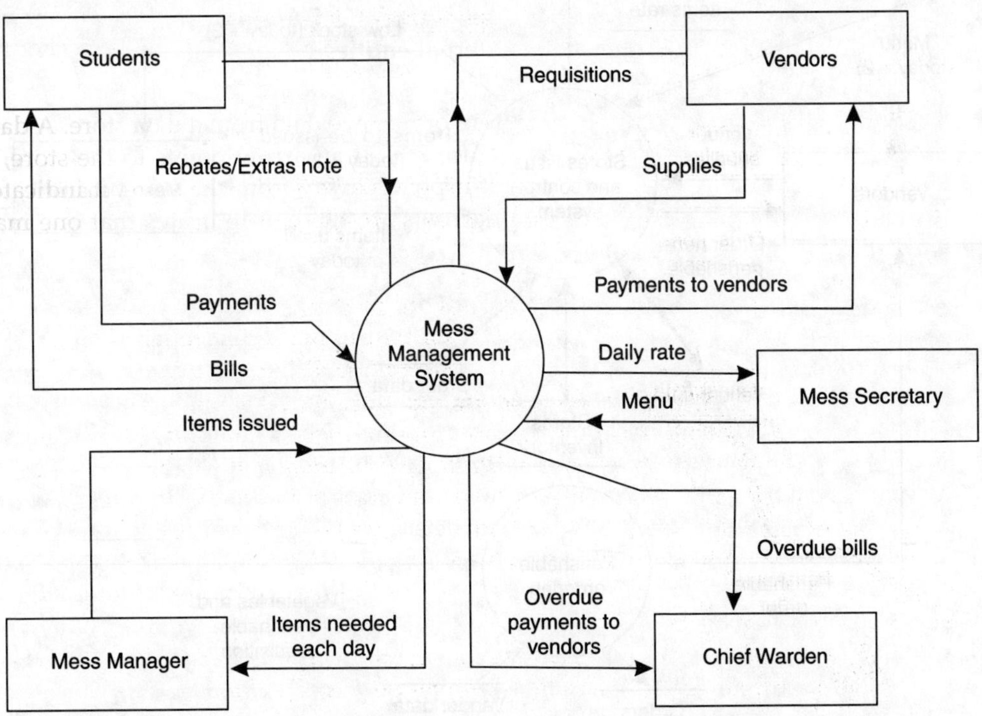

Figure 7.5 Context diagram for mess management system.

into major processes is drawn. This is called a top level DFD and is shown in Figure 7.6 (which is the same as Figure 6.2). This DFD can be further expanded by taking one of the circles and detailing the procedures. In the example being considered, the billing system

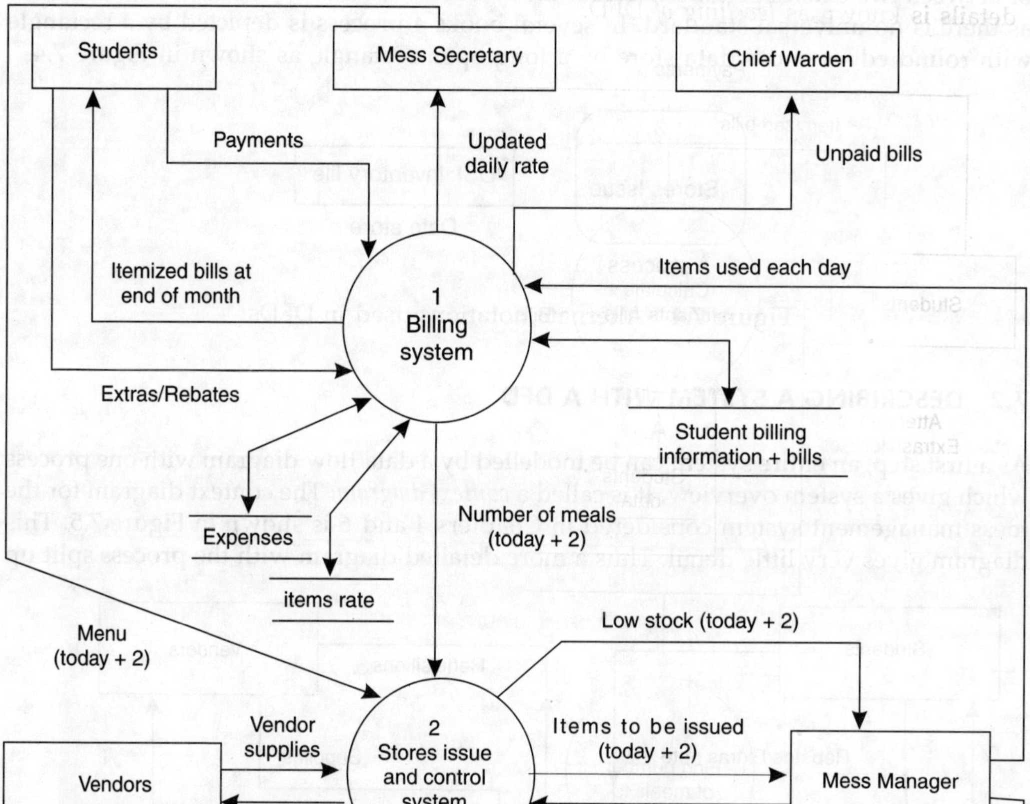

Figure 7.6 Data flow diagram for a hostel mess.

in the DFD of Figure 7.6 is expanded to the DFD of Figure 7.7. Observe that in detailing the DFD we still do not specify how the processing is done. We emphasize gross flow of data and specify what is to be done.

The successive expansion of a DFD from the context diagram to those giving more details is known as *levelling of* DFD.

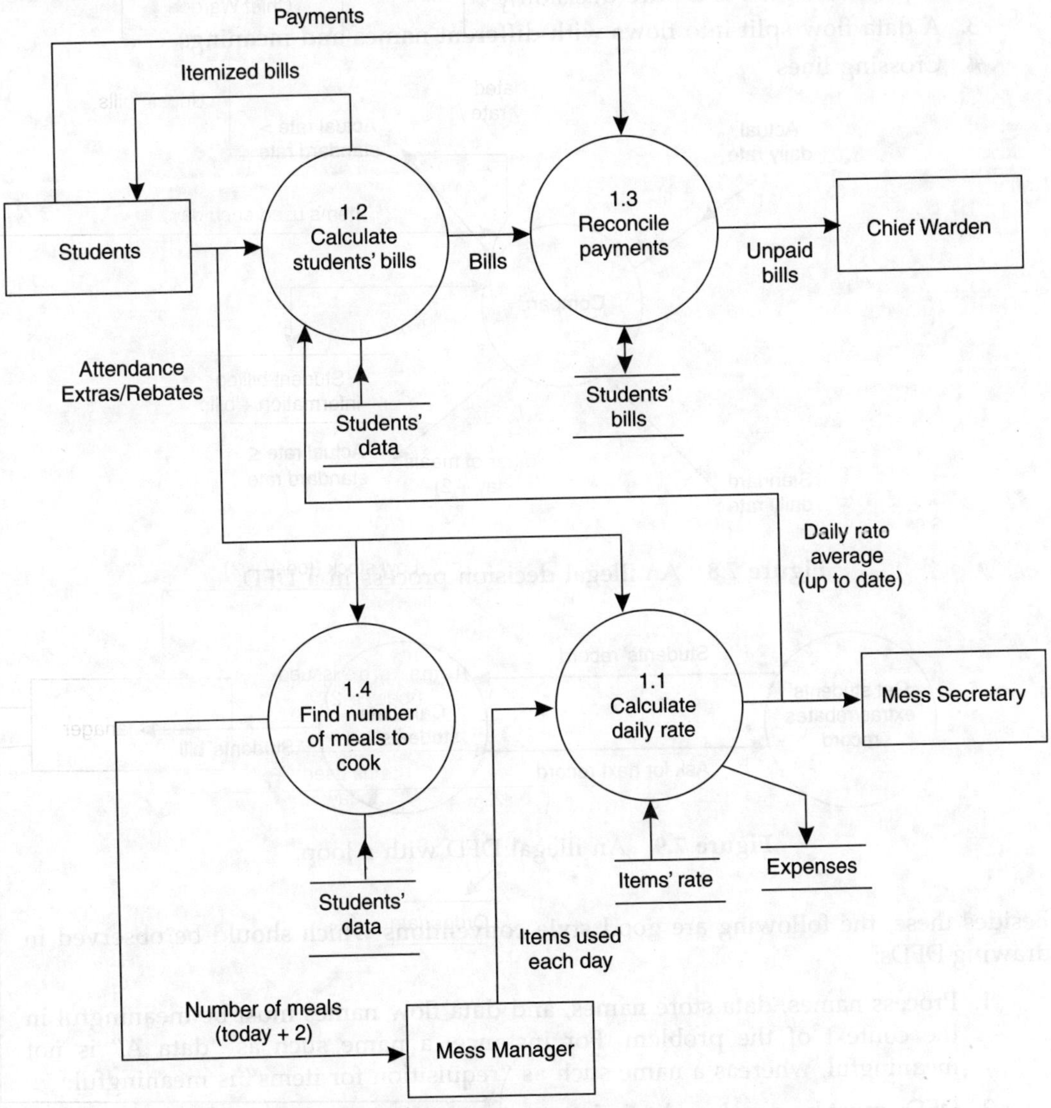

Figure 7.7 Expanded billing system process.

7.3 GOOD CONVENTIONS IN DEVELOPING DFDs

Data flow diagrams serve the dual purpose of specifying what data are needed for processing and as documentation of what procedures transform data. DFDs are different from flow charts. Flow charts give in detail how data is processed and include structures such as decision nodes and loops. DFDs do not have decision nodes and loops. For example, the DFDs given in Figures 7.8 and 7.9 are incorrect. A good data flow diagram *should not* have the following:

1. Loops
2. A process which is a pure decision
3. A data flow split into flows with different names and meanings
4. Crossing lines

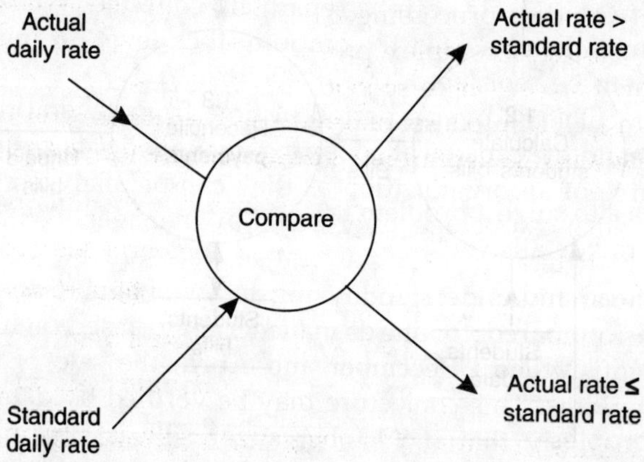

Figure 7.8 An illegal decision process in a DFD.

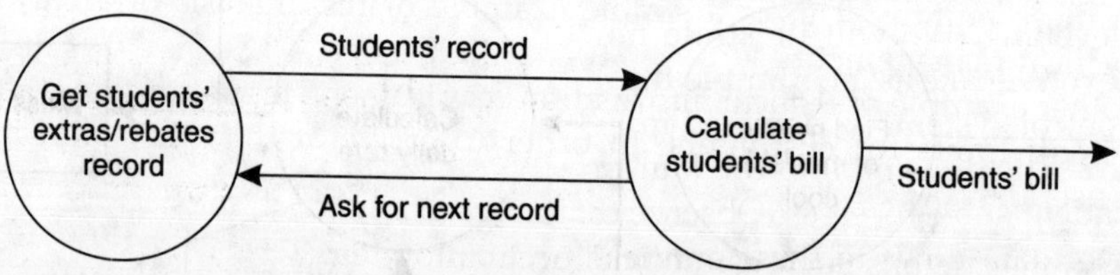

Figure 7.9 An illegal DFD with a loop.

Besides these, the following are good style conventions which should be observed in drawing DFDs:

1. Process names, data store names, and data flow names must be meaningful in the context of the problem. For instance, a name such as "data A" is not meaningful, whereas a name such as "requisition for items" is meaningful.
2. DFDs must be developed top down with lower levels giving more details.
3. Data should be conserved. A process cannot create new data. It can only transform input data to create output data. Similarly, any data retrieved from a store must have been stored in it earlier.
4. Data flows should not act as signals to activate or initiate processes.

7.4 LEVELLING OF DFDs

If a DFD contains all details, then it will have too many data flow lines. It will be mixed up and it will be difficult to understand all data flows and their processing. It is thus a good idea to start with an overview of the problem represented by a context DFD and expand the process appearing in it into an interconnected set of processes. This expanded DFD should specify the next level of detail and should be understandable at a glance. This implies that the number of processes in it should not be more than around seven. Each of the processes in this expanded DFD can in turn be expanded into another DFD. This method of specifying data processing is called a *top down method.* Such a method is an excellent way of solving a complex problem by breaking down the problem into smaller problems which are easier to solve. In the literature on DFD, the procedure of expanding a process in a DFD into a set of equivalent smaller processes is called *levelling*. The reasons why a DFD is levelled are:

1. It is a good idea to solve a problem top down, starting with an overview and then working out the details.
2. It is easier to read and understand levelled DFD (namely one with around seven processes) as compared to one large detailed DFD with a large number of processes. Each levelled DFD can normally be accommodated in one page.

7.4.1 Levelling Examples

We have already illustrated how a context diagram is expanded into a DFD which is in turn levelled (Figures 7.5–7.7). In Figure 7.10 we show how levelled diagrams are developed. This figure mainly illustrates mechanical details without specifying the logical context of levelling. We will illustrate now the idea of levelling within a meaningful logical context. We take as an example the mess management system. The process "stores issue and control system" is depicted in Figure 7.11. It is expanded into four processes in Figure 7.12. The division into the four processes is done by decomposing the process into independent sub-processes. Observe that the external entities remain the same, and the number of data flows to these entities is also same.

7.4.2 Levelling Rules

1. If a process numbered *p* in a DFD is levellled (i.e., expanded) into *n* processes, then each new process is numbered *p.1, p.2, . . ., p.n*, respectively.
2. All data flows entering a process must also enter its levelled equivalent DFD.
3. All data flows leaving a process must also leave its levelled DFD.
4. New data stores not referred to in a process may appear in the equivalent levelled DFD of that process. This implies that this store is local to the process and will not appear in the context diagram.
5. No external entities may appear in lower level DFDs. This is due to the fact that external entities are not part of a process.
6. The number of processes at each level is usually kept small (not exceeding 7). This is mainly to aid understanding of each level.

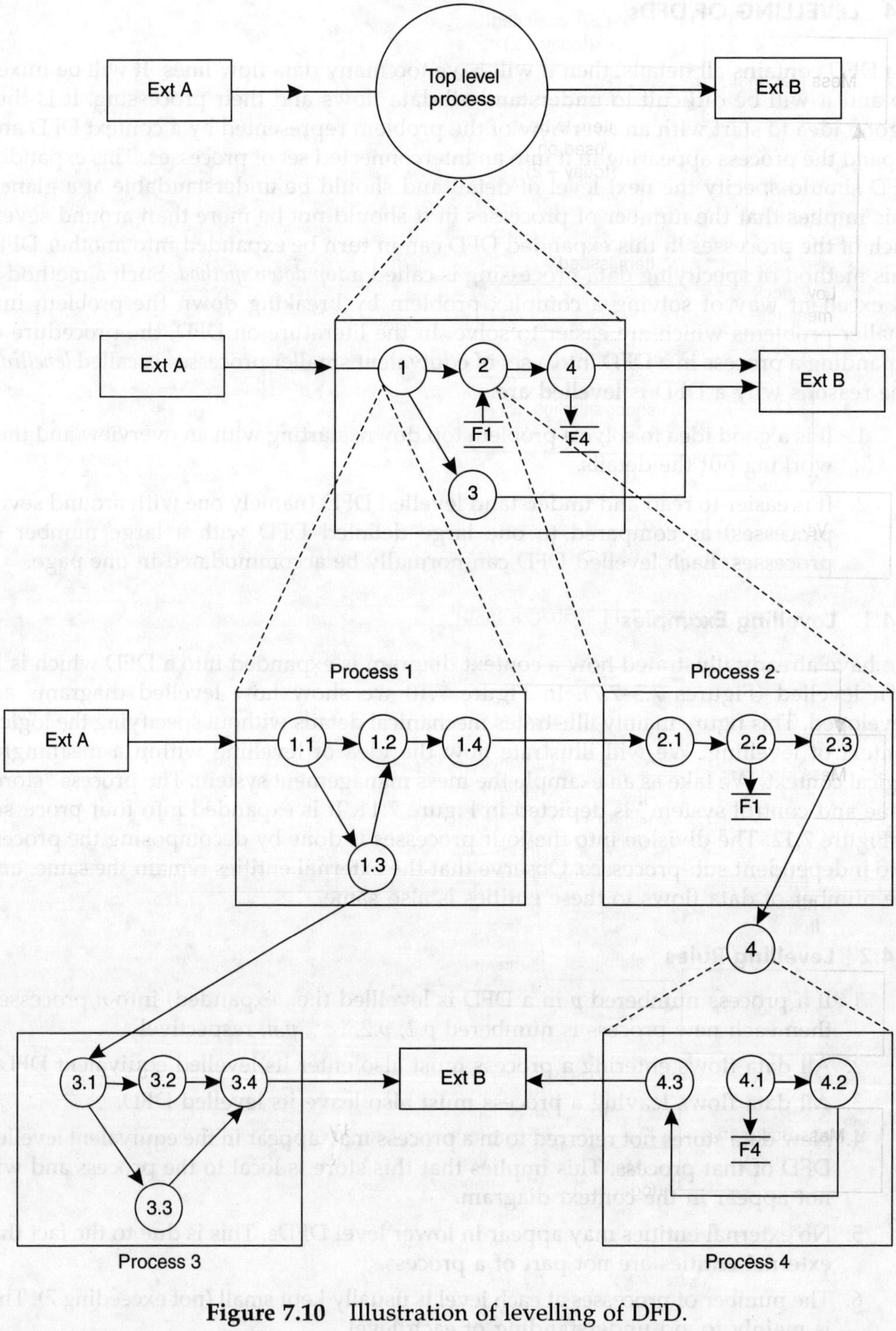

Figure 7.10 Illustration of levelling of DFD.

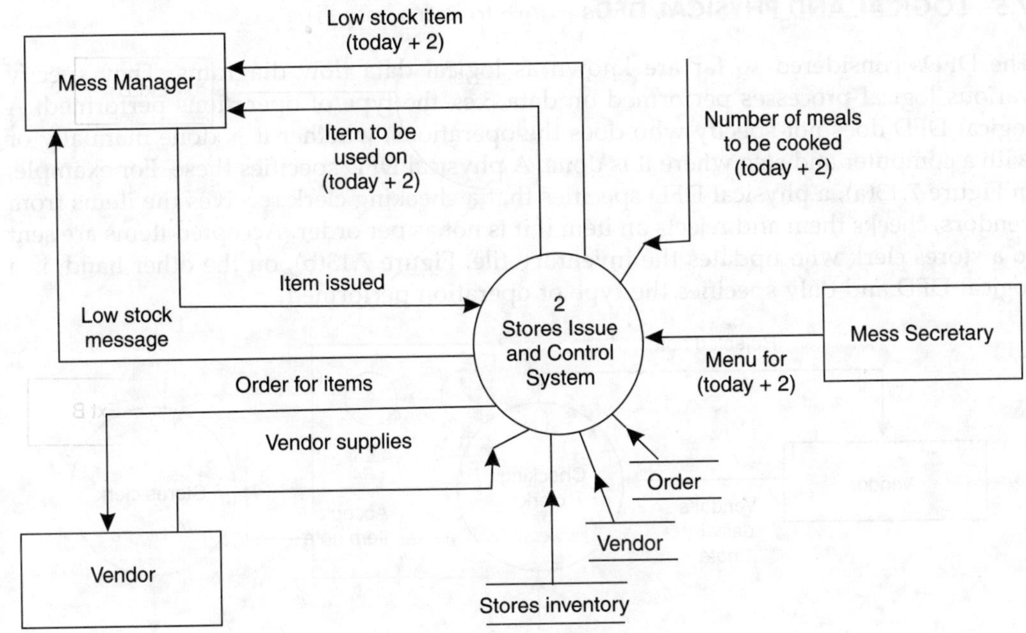

Figure 7.11 Stores issue and control system process.

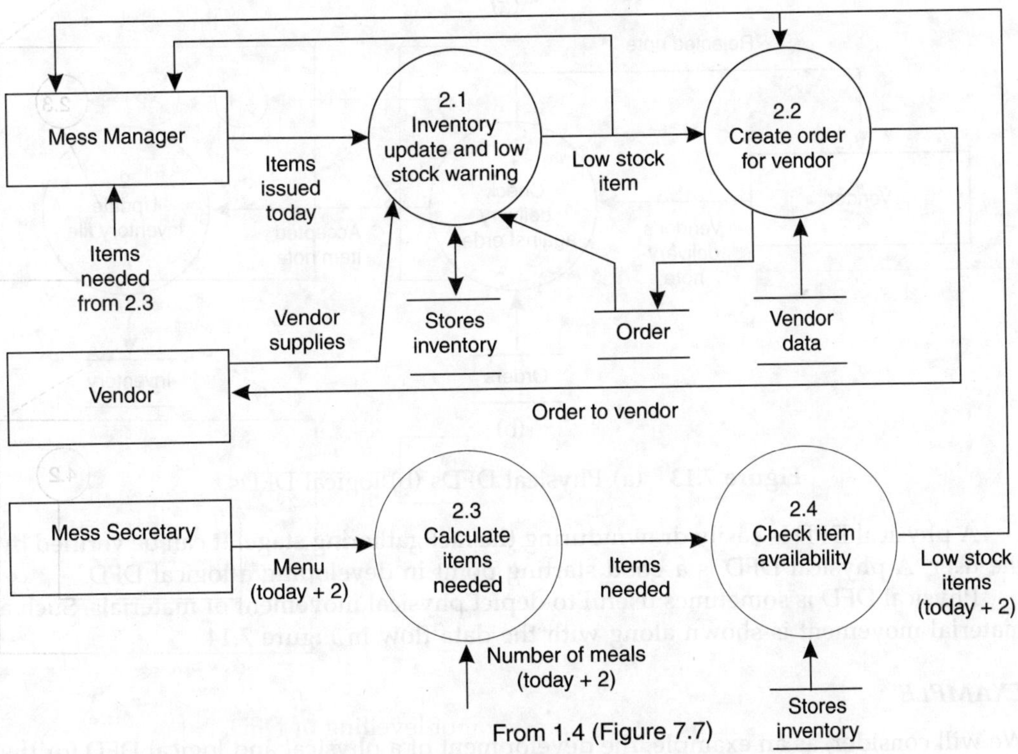

Figure 7.12 Levelled DFD stores issue and control system.

7.5 LOGICAL AND PHYSICAL DFDs

The DFDs considered so far are known as logical data flow diagrams. They specify various logical processes performed on data, i.e., the type of operations performed. A logical DFD does not specify who does the operations, whether it is done manually or with a computer and also where it is done. A physical DFD specifies these. For example, in Figure 7.13(a), a physical DFD specifies that a checking clerk receives the items from vendors, checks them and rejects an item if it is not as per order. Accepted items are sent to a stores clerk who updates the inventory file. Figure 7.13(b), on the other hand, is a logical DFD and only specifies the type of operation performed.

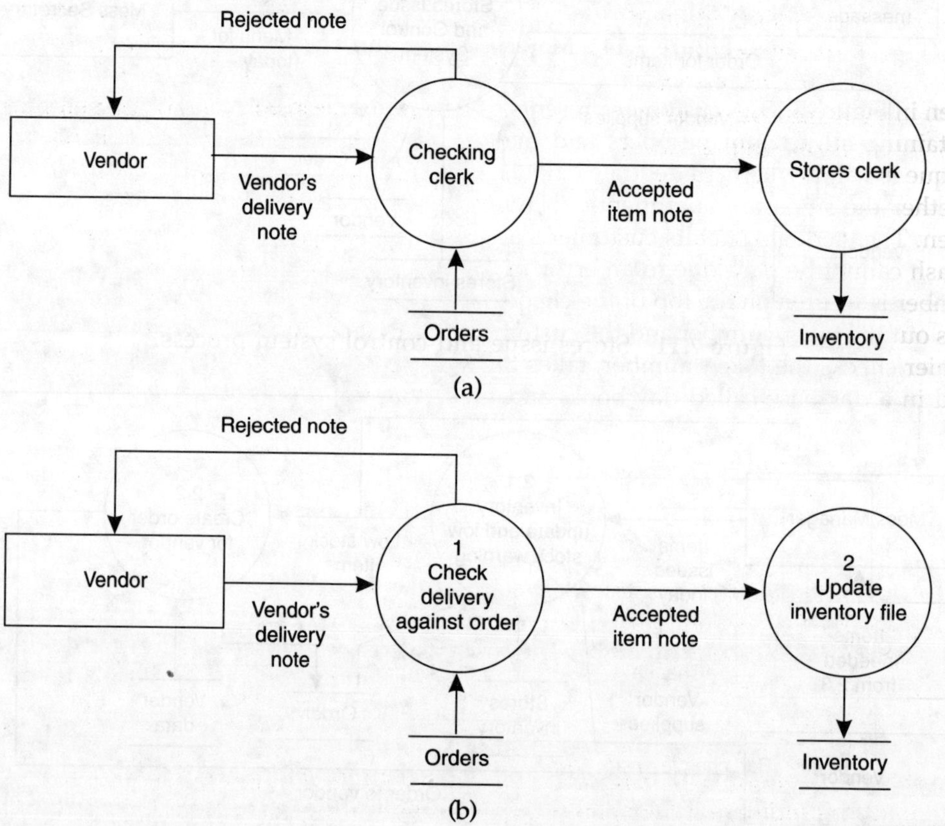

Figure 7.13 (a) Physical DFDs (b) logical DFDs.

A physical DFD is easily drawn during the fact gathering stage. It can be verified by the user. A physical DFD is a good starting point in developing a logical DFD.

Physical DFD is sometimes useful to depict physical movement of materials. Such a material movement is shown along with the data flow in Figure 7.14.

EXAMPLE

We will consider, as an example, the development of a physical and logical DFD for the process of getting a cheque encashed in a bank. The physical DFD for this operation is

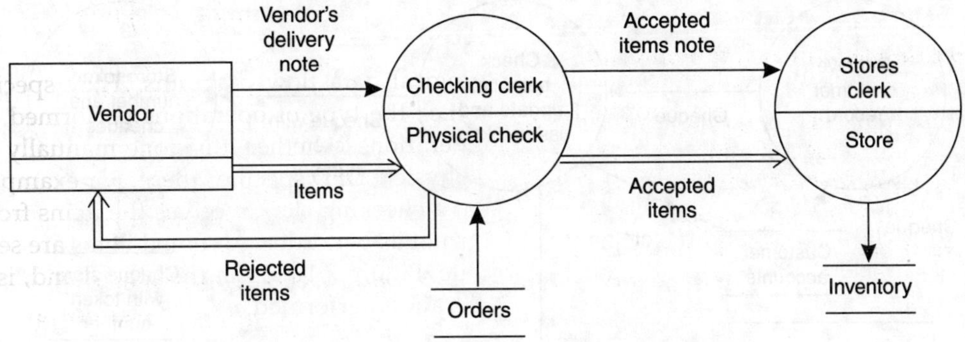

Figure 7.14 Materials flow shown on DFD.

given in Figure 7.15. A customer presents a cheque to a clerk. The clerk checks a database containing all account numbers and makes sure whether the account number in the cheque is valid, whether adequate balance is there in the account to pay the cheque, and whether the signature is authentic. Having done these, the clerk gives the customer a token. The clerk also debits customer's account by the amount specified on the cheque. If cash cannot be paid due to an error in the cheque, the cheque is returned. The token number is written on the top of the cheque and it is passed on to the cashier. The cashier calls out the token number and the customer goes to the cash counter with the token. The cashier checks the token number, takes the customer's signature, pays cash, enters cash paid in a database called day book, and files the cheque.

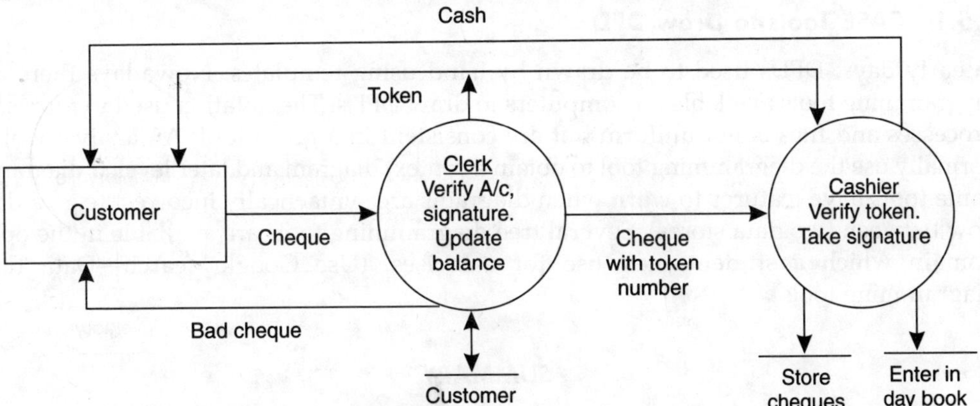

Figure 7.15 Physical DFD for encashing cheque.

The same diagram may now be converted to a logical DFD. In this conversion we take the functions performed at each step as same. Each process has a well-defined operation. Further, in this diagram we do not include details such as clerks/cashiers performing operations. The logical DFD corresponding to Figure 7.15 is shown in Figure 7.16.

If it is a teller system, then only Processes 1 and 2 will be there. Instead of token slip cash will be given to the customer by Process 2.

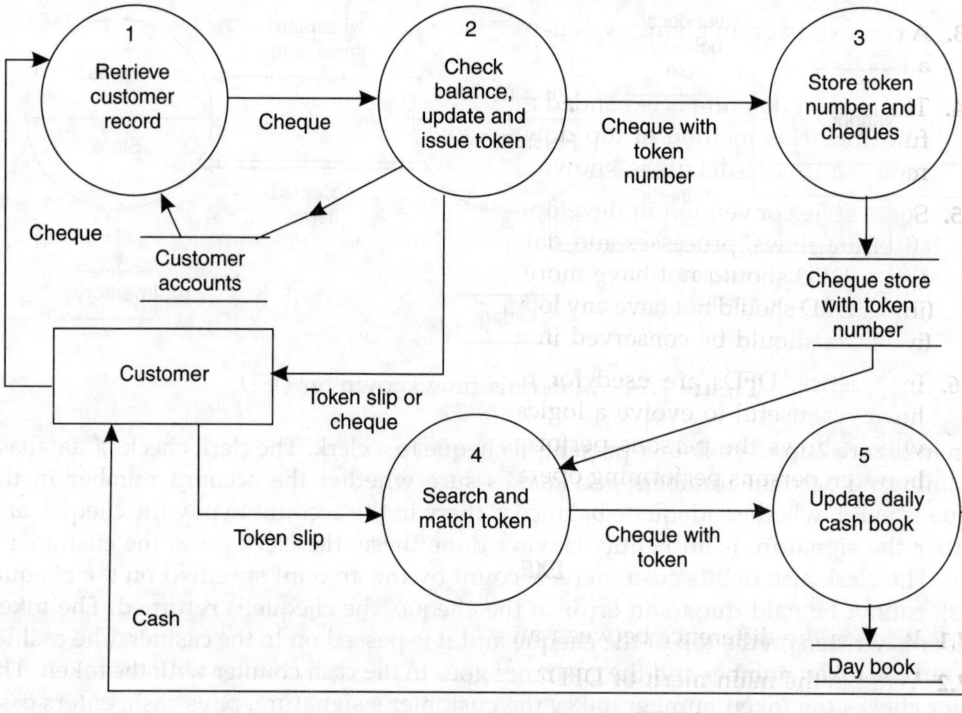

Figure 7.16 Logical DFD for encashing cheque.

7.5.1 CASE Tools to Draw DFD

In early days, DFDs used to be drawn by hand using templates. Nowadays there are diagramming tools available on computers to draw DFDs. The notation used to represent processes and files is not uniform but are consistent in a given tool. An analyst would normally use the diagramming tool to obtain a context diagram and later leveled diagrams. Some tools have features to warn when diagrams are syntactically incorrect (e.g., a data flow between two data stores). Several free diagramming tools are available in the open domain which a student may use for exercises. (Use Google search—Data flow diagramming tools).

SUMMARY

1. Data flow diagrams (DFDs) are used to depict the flow and transformation of data in an information processing system.
2. Symbols used in DFDs are a rectangle to depict an external entity, a circle to represent a process, two short parallel lines to represent a data store, and lines with arrows to depict data flows. The direction of an arrow shows the direction of the flow of data. Another notation uses a rectangle with round edges to depict a process and an open rectangle to represent a data store (Figure 7.4)

3. A context diagram giving system overview is used as the first step in developing a DFD.
4. The context diagram is expanded into a series of DFDs, each describing a specific function. This method of top down analysis and breaking down DFDs to give more and more details is known as *levelling*.
5. Some style convention in developing DFDs are:
   (i) Data flows, processes and data store must have meaningful names.
   (ii) A DFD should not have more than 7-9 processes.
   (iii) A DFD should not have any loops, crossing lines or pure decisions as processes.
   (iv) Data should be conserved in a DFD.
6. In practice, DFDs are used for representing logical processing of data. It is, however, useful to evolve a logical DFD after first developing a physical DFD which shows the persons performing various operations and how data flows between persons performing operations.

EXERCISES

7.1 What is the difference between an external entity and a process in a DFD?

7.2 What is the main merit of DFD?

7.3 What is the role of DFD as a documentation aid?

7.4 What is the difference between a source and a sink in a DFD?

7.5 Why is data flow between data stores not allowed in DFD?

7.6 Is data flow allowed between an external entity and a data store? If your answer is "no", explain why?

7.7 What is a context diagram?

7.8 What do you understand by levelling of DFD?

7.9 What is the main difference between a flow chart and a DFD?

7.10 What do you understand by top down development of a DFD?

7.11 What is a physical DFD?

7.12 What is the difference between a physical and a logical DFD?

7.13 In what way is a physical DFD useful?

7.14 What are the mistakes in each of the DFDs of Figure 7.17(a)–7.17(e)? Correct these mistakes.

7.15 Prepare physical and logical DFDs for the following activities:
   (i) Issuing out a book from a library
   (ii) Returning a book to a library
   (iii) Getting a ticket reserved for a train journey
   (iv) Buying an item from a super market
   (v) Getting your marksheet from a University office.

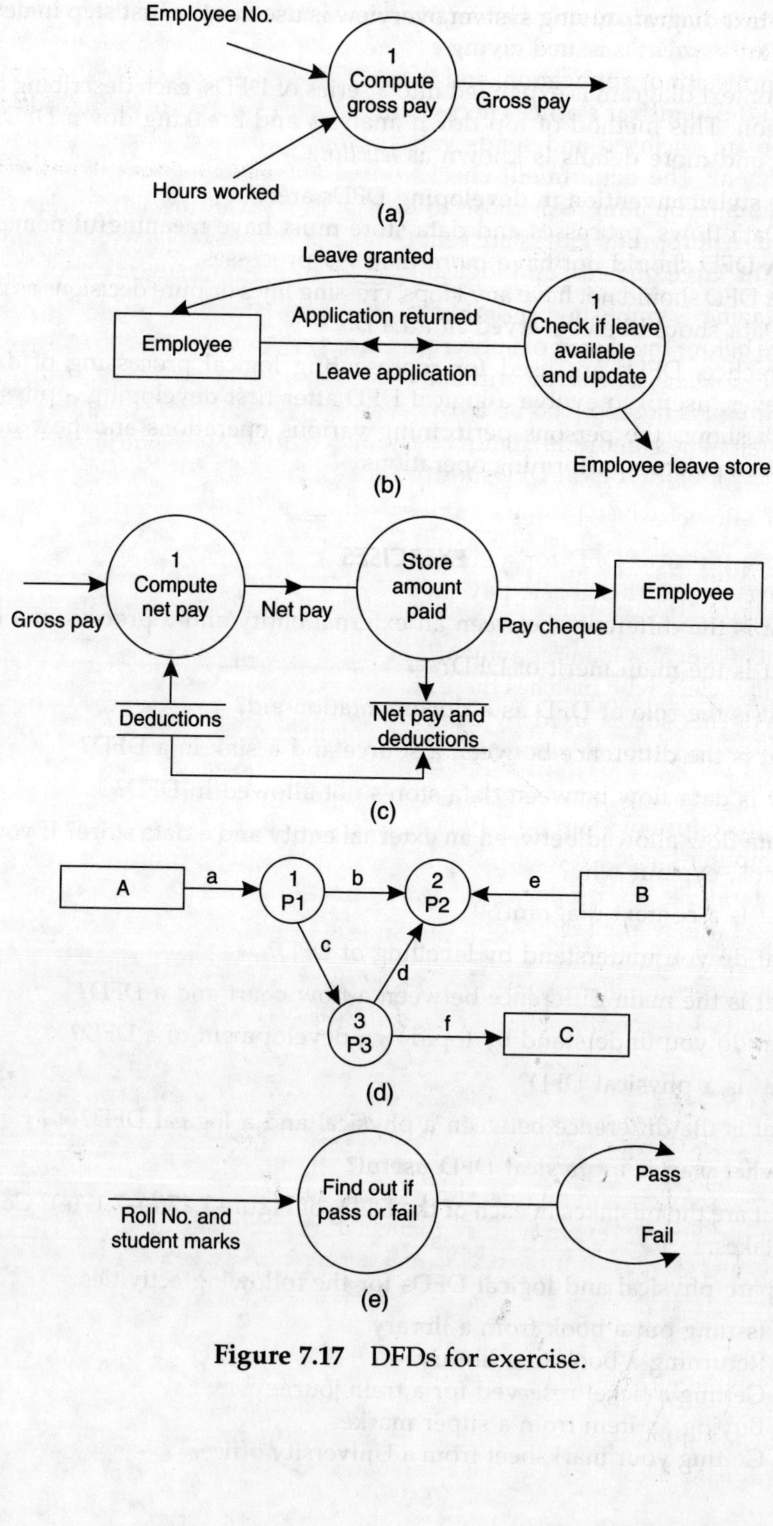

Figure 7.17 DFDs for exercise.

7.16 Admission procedure in a University is as follows:
An advertisement is issued giving essential qualifications for the course, the last date for receipt of application, and the fee to be enclosed with the application. A clerk in the Registrar's office checks the received applications to see if marksheet and fee are enclosed and sends valid applications to the concerned academic department. The department checks the application in detail and decides the applicants to be admitted, those to be put in the waiting list, and those to be rejected. Appropriate letters are sent to the Registrar's office which intimates the applicant. Give physical and logical DFDs corresponding to the above problem.

7.17 A magazine is published monthly and is sent by post to its subscribers. Two months before the expiry of subscription, a reminder is sent to the subscribers. If subscription is not received within a month, another reminder is sent. If renewal subscription is not received up to two weeks before the expiry of the subscription, the subscriber's name is removed from the mailing list and the subscriber informed. Obtain logical DFDs for this problem.

7.18 Obtain a flowchart for Exercise 7.17 and state in what way it differs from the DFD.

7.19 Obtain a physical DFD for a simple payroll system described below. A list of employees with their basic pay is sent to a clerk. He calculates the gross pay using standard allowances which are known for each pay slab. Deduction statements such as loan repayment, subscription to association, etc., are also sent to another clerk who matches these slips with the slips of gross pay and calculates net pay. This slip is used by a third clerk to write out pay cheques for each employee and sent to respective employees. The total pay bills paid are also computed.

7.20 If the procedure of Exercise 7.19 is to be computerized, obtain a logical DFD for the computer-based system.

8

Process Specifications

LEARNING GOALS

After reading this chapter a student should be able to:

1. Identify key words and control structures used in structured English.
2. Specify processes in data flow diagrams using structured English.

In Chapter 7, we saw how data flow diagrams are formulated to model information systems. In data flow diagrams processes are described briefly by what transformations they perform on the input data flow. In this chapter we will discuss in detail the methods used to describe processes.

8.1 PROCESS SPECIFICATION METHODS

Systems analysts find out the requirements to be met by a system by interviewing and looking at various documents in the organization. Requirements for processing are often imprecisely stated. For instance, consider the statement: "Give a discount of 5% if a customer pays advance or if the purchase is for ₹ 10,000 or more and the customer is a regular customer". What does one do if the purchase is for ₹ 20,000 and the customer is a first time customer and pays advance? If we apply the *or* connective on clauses "pays advance" and "if the purchase is for ₹ 10,000 or more" first and then apply the *and* connective on the clause the "customer is a regular customer" then we do not give discount. If we apply the *and* connective to clauses "if purchase is for ₹ 10,000 or more" and he is a "regular customer" and then apply the *or* connective with clause "pays advance" we give discount as the customer pays advance. Such an ambiguity is due to the possibility of different interpretations of sentences. Thus there is a need for a more precise statement which can be unambiguously transformed to computational procedures. Two main techniques are used to express computational procedures:

- Structured English
- Decision tables

(i) Structured English. It tries to express verbal statements in a more logical framework. It uses natural language along with the ideas of logic and block structuring used in computer languages. This makes the description more precise and understandable. Such a technique is useful if there are not too many conditions to be checked and the procedures are sequential and repetitive.

(ii) Decision tables. These are appropriate when a large number of conditions are to be checked in arriving at a set of actions. Decision tables have many other advantages for computer processing which we will explain in some detail in Chapter 9. The difference between the two methods is illustrated, taking the word statement given at the beginning of this section.

STATEMENT: Give a discount of 5% if the customer pays advance or if the purchase is for ₹ 10,000 or more and the customer is a regular customer. This can be written in structured English as

```
If Customer pays advance
  then
    Give 5% discount
  else
    if purchase amount ≥ 10,000
      then
        if customer is a regular customer
          then
            Give 5% discount
          else
            No discount
        endif
      else
        No dicount
    endif
  endif
```

In the structured English description the conditions are checked one after another. For each question the part after *then* states the action to be carried out if the answer to the question is "Yes". The part after *else* states the action to be carried out if the answer to the question is "No". Observe that the *then* and *else* parts for each *if* are properly paired. At the end, the *if* is paired with *endif* which indicates the end of the *if* statement. Indentation is used to aid the reader to see the correct grouping of answers. This style is closer to the way programs are written in a programming language.

The decision table (Table 8.1), on the other hand, is not programming language oriented. It is oriented towards the *specification* of what is to be done in a process. The way the decision table is interpreted is to read the rules. A "Y" entered for a rule indicates the answer as "yes" to the condition in the corresponding row, a "N" the answer as "no", and "-" means the answer is not relevant. Table 8.1 is interpreted as follows:

RULE 1: *If* advance payment made
then give 5% discount

Table 8.1 Decision Table for Discount Decision

Conditions	Rule 1	Rule 2	Rule 3	Rule 4
Advance payments made?	Y	N	N	N
Purchase amount ≥ 10,000?	–	Y	Y	N
Regular customer	–	Y	N	–
Actions				
Give 5% discount	X	X	–	–
No discount	–	–	X	X

RULE 2: *If* no advance payment is made and *if* the purchase amount ≥ 10,000 and the customer is a regular customer
then give 5% discount

RULE 3: *If* no advance payment is made and *if* the purchase amount ≥ 10,000 and customer is not a regular customer
then give no discount

RULE 4: *If* no advance payment is made and *if* the purchase amount is not ≥ 10,000
then give no discount

Observe that these lengthy rules are beautifully summarized by the decision table. In Chapter 9, we will describe how decision tables are obtained from a word statement. We now give another example of structured English and decision table specifications of a set of rules.

EXAMPLE 8.1

A University has the following rules for a student to qualify for a degree with Physics as the main subject and Mathematics as the subsidiary.

1. Marks should be 50% or more in Physics and 40% or more in Mathematics.
2. If marks in Physics are less than 50% then marks in Mathematics must be 50% or more. However, Physics marks must be at least 40%.
3. If marks in Mathematics are less than 40% but those in Physics are 60% or more then only examination in Mathematics has to be repeated.
4. In all other cases the student fails.

The above statements are expressed in structured English as follows:

```
if Physics marks ≥ 50% and Math marks ≥ 40%
  then
    Pass
  else
    if Physics marks ≥ 40% and Math marks ≥ 50%
      then
        Pass
```

```
        else
            Fail
        endif
endif
if Math marks < 40% and Physics marks ≥ 60%
    then
        Repeat Math exam
    else
        Fail
endif
```

A decision table equivalent of the above rules is given as Table 8.2.

Table 8.2 Decision Table for Passing Rules

Conditions	Rule 1	Rule 2	Rule 3	Rule 4
Physics marks	≥ 50%	≥ 40%	≥ 60%	ELSE
Math marks	≥ 40%	≥ 50%	< 40%	
Actions				
Pass candidate	X	X	–	–
Repeat Math	–	–	X	–
Fail candidate	–	–	–	X

Observe that in Table 8.2 the entries in the columns for rules are not Yes or No answers as in Table 8.1. Part of the condition appears in the rule. The table is interpreted as follows:

RULE 1: *If* Physics marks ≥ 50% **and** Math marks ≥ 40%
then Pass candidate

RULE 2: *If* Physics marks ≥ 40% **and** Math marks ≥ 50%
then Pass candidate

RULE 3: *If* Physics marks ≥ 60% **and** Math marks < 40%
then Repeat Math exam

RULE 4: In all other cases Fail candidate. (In other words, if none of the rules 1, 2, 3 is true *then* the rule marked ELSE holds.)

8.2 STRUCTURED ENGLISH

The methodology used in structured English description is similar to that used in block structured computer languages. The main difference is the use of English statements to specify actions. The syntax rules are also not very strict. The aim is to allow easy readability which aids documentation and maintenance and at the same time ensures some discipline in process description which will help the programmer. The following conventions are used in writing structured English process description.

(i) **Imperative sentences.** This consists of an imperative verb followed by operations to be performed on variables. For example:

Multiply gross price by discount rate
Store result in rebate
Subtract rebate from gross price and obtain net price

are three imperative sentences.

It is important to use verbs which are precise. Some notable examples are:

Find sum of marks
Check if marks ≥ 50

Some imprecise statements are:

Process student grades
Sample a few cases
Allow substantial discount

(ii) **Arithmetic and relational operations.** Common symbols used in arithmetic are used in structured English descriptions. Symbols used for arithmetic are:
+ add; – subtract; * multiply; / divide

Symbols used in relations are:
= equal to; > greater than; < less than; ≠ not equal to; ≥ greater than or equal to; ≤ less than or equal to.

Symbols used for logical operations are:

and, or, not

Symbols used as keyword are:

If, then, else, repeat, until, while, search, retrieve, do, case, for, of, endif, elseif, Read, Write.

We will illustrate the use of these symbols with some examples. A sequence of imperative statements can be enclosed optionally by curly brackets if it helps avoid ambiguity. For example:

{**Read** 'store issue note'.
Select item issued record from 'store issue note'
Cost of issued item = quantity * price of item
Balance in account = balance in account – cost of issued item}

(iii) **Decision structures**. There are two types of decision structures used in structured English statements. One of them is an **if then else** structure which is used to make a choice between two alternatives. Figure 8.1 shows such a decision structure as a flowchart. The corresponding structured English description is:

```
if condition
   then
      {Group Y statements}
   else
      {Group N statements}
   endif
```

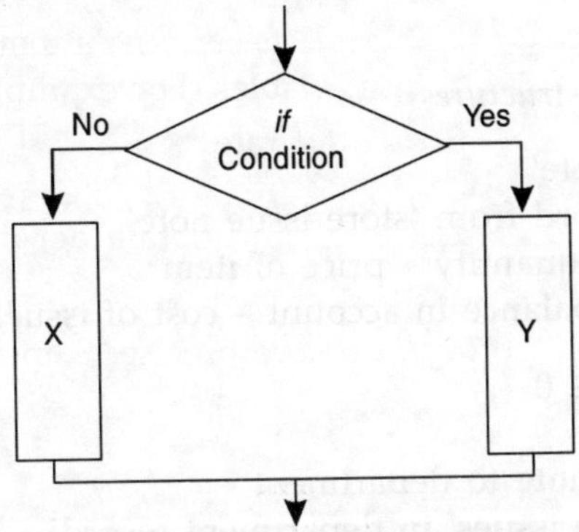

Figure 8.1 Illustration of **if then else** structure.

The second type of structure is to choose one out of a set of alternatives. Here a **case** keyword is used. Figure 8.2 shows a structure. The corresponding structured English description is:

case (variable)
(Variable = P) : {statements for alternative P}
(Variable = Q) : {statements for alternative Q}
(Variable = R) : {statements for alternative R}
(Variable = S) : {statements for alternative S}
None of the above: {statements for default case }
end case.

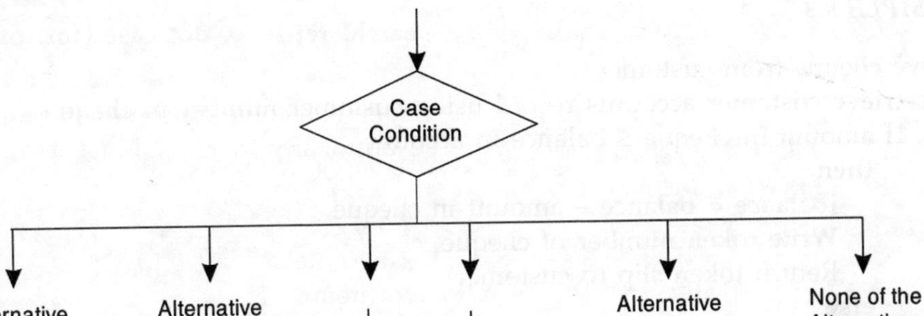

Figure 8.2 Illustration of case statement.

In this example, for each specified value of the **case** variable a different sequence of statements is executed. If the variable does not have a value within the specified values then the statements specified in 'None of the above' are to be executed. We give the following example to illustrate these structures:

EXAMPLE 8.2

Illustration of **if then else** *structures*

```
Receive 'store issue note'
Select item issued record from 'store issue note'
Cost of issued item = quantity * price of item
Balance in account = balance in account – cost of issued item.

If balance in account ≤ 0
  then
    {Issue exception note to department
    Mark 'stop future issues' in department record}
  else
    if balance in account < minimum balance
      then
        Issue item to department
        Issue warning note to department
      else
        Issue item to department
    endif
  endif
```

We now give another example corresponding to processes 1 and 2 of DFD given in Figure 7.16.

EXAMPLE 8.3

```
Receive cheque from customer
  Retrieve customer accounts record using customer number in cheque
    If amount in cheque ≤ balance in account
      then
        {Balance = balance – amount in cheque
        Write token number of cheque
        Return token slip to customer}
      else
        {Write "insufficient balance" in cheque
        Return cheque to customer}
    endif
```

We illustrate the use of **case** with the following example.

EXAMPLE 8.4

A company makes three products which are codified as class A, class B and class C. On class A items for purchase above ₹5000, 10% discount is given. For class B for purchase above ₹8000, a discount of 5% is given. On class C on a purchase of ₹10,000 and above, 4% discount is given.

The above statements are expressed in structured English as follows:

```
Case (Product class)
  (Product Class = A)   :  {if Purchase > 5000
                            then
                               Discount = 10%}
  (Product Class = B)   :  {if Purchase > 8000
                            then
                               Discount = 5%}
  (Product Class = C)   :  {if Purchase > 10,000
                            then
                               Discount = 4%}
  None of the above     :  Discount = 0
```

(iv) Repetition. One of the most powerful techniques used in computer procedures is repetition, i.e., a sequence of statements are carried out many times. The number of times the repetition is to take place is governed by the type of structure used in the description. If a sequence of statements are to be repeated a specified fixed number of times, a **for** structure is used. For example, if each student in a class takes 5 subjects and the marks in each subject are represented by Marks (subject) then the following **for** loop computes total marks of the student:

```
Total marks = 0
  for subject = 1 to subject = 5 do
    total marks = total marks + marks (subject);
    write student roll number, total marks
  endfor
```

Assume we want to find out the total marks of all students in the class and each student's record has the form:

Student roll number, marks (1), marks (2), marks (3), marks (4), marks (5).

If the number of students in the class is not known then we can write the following procedure:

```
while there are more student records do
  Read student record
  Total marks = 0
  for subject = 1 to subject = 5 do
    total marks = total marks + marks (subject)
  endfor
  write student roll number, total marks
endwhile
```

The set of statements between **while** and **endwhile** are repeated as long as there are student records to be processed. Flowcharts showing **for** loop and **while** loop are given in Figures 8.3 and 8.4, respectively. Using these basic structures it is possible to specify fairly complex procedures. The point to remember is to make procedures simple so that any one can easily read and understand the procedure.

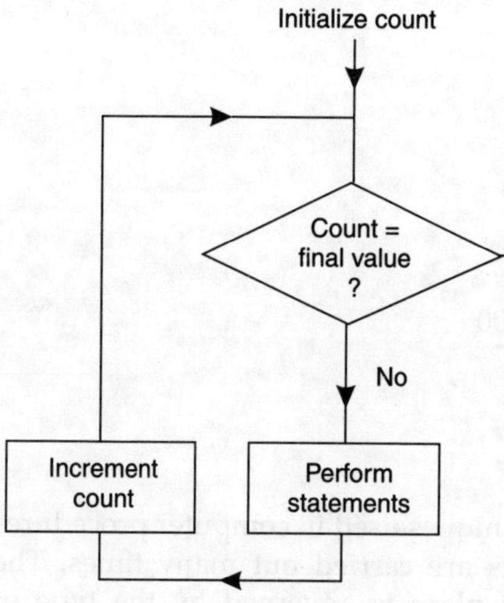

Figure 8.3 Illustration of a **for** loop.

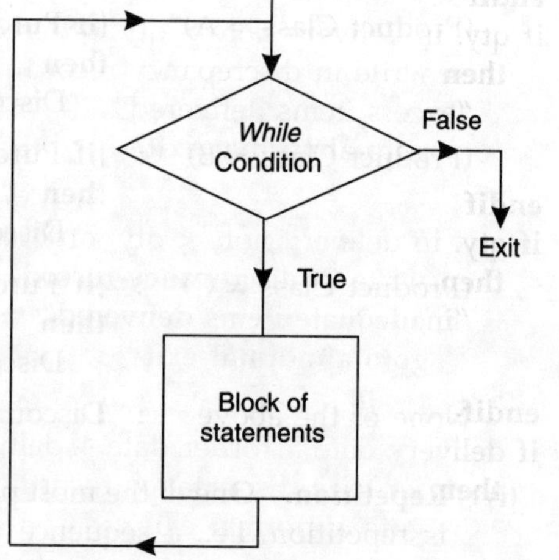

Figure 8.4 Illustration of a **while** loop structure.

8.3 SOME EXAMPLES OF PROCESS SPECIFICATION

In Chapter 5, we obtained the specification of requirements of an office receiving goods from a vendor. A narrative of the procedure followed by the receiving office was given in Section 5.1. A more precise statement of processing rules in structured English is now given:

EXAMPLE 8.5

```
while there are delivery notes do
for each delivery note
   search order file using order number from delivery note
   if search fails
      then write in discrepancy record
         "order not found"
            goto abnormal exit
   endif
   if search succeeds
      then retrieve order record corresponding to delivery note
   endif
   if vendor code in delivery note ≠ vendor code in order record
      then write in discrepancy record
         "vendor code incorrect"
            goto abnormal exit
   endif
   if item code in delivery note ≠ item code in order record
      then write in discrepancy record
         "item code incorrect"
            goto abnormal exit
```

endif
if qty. in delivery note > qty. ordered in order record
then write in discrepancy record
"excess items delivered"
goto abnormal exit
endif
if qty. in delivery note < qty. ordered in order record
then write in discrepancy record
"inadequate items delivered"
goto abnormal exit
endif
if delivery date > (order date + delivery time)
then write in discrepancy record "late delivery"
goto abnormal exit
endif
if delivery date < (order date + delivery time)
then write in discrepancy record "early delivery"
goto abnormal exit
endif
write in items received note
< Order no., vendor code, vendor name, vendor address,
item code, item name, delivery date, qty. supplied, unit >
and forward to Inspection office. **goto exit.**
abnormal exit:
Present discrepancy note to purchase office to initiate action.
exit: continue
endwhile.

The structured English procedure for updating the inventory in the stores process of Figure 5.8 is given as Example 8.6.

EXAMPLE 8.6

Update Inventory.
for each item accepted record **do**
Search inventory file using item code
if successful
then retrieve inventory record
qty. in stock ← qty. in stock + qty. in item accepted record
write updated record in inventory file
else create new record in inventory file and enter accepted item in it
endif
endfor

In Examples 8.7 we give a structured English description of the stores process of Figure 5.8

EXAMPLE 8.7

```
Requisition of items from store
  for each requisition
        search inventory file if item code available
    if item code found
          retrieve item record
            if qty. requisitioned ≤ qty in stock
                then issue item note
                      qty. in stock = qty in stock – qty issued
                  if qty. in stock ≤ reorder level
                      then send reorder note to purchase dept.
                else enter record in back order file
                  send out of stock note to customer
                  send reorder note to purchase dept.
                endif
  endfor
```

In the above procedure we have not specified what is to be done with the records in the back order file. Item in back order file should be automatically issued to customer who requisitioned it when the item arrives in the store. The procedure for this is given in Example 8.8.

EXAMPLE 8.8

```
  while records are left in back order file
    Retrieve record
      Search inventory file for item code and retrieve it
      if qty. in back order ≤ qty. in inventory
        issue item to customer
      else send message to customer to wait
      endif
  endwhile
```

The periodicity of checking back order file has to be decided by management. It is usually done every day. In Example 8.7 if item code not found an error message must be printed.

Processing rules for the other processes in Chapter 5 can also be expressed in structured English. This is left as an exercise to the reader.

SUMMARY

1. Procedures are used to transform input data into processed results.
2. Procedure can be described in natural English. Such a description is often ambiguous due to impreciseness of natural English.
3. Structured English attempts to describe verbal statements made in natural English more precisely.
4. Precision is obtained by using ideas of logic and block structuring as in programming languages.

5. Decision tables are more appropriate when a large number of conditions are to be checked in arriving at a set of actions.
6. In structured English a number of key words are used for representing commonly used operations. Some important keywords are; **search, retrieve, if, then, else, do, case, while, for.**
7. The important control structures used to describe a process are: Selection structure (**if then else** or **case**), loop structure (**for, while** and **repeat until**).

EXERCISES

8.1 A bank has the following policy on deposits: On deposits of ₹ 5000 and above and for three years or above the interest is 12%. On the same deposit for a period less than 3 years it is 10%. On deposits below ₹ 5000 the interest is 8% regardless of the period of deposit. Write the above process using

i. Structured English

ii. A decision table

8.2 An organization maintains an employee file in which each record has the following data:

(Employee No., employee name, employee gross pay).

It has been decided to increase the pay as per the following formula:

For pay of ₹1000 or less increase 15%

For pay of more than ₹1000 but up to ₹2500 increase 10%.

For pay over ₹2500 increase 5%.

i. Write a structured English processing rule corresponding to the above policy

ii. Express the policy as a decision table.

8.3 An offshore gas company bills its customer according to the following rate schedule:

First 500 litres ₹10 (flat)

Next 300 litres ₹1.25 per 100 litres

Next 30,000 litres ₹1.20 per 100 litres

Next 1,00,000 litres ₹1.10 per 100 litres

Above this ₹1.00 per 100 litres.

The input record has customer identification, name and address, meter reading, past and present. Write a structured English procedure to obtain a bill for the customer.

8.4 Develop the processing rules in structured English for the stores issue and control system shown in Figure 7.6.

8.5 Obtain the processing rules in structured English for the accounting process shown in Figure 5.10.

8.6 Develop the processing rules in structured English for the purchase process of Figure 5.9.

8.7 State the processing rules in structured English for the various processes in the data flow diagram for the hostel (Figure 7.11, Figure 7.12).

8.8 Specify processing rules in structured English for the Billing System shown in Figure 7.6 and Figure 7.7.

9

Decision Tables

LEARNING GOALS

After reading this chapter a student should be able to:

1. Obtain decision tables to represent complex business rules.
2. Detect incompleteness, ambiguity and contradictions in decision table representation of business rules.
3. Simplify complex business rules by removing redundant conditions and rules by using decision tables.

In Chapter 8, we discussed how structured English is used to represent data processing procedures. Structured English is *procedural.* In other words, it describes in detail *how data* would be processed. The details of how data would be processed do not concern most managers who use information systems. They are in fact interested in what business rules are used for processing data. Such specification of rules is *non-procedural.* Decision tables are non-procedural specification of decision rules. They have been used extensively in systematizing design of information systems. They have also been used for communicating and documenting complex business rules. The format of decision tables allows systematic validation of specifications. Algorithms have also been developed to automatically convert decision tables to computer programs.

We have already used decision tables informally in Chapters 5 and 8. In this chapter we formally introduce decision tables and discuss in some detail how they can be checked for incompleteness of specifications, inconsistency in specifications and redundancy in specifications.

9.1 DECISION TABLE TERMINOLOGY AND DEVELOPMENT

A decision table defines a logical procedure by means of a set of *conditions* and related *actions.* We will discuss how a written description of a procedure (commonly known as specification of business rules) is expressed in the form of a decision table. This is best done by using a set of illustrative examples.

EXAMPLE 9.1

A bank uses the following rules to classify new accounts. If a depositor's age is 21 or above and if the deposit is ₹100 or more, classify the account (type) as A. If the depositor is under 21 and the deposit is ₹100 or more, classify it as account B. If the depositor is 21 or over and the deposit is below ₹100 classify it as account C. If the depositor is under 21 and the deposit is below ₹100, do not open an account.

In order to obtain a decision table corresponding to the above problem statement, one has to first isolate in each sentence a part which spells out the conditions, and another part which spells out actions which are to be performed, based on the results obtained by testing the conditions. For the above problem the relevant condition clauses and action clauses can be stated as follows:

RULE 1: **if** a depositor's age is 21 or above (condition 1) and
if the deposit is ₹100 or more (condition 2)
then classify the account as type A (action 1).

RULE 2: **if** the depositor's age is under 21 (condition 3) and
if the deposit is ₹100 or more (condition 2)
then classify the account as type B (action 2).

RULE 3: **if** the depositor's age is 21 or over (condition 1) and
if the deposit is below ₹100 (condition 4)
then classify the account as type C (action 3).

RULE 4: **if** the deposit is below ₹100 (condition 4) and
if the depositor's age is under 21 (condition 3)
then do not open an account (action 4).

These conditions can be written as

Condition 1: Depositor's age ≥ 21
Condition 2: Deposit ≥ ₹100
Condition 3: Depositor's age < 21
Condition 4: Deposit < ₹100

Observe that conditions 1 and 3 are not independent; they are complementary. In other words, if the answer to the question "Depositor's age ≥ 21" is No, the answer to the question "Depositor's age < 21" is automatically Yes. Similarly, conditions 2 and 4 are complementary. Using this, the rules may be expressed in a concise tabular form as shown in Table 9.1. The notation is explained in Table 9.2. A decision table is divided into four parts by horizontal and vertical double lines. The part above the horizontal double lines pertains to the conditions being tested (i.e., the questions asked) and the outcomes of these tests (i.e., the answers to these questions). The part to the left of the vertical double lines in the top portion of the table (top left quadrant) lists the conditions and is known as the *condition stub.* The part to right of the double lines lists the answers to these questions, namely, Yes or No (which are usually abbreviated as Y and N). The top right quadrant of the table is said to contain the *condition entries*. Each row in the top half of the table listing a question and its answer is called a *condition row.*

The bottom half of the decision table, namely, the part below the horizontal double lines, lists the actions to be performed. The left bottom half quadrant is called the *action*

Table 9.1 Decision Table for Bank Account Classification

	Rule 1	*Rule 2*	*Rule 3*	*Rule 4*
Condition C1: Depositor's age ≥ 21	Yes	No	Yes	No
Condition C2: Deposit ≥ ₹ 100	Yes	Yes	No	No
Action 1: Classify as account A	X	–	–	–
Action 2: Classify as account B	–	X	–	–
Action 3: Classify as account C	–	–	X	–
Action 4: Do not open account	–	–	–	X

Table 9.2 Notation Used in Decision Tables

			R1 R2 Rn	
Condition stub	C1 C2 ⋮ Ck	Rule	Condition row	Condition entries
Action stub	A1 A2 ⋮ Am		Action row	Action entries

stub and in it the actions to be carried out are listed. The right bottom quadrant has crosses (X) and dashes (–) entered. A cross indicates that the action listed in that row has to be performed and a dash indicates that the action is not performed. These are called *action entries*. Each row in this half of the table is called an *action row*. Vertical columns to the right of the vertical double lines are called *rules*. A rule is interpreted by using the condition listed in the condition stub and the answers which are listed as condition entries. The corresponding actions are obtained from the action stub and entries. For instance, Rule 3 (Table 9.1) is interpreted as follows:

If depositor's age is ≥ 21 and **if** the deposit is < ₹ 100 **then** classify it as account C. Observe the AND connective between condition clauses. This connective is assumed to be present in the decision table. It is also assumed that the order in which the questions are asked in a rule is arbitrary. Thus Rule 3 may also be written as:

If the deposit is < ₹ 100 and **if** the depositor's age is ≥ 21 **then** classify it as account C.

The order of appearance of the rules in the decision table is also arbitrary. For example, columns 1 and 4 in Table 9.1 may be interchanged and the meaning of the table does not change. The order of appearances of actions, however, is important. Actions are performed from top to bottom in the order in which they appear.

We will now illustrate the formulation of decision tables with a few more examples.

EXAMPLE 9.2

Applications for admission to an extension course are screened using the following rules: For admission, a candidate should be sponsored by his employer and he should possess prescribed minimum academic qualifications. If his fee is also paid, then he is sent a letter of admission. If his fee is not paid, then a letter of provisional admission is sent. In all other cases a letter of regret is sent. To arrive at a decision table, we first isolate the conditions and actions. The condition to be tested are:

Condition 1: Is the applicant sponsored by an employer?
Condition 2: Does the applicant possess the prescribed minimum academic qualifications?
Condition 3: Is the fee paid?

The action is to be taken are:

1. Send letter of admission
2. Send letter of provisional admission
3. Send regret letter

Having isolated the conditions and the actions, the rules are expressed in the form of a decision table as shown in Table 9.3.

Table 9.3 Decision Table for Admission

	R1	R2	ELSE
C1: Is the applicant sponsored?	Y	Y	
C2: Does he possess prescribed minimum qualifications?	Y	Y	
C3: Is the fee paid?	Y	N	
A1: Send letter of admission	X	–	–
A2: Send letter of provisional admission	–	X	–
A3: Send regret letter	–	–	X

Observe here a rule called "ELSE". This rule is used to specify all cases not explicitly covered by Rules 1 and 2. This corresponds to the statement "in all other cases send regret letter". The ELSE rule is a useful device to make a table complete, that is, to cover all eventualities not explicitly stated in the word statement. Therefore, it should be used carefully.

EXAMPLE 9.3

A policy to be followed in a stores inventory system is stated as follows:

1. If the quantity of an item ordered by a customer is available in the store, then it is shipped. The quantity of the specified item remaining in the store is checked against the reorder level. If it is below the reorder level then a reorder procedure is initiated.

2. If the quantity ordered by the customer is greater than that in stock, he is asked whether he would be willing to accept partial shipment. If he is willing, then the available quantity is shipped, reorder is initiated, and quantity in stock is set to zero. The quantity to be shipped later is entered in a back order file. If the customer does not accept partial shipment, then nothing is shipped and his entire order is entered in the back-order file and reorder is initiated. The relevant conditions and actions in this example are:

Conditions:

C1: Is the quantity ordered ≤ quantity in stock?
C2: Is the (quantity in stock – quantity ordered) ≤ reorder level?
C3: Would the customer accept partial shipment?

Actions:

A1: Set quantity shipped = quantity ordered
A2: Set quantity shipped = quantity in stock
A3: Set quantity shipped = 0
A4: Set quantity in stock = 0
A5: Enter (quantity ordered – quantity shipped) in the backorder file.
A6: Initiate reorder procedure
A7: Quantity in stock ← quantity in stock – quantity shipped

The decision table is shown as Table 9.4. Observe the entry marked by dash in the table. This is called a dash or a "don't care" entry. The dash entry is used whenever the answer to a question is irrelevant in arriving at an action. The question "Is partial shipment acceptable?" is irrelevant in rules R1 and R2 as the quantity ordered is less than or equal to the quantity in stock. In rules R3 and R4, as the quantity ordered is greater than that in stock, an automatic reordering action is required.

Table 9.4 Decision Table Showing Shipping Rules

	R1	R2	R3	R4
C1: Quantity ordered ≤ qty. in stock?	Y	Y	N	N
C2: (Qty. in stock – qty. ordered) ≤ reorder level.	N	Y	–	–
C3: Is partial shipment acceptable?	–	–	Y	N
A1: Set qty. shipped = qty. ordered	X	X	–	–
A2: Set qty. shipped = qty. in stock	–	–	X	–
A3: Set qty. shipped = 0	–	–	–	X
A4: Set qty. in stock = 0	–	–	X	–
A5: Enter (qty. ordered – qty. shipped) in back order file	–	–	X	X
A6: Initiate reorder procedure	–	X	X	X
A7: Quantity in stock ← qty. in stock – qty. shipped	X	X	–	–

It should also be observed that some rules in this example require more than one action to be performed. Further, action A5 depends on actions A1, A2 and A3. In interpreting a rule, it is assumed that the actions against which crosses are placed are executed in the order in which they appear from top to bottom in the action stub.

9.2 EXTENDED ENTRY DECISION TABLES

So far we have considered examples in which the questions asked have only Yes or No as answer. It was thus logical to separate the questions from the answers by entering the questions in the condition stub and the answers in the condition entry part of the decision table. However, there are a number of problems in which a question can have multiple answers and it is clearly and concisely expressed with the questions being extended into the concerned entry part of the decision table. We will now consider an example of such a decision table.

EXAMPLE 9.4

A company markets two products to three types of customers. It has the following discount policy:

The products are bulbs and fans, and the customers are classified as retailers, distributors and government agencies. If the order is from a retailer for amounts up to ₹500, it allows 5% discount. If it is from a distributor, 7.5% discount is given. On retail orders exceeding ₹500, 7.5% discount is allowed. For the same order from a distributor, 10% discount is given. In all the above cases a flat discount of 6% is given to government agencies. The above policies apply only to orders for bulbs. A flat discount of 5% is given on orders for fans regardless of the amount of purchase or the customer classification.

Analyzing the word statement of the problem we see that the action, namely, the percent discount to be allowed, is dependent on three conditions:

1. The type of product
2. The type of customer
3. The amount of the order

The two products, namely, bulbs and fans, may be coded respectively as 1 and 2. Similarly, the three types of customers, namely, retailers, distributors, and government agencies, could be coded respectively as A, B and C. Using these codes the policy may be formulated as a decision table (Table 9.5).

Table 9.5 Decision Table Showing Discount Policy

	R1	R2	R3	R4	R5	R6
C1: Product Code	1	1	1	1	1	2
C2: Customer Code	A	B	A	B	C	–
C3: Order Amount	≤ 500	≤ 500	> 500	> 500	–	–
Discount =	5%	7.5%	7.5%	10%	6%	5%

The rules may be interpreted with the help of Table 9.5. For instance, rule R4 states that: If the product code equals 1 (namely, if the product is bulb) and the customer code equals B (namely, if the customer is a distributor) and if the order amount is greater than ₹500, then allow 10% discount.

It should be observed that each question is formulated by combining the statement in the condition stub with that in the condition entry of the decision table. Such a table is called an *extended entry decision table*. This may be contrasted with the tables in the previous examples where questions are written in the condition stub and their answers in the condition entry part of the decision table. Such tables are called *limited entry decision tables*.

It should be noted that in an extended entry decision table more than one question is asked in a condition row of the table. This enables the table to be expressed in a condensed form. The same word statement can also be expressed as a limited entry decision table as shown in Table 9.6. Observe that the number of condition rows are more as compared to Table 9.5.

Table 9.6 Limited Entry Decision Table Corresponding to Table 9.5

	R1	R2	R3	R4	R5	R6
C1: Product code = 1?	Y	Y	Y	Y	Y	N
C2: Customer code = A?	Y	–	Y	–	–	–
C3: Customer code = B?	–	Y	–	Y	–	–
C4: Customer code = C?	–	–	–	–	Y	–
C5: Order amount ≤ 500	Y	Y	N	N	–	–
A1: Discount 5%	X	–	–	–	–	X
A2: Discount 6%	–	–	–	–	X	–
A3: Discount 7.5%	–	X	X	–	–	–
A4: Discount 10%	–	–	–	X	–	–

Table 9.5 could also be shown in another form, with some questions extending to the entry part and others limited to the stub. This is illustrated in Table 9.7. Such a table is called *mixed entry decision table*.

Table 9.7 Mixed Entry Decision Table Corresponding to Table 9.5

	R1	R2	R3	R4	R5	R6
C1: Product code = 1?	Y	Y	Y	Y	Y	N
C2: Customer code	A	B	A	B	C	–
C3: Order amount ≤ 500	Y	Y	N	N	–	–
Discount	5%	7.5%	7.5%	10%	6%	5%

The choice of the form in which a decision table is formulated depends on the problem, available software system, need to standardise, debugging, ease of communication with lay persons, etc. We will now discuss a more complex example which is formulated as a set of decision tables rather than as a single table.

EXAMPLE 9.5

A set of rules in a public service for promotion is as follows: The service has six salary points. Provided a candidate's conduct, diligence and efficiency are considered satisfactory, and he has spent one year as a Class I officer, and has satisfactorily passed the departmental test, he advances to the next higher salary point from points 1 or 2. If he is in a higher salary point, then if his conduct, diligence and efficiency are considered satisfactory, and if one year has elapsed since his last increment and he has satisfactorily completed a departmental course, then he advances to the next higher salary point.

Analyzing this word statement we isolate the following conditions which affect the action to be taken, namely, to promote or not to promote:

1. Conduct
2. Diligence
3. Efficiency
4. Duration of service as Class I officer
5. Result of departmental test
6. Current salary point
7. Time elapsed since last increment
8. Completion of departmental course.

One may attempt to obtain a single decision table in which all the eight conditions are used. It will, however, be error prone and less readable. We will break up the procedure into a set of linked decision tables. On reading the statement we see that every action requires testing conditions 1, 2 and 3 in the list, namely, the candidate's conduct, diligence and efficiency. It is therefore advisable to do the testing at the beginning. Further, if a candidate is already in the highest salary point (namely, salary point 6) the promotion policy is not applicable to him. We will thus have these four conditions tested in one decision table at the beginning of the procedure.

If a candidate is found eligible for promotion, then the promotion rules depend on whether his salary point is less than or equal to 2 or greater than 2. The next table could thus test the salary point and if it is less than or equal to 2, test conditions 4 and 5 (given in the list) and arrive at an action. If the salary point is above 2, then one may proceed to another table which tests conditions 7 and 8. This analysis leads to the three decision tables given in Table 9.8.

Table 9.8 A Set of Linked Decision Tables

Decision Table 1

Salary point = 6	N	ELSE
Conduct OK?	Y	
Diligence OK?	Y	
Efficiency OK?	Y	
GOTO Decision Table 2	X	–
No promotion	–	X

Decision Table 2

	R1	R2	R3	R4
Salary point > 2	N	N	N	Y
1 Year as Class I officer?	Y	N	–	–
Departmental test passed?	Y	–	N	–
Advance to next salary point	X	–	–	–
No promotion	–	X	X	–
GOTO Decision Table 3	–	–	–	X

Decision Table 3

Completed departmental course?	Y	ELSE
1 year since last increment?	Y	
Advance to next salary point	X	–
No promotion	–	X

We will now see how this set of linked decision tables can be shown in the form of a chart to graphically depict the linkages between different decision tables. Assume that an employee record is given as in Table 9.9.

Table 9.9 Employee Record Format

Employee's identity number	Name	Conduct	Diligence	Efficiency	Salary point	Years as Class I officer	Dept. test	Dept. course	Date of last increment

This has all the requisite data to arrive at a decision. The employee file will consist of a set of such records. Given an employee file, the decision table-oriented chart of Figure 9.1 can be used to implement the given rules.

9.3 ESTABLISHING THE LOGICAL CORRECTNESS OF DECISION TABLES

When a decision table is formulated, it is essential to check if it is correct. One of the great advantages of using decision tables is that it is reasonably simple to check whether the table is correct. In this section we will show how it can be done. Consider the decision table given in Table 9.10.

Table 9.10 Decision Table with Dash Entries

	R1	R2
C1: x > 60	Y	–
C2: x < 40	–	Y
A1	X	–
A2	–	X

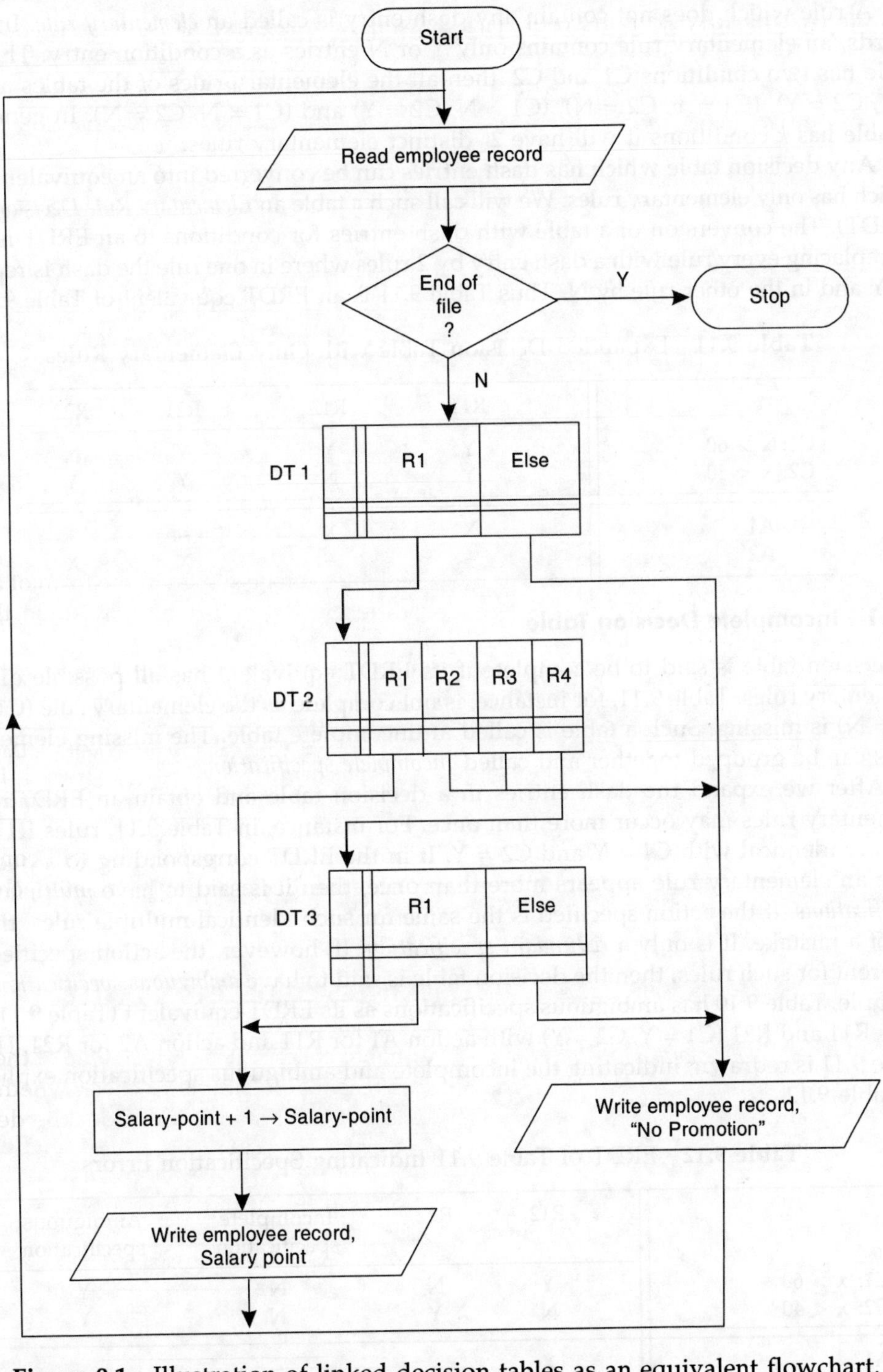

Figure 9.1 Illustration of linked decision tables as an equivalent flowchart.

A rule which does not contain any dash entry is called an *elementary rule.* In other words, an elementary rule contains only Y or N entries as a condition entry. Thus if a table has two conditions C1 and C2, then all the elementary rules of the tables are (C1 = Y, C2 = Y), (C1 = Y, C2 = N), (C1 = N, C2 = Y) and (C1 = N, C2 = N). In general, if a table has k conditions it will have 2^k distinct elementary rules.

Any decision table which has dash entries can be converted into an equivalent table which has only elementary rules. We will call such a table an *Elementary Rule Decision Table* (ERDT). The conversion of a table with dash entries for conditions to an ERDT is done by replacing every rule with a dash entry by 2 rules where in one rule the dash is replaced by Y and in the other rule by N. Thus Table 9.11 is an ERDT equivalent of Table 9.10.

Table 9.11 Expanded Decision Table with Only Elementary Rules

	R11	R12	R21	R22
C1: x > 60	Y	Y	Y	N
C2: x < 40	Y	N	Y	Y
A1	X	X	–	–
A2	–	–	X	X

9.3.1 Incomplete Decision Table

A decision table is said to be complete if its ERDT equivalent has all possible distinct elementary rules. Table 9.11, for instance, is not complete as the elementary rule (C1 = N, C2 = N) is missing. Such a table is called an incomplete table. The missing elementary rules can be grouped together and called *incomplete specification*.

After we expand the dash entries in a decision table and obtain an ERDT, some elementary rules may occur more than once. For instance, in Table 9.11, rules R11 and R21 are identical with C1 = Y and C2 = Y. If in the ERDT corresponding to a decision table an elementary rule appears more than once, then it is said to have *multiplicity of specifications.* If the action specified is the same for such identical multiple rules, then it is not a mistake. It is only a *redundant specification.* If, however, the action specified are different for such rules, then the decision table is said to have *ambiguous specification.* For example, Table 9.10 has ambiguous specifications as its ERDT equivalent (Table 9.11) has rules R11 and R21 (C1 = Y, C2 = Y) with action A1 for R11 and action A2 for R21. Hence Table 9.11 is redrawn, indicating the incomplete and ambiguous specification explicitly, as Table 9.12.

Table 9.12 ERDT of Table 9.11 Indicating Specification Errors

	R12	R22	Incomplete specification	Ambiguous specification
C1: x > 60	Y	N	N	Y
C2: x < 40	N	Y	N	Y
A1	X	–	–	–
A2	–	X	–	–
Error	–	–	X	–

Consider the ambiguous specification of Table 9.12. The rule C1 = Y, C2 = Y is satisfied if $x > 60$ and $x < 40$. No value of x exists which can satisfy both these conditions. Thus the ambiguous specification *is logically impossible.* A decision table is said to have a *logically impossible ambiguity* if it contains ambiguous specification such that no values can be assigned to the variables in the condition stub of the decision table which satisfies that specification. If values can be assigned to the variables in the condition stub of the decision table which satisfy the ambiguous specification, then the decision table is said to contain a *real ambiguity*. A logically impossible ambiguity is not an error, whereas a real ambiguity is an error.

Consider the incomplete specification of Table 9.12, namely, C1 = N, C2 = N. This rule is satisfied if $x \le 60$ and $x \ge 40$. A value of x (say, $x = 50$) satisfies these inequalities. Thus the incomplete specification is a real *incompleteness.* It is an error. If no values can be assigned to the variables in the condition stub which satisfies an incomplete specification, then the *incompleteness is logically impossible.* A decision table has a *logical error* whenever it has real ambiguity or real incompleteness. Such tables should be corrected by the systems analyst at the formulation stage.

9.4 USE OF KARNAUGH MAPS TO DETECT LOGICAL ERRORS IN DECISION TABLES

Incomplete and ambiguous specifications in a decision table can be perceived quickly if the table is mapped onto a two-dimensional diagram called a *Karnaugh map* (named after its inventor Karnaugh and abbreviated as K-map). A map of two conditions is shown in Figure 9.2. Each square in the map corresponds to one elementary rule. The square marked 0 corresponds to the rule (C2 = N, C1 = N), the one marked 1 to (C2 = N, C1 = Y), and so on. In Figure 9.3, we have mapped Table 9.11. Observe that an action or actions correspond to each elementary rule in the decision table and is entered in the corresponding square of the K-map. By inspection of the map it is obvious that no action is specified for (C1 = N, C2 = N) and that the rule (C1 = Y, C2 = Y) is ambiguous due to two action entries in the corresponding square. Besides providing a quick check, a K-map is also useful in reducing redundancies in a decision table.

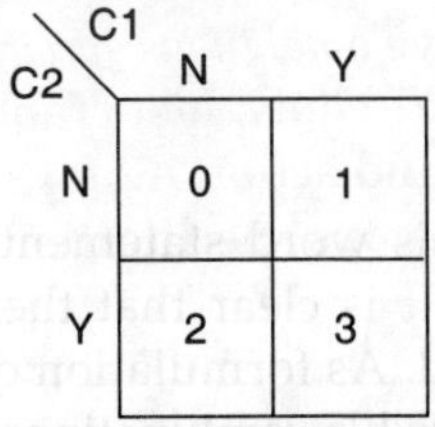

Figure 9.2 A two-variable Karnaugh map.

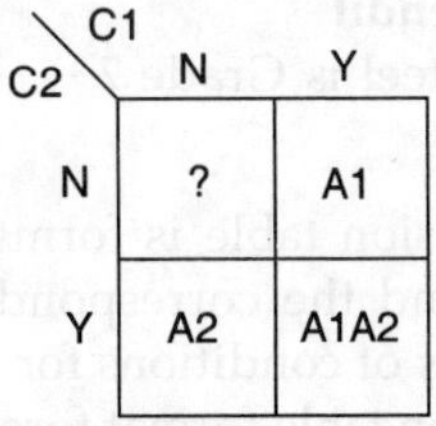

Figure 9.3 Mapping a decision table onto a Karnaugh map.

A K-map for a decision table with three variables is shown in Figure 9.4. The square marked A corresponds to C1 = N, C2 = Y, C3 = Y. It should be noted that any two physically adjacent squares in the map will not have more than one condition changing from N to Y or Y to N. For example, the square marked B has C1 = Y, C2 = Y and C3 = Y. Only C1 is different for squares marked A and B. The squares marked P and Q also are "neighbours" in the sense that in their case, only one condition value (C1) is different.

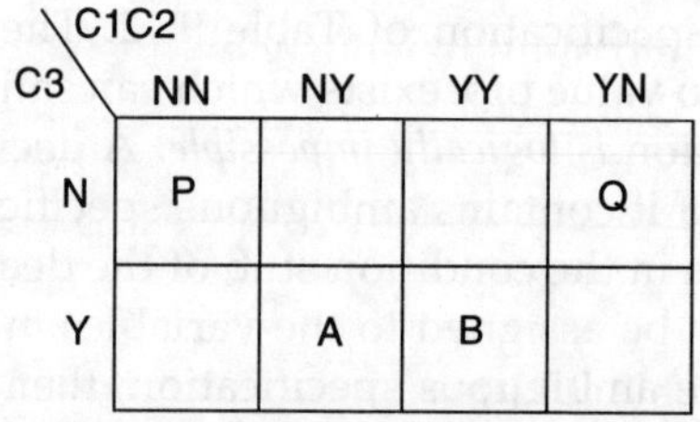

Figure 9.4 A three-variable Karnaugh map.

We now take another example to illustrate detection of incomplete specification.

EXAMPLE 9.6

A certain steel is graded according to the results of three tests given below:

1. Carbon content < 0.7
2. Rockwell hardness > 50
3. Tensile strength > 30,000 kg/cm^2

The steel is graded 10 if it passes all 3 tests, 9 if it passes only tests 1 and 2, 8 if it passes only test 1, and 7 if it passes none of the tests.

If one represents the above problem statement in structured English (as shown below), then the incomplete nature of the statement will not be obvious.

STRUCTURED ENGLISH STATEMENT FOR EXAMPLE 9.6

```
if Carbon content < 0.7
   then if Rockwell hardness > 50
          then if Tensile strength > 30,000
                 then steel is Grade 10
                 else steel is Grade 9
                 endif
          else steel is Grade 8
          endif
   else steel is Grade 7
endif
```

If a decision table is formulated corresponding to this word statement we obtain Table 9.13 and the corresponding K-map of Figure 9.5. It is clear that there are four combinations of conditions for which no action is specified. As formulation of problems in the decision table format forces analyst to exhaust all possible combinations of answers to questions, any incompleteness in the specifications can be pinpointed. The analyst must now get back to the person who gave the rules to get a clarification.

Table 9.13 Decision Table for Example 9.6

C1: Carbon content < 0.7	Y	Y	Y	N	Y	N	N	N
C2: Rockwell hardness > 50	Y	Y	N	N	N	Y	Y	N
C3: Tensile strength > 30,000 kg/cm^2	Y	N	N	N	Y	Y	N	Y
Grade	10	9	8	7	?	?	?	?

C3 \ C1C2	NN	NY	YY	YN
N	7	?	9	8
Y	?	?	10	?

Figure 9.5 Table 9.13 mapped onto a Karnaugh map.

EXAMPLE 9.7

Consider a decision table depicting the actions to be taken in an instalment payment system shown in Table 9.14. There are five distinct action sets (i.e., combination of actions). These action sets are marked onto a K-map in Figure 9.6.

Table 9.14 Decision Table for Instalment Payment System

	R1	R2	R3	R4	R5	R6
C1: Payment in current month > minimum specified payment.	Y	N	N	–	–	–
C2: Payment in current month > 0?	–	Y	Y	–	N	N
C3: Any payment in last 3 months?	–	–	–	N	Y	Y
C4: Actual arrears > 3 (minimum specified payment per month)	–	Y	N	Y	N	Y
A1: Send letter A	X	–	–	–	–	–
A2: Send letter B	–	X	–	–	–	–
A3: Send letter C	–	–	X	–	–	–
A4: Send letter D	–	–	–	X	–	X
A5: Send letter E	–	–	–	–	X	–

C3C4 \ C1C2	NN	NY	YY	YN
NN		A3	A1	* A1
NY	A4	A2A4	A1A4	* A1A4
YY	A4	A2	A1	* A1A4
YN	A5	A3	A1	* A1A5

Figure 9.6 Karnaugh map for Table 9.14.

Observe from the K-map (Figure 9.6) that the table is incomplete as no action is specified for C1 = N, C2 = N, C3 = N and C4 = N. It is also ambiguous as 5 boxes in the K-map have more than one action specified. Thus this K-map needs further examination to remove incompleteness and ambiguities. In order to do this we examine the conditions in detail.

Observe that an answer of Y to question C1 and simultaneously an answer of N to C2 is logically impossible. This can be checked by the following method. Let the payment in the current month be represented by x and the payment in the last three months by y. Let z represent the actual arrears and m the minimum specified payment. Thus conditions C1–C4 of the decision table may be represented by the inequalities.

C1: $x > m$
C2: $x > 0$
C3: $y > 0$
C4: $z > 3\,m$

The minimum specified payment m must be greater than zero. It is clear that C3 and C4 are independent and also not dependent on C1 and C2. In other words, answers to C3, C4 are unrelated and also not related to answers for C1, C2. To check logical incompleteness and consistency, we have to pay attention only to conditions C1 and C2. The four possible combinations of values for C1 and C2 are: YY, YN, NY and NN. In terms of the inequalities we have:

$$\left.\begin{array}{lll} \text{C1:} & \text{Y:} & x > m \\ \text{C2:} & \text{Y:} & x > 0 \end{array}\right\} \text{ logically possible as } m > 0$$

$$\left.\begin{array}{lll} \text{C1:} & \text{Y:} & x > m \\ \text{C2:} & \text{N:} & x \le 0 \end{array}\right\} \text{ logically impossible as } m > 0$$

$$\left.\begin{array}{lll} \text{C1:} & \text{N:} & x \le m \\ \text{C2:} & \text{Y:} & x > 0 \end{array}\right\} \text{ logically possible as } m > 0$$

$$\left.\begin{array}{lll} \text{C1:} & \text{N:} & x \le m \\ \text{C2:} & \text{N:} & x \le 0 \end{array}\right\} \text{ logically possible as } m > 0$$

Thus all rules in the decision table with C1 = Y and C2 = N (regardless of the values of C3 and C4) are logically impossible. Thus there are four *logically impossible rules* in this decision table. This fact is illustrated in the Karnaugh map of Figure 9.6 by marking the boxes, for which C1 = Y and C2 = N, with a star (*).

In general, when the conditions in a decision table are specified as linear inequalities in reals or integers, the problem of finding logically impossible rules is equivalent to finding whether a set of inequalities are consistent. Algorithms exist to decide this. As most inequalities occurring in the context of data processing are simple inequalities in integers or reals, such algorithms are also applicable.

9.4.1 Incomplete Specification

Continuing with our example, we see by inspection of the K-map (Figure 9.6) that no action is specified when C1 = N, C2 = N, C3 = N, C4 = N. As this rule is logically possible and no action is specified this decision table is *incomplete*.

9.4.2 Logical Contradiction in Specification

By inspection of the K-map (Figure 9.6), we see that when C1 = N, C2 = Y, C3 = N, C4 = Y and C1 = Y, C2 = Y, C3 = N, and C4 = Y, two contradictory actions are specified.

Both these rules are logically possible. Thus there is a *real contradiction* in the decision table. After this analysis we see that the decision table has logical errors as it is incomplete and has contradictory specifications. The analyst points this out to the user. If the user specifies action A4 for the cases

C1 = N, C2 = N, C3 = N, C4 = N
C1 = N, C2 = Y, C3 = N, C4 = Y
C1 = Y, C2 = Y, C3 = N, C4 = Y

then the corrected decision table for instalment payment is as obtained in Table 9.15 and the corresponding K-map as in Figure 9.7.

Table 9.15 Corrected Decision Table for Instalment Payment

	R1	R2	R3	R4	R5	R6	R7	R8	R9	R10	R11	R12
C1	Y	Y	Y	N	N	N	Y	N	N	N	N	N
C2	Y	Y	Y	Y	Y	Y	Y	Y	N	N	N	N
C3	N	Y	Y	Y	N	Y	N	N	N	N	Y	Y
C4	N	Y	N	Y	N	N	Y	Y	Y	N	Y	N
Action	A1	A1	A1	A2	A3	A3	A4	A4	A4	A4	A4	A5

C3C4 \ C1C2	NN	NY	YY	YN
NN	A4	A3	A1	X
NY	A4	A4	A4	X
YY	A4	A2	A1	X
YN	A5	A3	A1	X

Figure 9.7 Karnaugh map for Decision Table 9.15.

9.5 ELIMINATING REDUNDANT SPECIFICATIONS

After removing contradictions and incomplete specifications, we obtained Table 9.15. It has 12 rules. The question which arises is: "Can the number of rules be reduced?" By inspection of R2 and R3 we see that they both specify action A1 and have C1 = C2 = C3 = Y and C4 = Y for R2 and C4 = N for R3. In fact the rules can be merged into one with C1 = C2 = C3 = Y and C4 = —. This can be seen by expressing rules R2 and R3 as

$$R2 = C1 \cdot C2 \cdot C3 \cdot C4 \tag{9.1}$$

$$R3 = C1 \cdot C2 \cdot C3 \cdot \overline{C4} \tag{9.2}$$

and Action A1 as

$$\begin{aligned} A1 &= R2 \text{ or } R3 && (9.3)\\ &= C1 \cdot C2 \cdot C3 \cdot C4 + C1 \cdot C2 \cdot C3 \cdot \overline{C4} && (9.4)\\ &= C1 \cdot C2 \cdot C3 \cdot (C4 + \overline{C4}) && (9.5)\\ &= C1 \cdot C2 \cdot C3 && (9.6) \end{aligned}$$

We have expressed action A1 by a Boolean expression. The variables C1, C2, C3, C4 are two valued variables (having values Yes or No). The symbol . (dot) denotes a boolean AND operator, + the OR operator, and — (bar) above a variable, a NOT operator. The following identities of Boolean algebra.

$$A + \overline{A} = 1 \quad (9.7)$$

$$1 \cdot A = A \quad (9.8)$$

have been used in Eqs. (9.5) and (9.6). These two identities are used repeatedly to reduce rules in a decision table, e.g., to reduce Table 9.15 to 9.16. Observe that Table 9.16 has eight rules.

Table 9.16 Reduced Table 9.15 with Smaller Number of Rules

C1	Y	Y	N	N	–	N	N	N
C2	Y	Y	Y	Y	Y	N	N	N
C3	N	Y	Y	–	N	N	–	Y
C4	N	–	Y	N	Y	–	Y	N
	A1	A1	A2	A3	A4	A4	A4	A5

As pointed out earlier, K-maps are designed to bring out possibilities of eliminating variables in Boolean expressions by repeated application of the equation

$$A \cdot B + A \cdot \overline{B} = A \cdot (B + \overline{B}) = A \cdot 1 = A \quad (9.9)$$

Consider the K-map of Figure 9.8. Observe that the values of AB are listed in the K-map in the order NN, NY, YY and YN. This choice is made to facilitate the application of the reduction procedure given in Eq. (9.9). Two physically adjacent squares in the K-map represent Boolean terms which are identical except for one variable. For example, the two squares marked P in Figure 9.8 represent $\overline{A} \cdot B \cdot \overline{C}$ and $A \cdot B \cdot \overline{C}$ respectively.

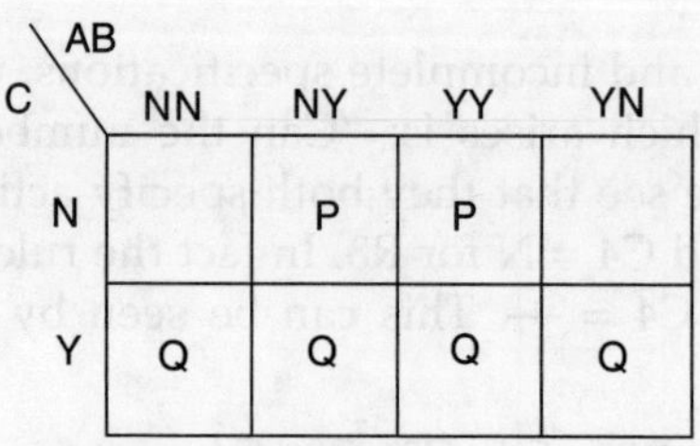

Figure 9.8 Three-variable Karnaugh map showing elimination of variables.

These terms have the common variable $B \cdot \overline{C}$ and the variable A occurs in its complement and true form, respectively. Thus

$$\begin{aligned} P &= \overline{A} \cdot B \cdot \overline{C} + A \cdot B \cdot \overline{C} \\ &= (\overline{A} + A) \cdot B \cdot \overline{C} \\ &= B \cdot \overline{C} \end{aligned}$$

The first and the last squares in row 1 in the K-map of Figure 9.8 are *logically adjacent* because the term for the first box is $\overline{A} \cdot \overline{B} \cdot \overline{C}$ and that for the last box is $A \cdot \overline{B} \cdot \overline{C}$. Thus $\overline{B} \cdot \overline{C}$ is common and the variable A appears in its true form in one term and in its complement form in the other. Therefore,

$$\overline{A} \cdot \overline{B} \cdot \overline{C} + A \cdot \overline{B} \cdot \overline{C} = \overline{B} \cdot \overline{C}$$

It may also be shown that the first and last squares of row 2 in the K-map are logically adjacent. Physical adjacency of these logically adjacent boxes in the K-map may also be assured by imagining the map to be wrapped around a cylinder.

In the above example we combined two adjacent squares to eliminate one variable. The next possible combination is a set of four adjacent squares with the same action specified to eliminate two variables. Referring to Figure 9.8, we see that

$$\begin{aligned} Q &= \overline{A} \cdot \overline{B} \cdot C + \overline{A} \cdot B \cdot C + A \cdot B \cdot C + A \cdot \overline{B} \cdot C \\ &= (\overline{A} \cdot \overline{B} + \overline{A} \cdot B) \cdot C + (A \cdot B + A \cdot \overline{B}) \cdot C \\ &= \overline{A} \cdot (\overline{B} + B) \cdot C + A \cdot (B + \overline{B}) \cdot C \\ &= \overline{A} \cdot C + A \cdot C \\ &= (\overline{A} + A) \cdot C \\ &= C \end{aligned}$$

From the K-map of Figure 9.9, four adjacent squares with same action can be combined again to eliminate two variables as

$$\begin{aligned} R &= \overline{A} \cdot B \cdot \overline{C} + A \cdot B \cdot \overline{C} + \overline{A} \cdot B \cdot C + A \cdot B \cdot C \\ &= (\overline{A} \cdot B + A \cdot B) \cdot \overline{C} + (\overline{A} \cdot B + A \cdot B) \cdot C \\ &= (\overline{A} + A) \cdot B \cdot \overline{C} + (\overline{A} + A) \cdot B \cdot C \\ &= B \cdot \overline{C} + B \cdot C \\ &= B \cdot (\overline{C} + C) \\ &= B \end{aligned}$$

C \ AB	NN	NY	YY	YN
N		R	R	
Y		R	R	

Figure 9.9 Three-variable Karnaugh map showing combination or four adjacent squares.

This elimination can be done by inspection after some practice by observing that in this case when we see four adjacent squares B is always Y, whereas A and C take both N and Y values.

Refer to the K-map for four variables shown in Figure 9.7. The top and bottom rows of this map are also adjacent. Thus the map may be visualized as being wrapped around a cylinder and the cylinder again wrapped around to form a toroid. The principle applied for minimization is the same as that for the three-variable map. Hence for the K-map of Figure 9.7 we can write

$$\begin{aligned} A3 &= \overline{C1} \cdot C2 \cdot \overline{C3} \cdot \overline{C4} + \overline{C1} \cdot C2 \cdot C3 \cdot \overline{C4} \\ &= \overline{C1} \cdot C2 \cdot \overline{C4} \cdot (\overline{C3} + C3) \\ &= \overline{C1} \cdot C2 \cdot \overline{C4} \end{aligned}$$

In the map of Figure 9.7 we have marked the last column as logically impossible rules. This implies that these cases will not arise. Such impossible rules can be used optionally for any action if it helps in minimizing the logical expression for that action. Thus we can combine boxes for A4 as shown in Figure 9.10.

$$A4 = \overline{C3} \cdot C4 + \overline{C3} \cdot \overline{C4} \cdot \overline{C2} + C3 \cdot C4 \cdot \overline{C2}$$

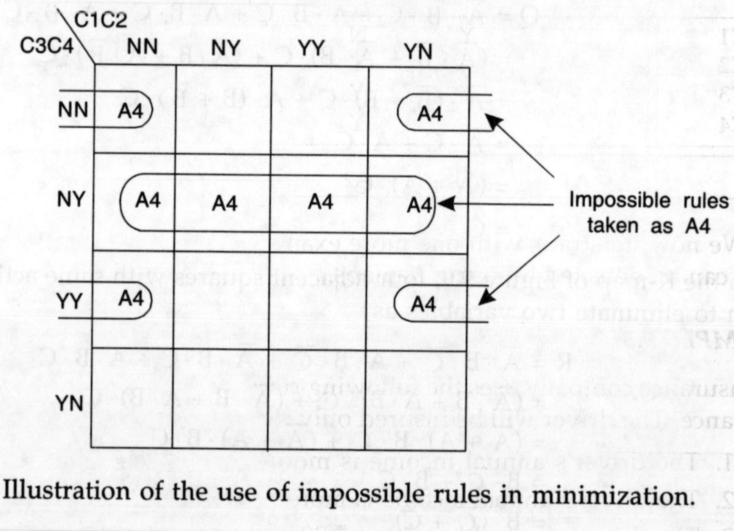

Figure 9.10 Illustration of the use of impossible rules in minimization.

Figure 9.11 shows the use of impossible rules to minimize the expression for A1. By inspection of K-map of Figure 9.11, we have

$$A1 = C1 \cdot \overline{C4} + C1 \cdot C3$$

From K-map of Figure 9.7,

$$A2 = \overline{C1} \cdot C2 \cdot C3 \cdot C4, \qquad A5 = \overline{C2} \cdot C3 \cdot \overline{C4}$$

Thus the decision table of Table 9.15 (which was mapped on the K-map of Figure 9.7) can be rewritten as shown in Table 9.17. Observe that this table also has eight rules (as in

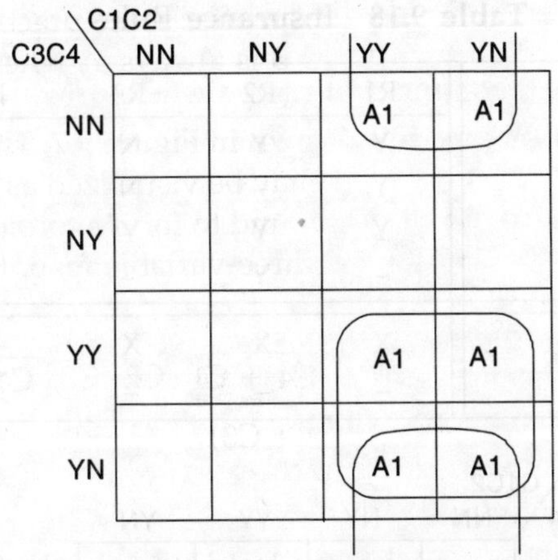

Figure 9.11 Illustration of minimization of rules.

Table 9.16) but there are more dash entries, which implies that there would be less testing of conditions to arrive at actions.

Table 9.17 Reduced Decision Table 9.15

C1	Y	Y	N	N	–	–	–	–
C2	–	–	Y	Y	–	N	N	N
C3	–	Y	Y	–	N	N	Y	Y
C4	N	–	Y	N	Y	N	Y	N
	A1	A1	A2	A3	A4	A4	A4	A5

We now illustrate, with one more example, how redundancy in the specification of rules can be detected by using K-maps.

EXAMPLE 9.8

An insurance company uses the following rules to determine the eligibility of a driver for insurance. The driver will be insured only if any of the following conditions is satisfied:

1. The driver's annual income is more than ₹ 20,000 and he is a married male.
2. The driver's annual income is more than ₹ 20,000 and he is married and over 30.
3. The driver's annual income is ₹ 20,000 or less and the driver is a married female.
4. The driver is a male and over 30
5. The driver is married and is 30 or below

In all other cases the driver is not insured.

The above rules may be expressed as a decision table (Table 9.18). Table 9.18 is mapped onto a K-map of Figure 9.12. Rule 1 implies that C4 can be either Y or N.

The action corresponding to this rule is depicted as A1, and entered in the two boxes C1C2C3C4 = YYYY and C1C2C3C4 = YYYN in the K-map.

$$R1 = C1 \cdot C2 \cdot C3 \cdot C4 + C1 \cdot C2 \cdot C3 \cdot \overline{C4}$$

Table 9.18 Insurance Rules

	R1	R2	R3	R4	R5	
C1: Annual income > ₹20,000	Y	Y	N	–	–	ELSE
C2: Male	Y	–	N	Y	–	
C3: Married	Y	Y	Y	–	Y	
C4: Age > 30	–	Y	–	Y	N	
A1: Insure	X	X	X	X	X	–
A2: Do not insure	–	–	–	–	–	X

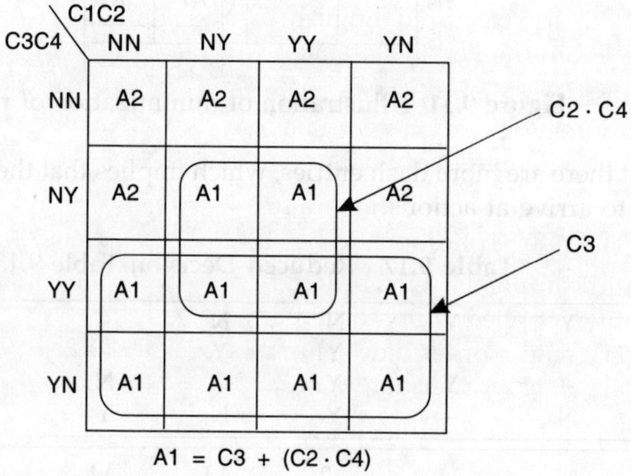

Figure 9.12 Karnaugh map for Table 9.18.

Expanding other rules also we get:

$$R2 = C1 \cdot C2 \cdot C3 \cdot C4 + C1 \cdot \overline{C2} \cdot C3 \cdot C4 = \text{Action A1}$$

$$R3 = \overline{C1} \cdot \overline{C2} \cdot C3 \cdot C4 + \overline{C1} \cdot \overline{C2} \cdot C3 \cdot \overline{C4} = \text{Action A1}$$

$$R4 = \overline{C1} \cdot C2 \cdot \overline{C3} \cdot C4 + C1 \cdot C2 \cdot C3 \cdot C4 + C1 \cdot C2 \cdot \overline{C3} \cdot C4 + \overline{C1} \cdot C2 \cdot C3 \cdot C4 = \text{Action A1}$$

$$R5 = \overline{C1} \cdot \overline{C2} \cdot C3 \cdot \overline{C4} + \overline{C1} \cdot C2 \cdot C3 \cdot \overline{C4} + C1 \cdot \overline{C2} \cdot C3 \cdot \overline{C4} + C1 \cdot C2 \cdot C3 \cdot \overline{C4} = \text{Action A1}$$

Entering these in K-map we get Figure 9.12. Combining the terms in the K-map, the Boolean expression for the action "Insure" may be written as

$$A1 = C3 + (C2 \cdot C4)$$

The above Boolean expression is depicted as a decision table (Table 9.19). In other words, the original set of rules have a number of redundancies which have been eliminated by using Boolean algebra. The reduced set of rules may be stated in words. "Insure if the driver is married or the driver is male and over 30. In all other cases do not insure the driver". Observe that the annual income is irrelevant.

Table 9.19 Reduced Insurance Rules

C2: Male	–	Y	ELSE
C3: Married	Y	–	
C4: Age > 30	–	Y	
Insure	X	X	–
Do not insure	–	–	X

EXAMPLE 9.9

In Chapter 5 we obtained the descriptive statement of rules to be followed when a receiving office gets goods from a vendor. The verbal description was given as a more precise statement in structured English in Chapter 8. Here we specify the rule as a decision table. From the verbal statement we isolate the conditions to be checked as:

C1: Order no. in delivery note = order no. in order file
C2: Item code in delivery note = item code in order file
C3: Vendor code in delivery note = vendor code in order file

If any of these conditions fails, then no further check is needed. We can thus put them in one table. The other conditions to be checked are:

C4: Quantity supplied = Qty. ordered
C5: Quantity supplied > Qty. ordered
C6: Delivery date = order date + number of days allowed for delivery
C7: Delivery date > order date + number of days allowed for delivery

We can specify the rules using the decision tables (Tables 9.20 and 9.21). Using the techniques already employed we will now check the specifications for logical errors and redundancies. The decision table of the receiving process (Table 9.20) is mapped on a K-map in Figure 9.13.

Table 9.20 Receiving Process: Decision Table for Checking Items Received against Order. (**For each** delivery note **do** Table: Receiving Process)

C1: Order no. in delivery note = order no. in order file?	N	Y	Y	Y	Y
C2: Item code in delivery note = Item code in order file	–	N	N	Y	Y
C3: Vendor code in delivery note = Vendor code in order file	–	Y	N	N	Y
Write in discrepancy record "Order not found"	X	–	–	–	
Write in discrepancy record "Incorrect item code"	–	X	X	–	–
Write in discrepancy record "Incorrect vendor code"	–	–	X	X	–
Store discrepancy record in discrepancy file	X	X	X	X	–
Intimate discrepancy to Purchase Officer	X	X	X	X	–
Go to Table 9.21 Quantity and Date check	–	–	–	–	X
Exit	X	X	X	X	–
	A1	A2	A3	A4	A5

Table 9.21 Quantity and Date Check

C4: Quantity supplied = Qty. ordered	Y	N	N	N	N	N	N	Y	Y
C5: Quantity supplied > Qty. ordered	–	Y	N	Y	Y	N	N	–	–
C6: Delivery date = order date + no. of delivery days	Y	Y	Y	N	N	N	N	N	N
C7: Delivery date > order date + no. of delivery days	–	–	–	Y	N	Y	N	N	Y
Write in discrepancy record "Excess qty. received" Excess = qty. supplied – qty. ordered.	–	X	–	X	X	–	–	–	–
Write in discrepancy record "Deficient quantity received". Deficiency = qty. ordered – qty. supplied.	–	–	X	–	–	X	X	–	–
Write in discrepancy record "Late delivery" Late days = delivery date – (order date + no. of delivery days)	–	–	–	X	–	X	–	–	X
Write in discrepancy record "Early delivery" Early days = (order date + no. of delivery days – delivery date)	–	–	–	–	X	–	X	X	–
Store discrepancy record in discrepancy file	–	X	X	X	X	X	X	X	X
Intimate discrepancy to Purchase Officer	–	X	X	X	X	X	X	X	X
Send delivery note to inspection	X	–	–	–	–	–	–	–	–
Exit	X	X	X	X	X	X	X	X	X
	A1	A2	A3	A4	A5	A6	A7	A8	A9

C3 \ C1C2	NN	NY	YY	YN
N	A1	A1	A4	A3
Y	A1	A1	A5	A2

Figure 9.13 Karnaugh map for receiving process.

The table is complete. When C1 = N, then C2 and C3 cannot really be tested. Thus the action taken is identical for all four cases with C1 = N. This is reflected in the K-map. The decision table for quantity and date check (of Table 9.21) is entered in the K-map of Figure 9.14. Observe that an answer of Y to C4 and simultaneously an answer of Y to C5 is logically impossible. Similarly, an answer of Y to C6 and simultaneously an answer of Y to C7 is impossible. This is shown by crossing out the squares corresponding to these answers in the K-map. We see that there are no boxes in K-map with a blank or duplicate entry. Thus the decision table is complete and has no ambiguity. The action A1 has three

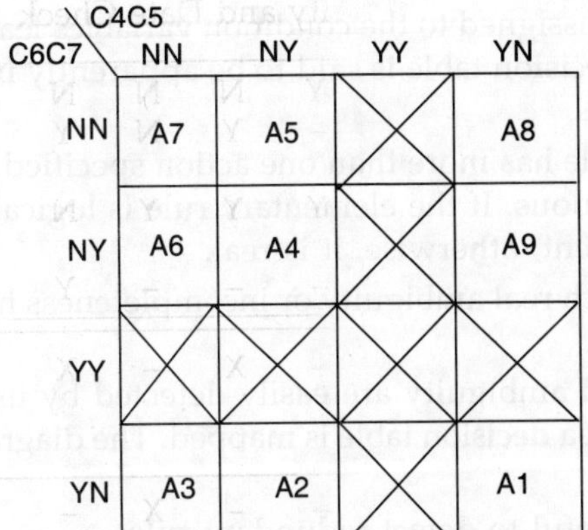

Figure 9.14 Karnaugh map for quantity and date check.

neighbouring boxes in the K-map, which are 'impossible'. Thus they can be used to simplify the rule for A1. Thus the rule for A1 is

$$C4 = Y, C5 = —, C6 = Y \text{ and } C7 = —, \text{ i.e., } C4 \cdot C6$$

Similarly, action A9 has three 'impossible' neighbours. Thus the rule for A9 is

$$C4 = Y, C5 = —, C6 = — \text{ and } C7 = Y, \text{ i.e., } C4 \cdot C7$$

Actions A2 and A4 may also be represented respectively by: A2 : $C5 \cdot C6$ and A4 : $C5 \cdot C7$.

With this analysis we can be reasonably certain that there are no logical errors in the formulation of rules.

SUMMARY

1. Decision tables are used to specify complex business rules. They are non-procedural in contrast with structured English which is a procedural specification.
2. Decision tables are particularly useful for documenting complex business rules with numerous conditions.
3. In order to obtain a decision table from a specification statement, we first isolate in each sentence a part which specifies the conditions and another part which specifies the actions to be performed, based on the result of testing the conditions. Conditions are then grouped in one or more table(s) and the rules are formulated.
4. It is not advisable to have more than four or five conditions per decision table. If there are many conditions they are grouped and a set of tables are formulated and linked.
5. Limited Entry Decision Tables (LEDT) use only Y or N answers to conditions.
6. In Extended Entry Decision Tables (EEDT) conditions are formed by combining the statement in the condition stub with that in the condition entry part of the decision table. EEDTs are more concise.
7. A decision table is said to be complete if it has all possible distinct elementary rules. If any rule is missing it is incomplete.

8. If no values can be assigned to the condition variables leading to the missing rule or rules then the decision table is said to be apparently incomplete. Otherwise it is really incomplete.
9. If an elementary rule has more than one action specified, then the decision table is said to be ambiguous. If the elementary rule is logically impossible, then the ambiguity is apparent; otherwise, it is real.
10. A decision table with real ambiguity or incompleteness has a logical error which should be corrected.
11. Incompleteness and ambiguity are easily detected by using a two-dimensional diagram onto which a decision table is mapped. The diagram is called a Karnaugh map (K-map).
12. K-maps are also useful to detect redundant rules.
13. Using basic ideas of Boolean algebra and K-maps, unnecessary conditions and unnecessary rules are eliminated. The number of conditions to be tested in each rule can also be minimized by using K-maps.
14. Pre-processors are available to automatically convert decision tables to programs in a programming language.

EXERCISES

9.1 The policy followed by a company to process customer orders is given by the following rules:

(i) If the customer order ≤ that in stock and his credit is OK, supply his requirement.

(ii) If the customer credit is not OK do not supply. Send him an intimation.

(iii) If the customer credit is OK but items in stock are less than his order, supply what is in stock. Enter the balance to be sent in a back-order file.

Obtain a decision table for the above policy.

9.2 Obtain a decision table to decide whether on a given date an employee in an organization has completed one year's service or not.

9.3 Obtain a decision table for Exercise 8.3.

9.4 Obtain a decision table for Exercise 8.4.

9.5 Obtain a decision table for Exercise 8.5.

9.6 Obtain a decision table for Exercise 8.6.

9.7 Obtain a decision table for an automatic stamp vending machine with the following specifications:

(i) To dispense 20, 15, 10 or 5 paise stamp

(ii) To accept 50, 25, 10 or 5 paise coin

(iii) Do not return change if it is necessary to return more than two coins of the same denomination. In such a case return the customer's coin and turn on "no change" light.

The machine should dispense a stamp, the right amount of change, "no stamp available", "no change available" signals, etc.

9.8 Obtain a decision table to be used by a person to enter the office of a manager. The conditions to be checked are:

Door open? Ring sign on? Enter sign on? Door locked?

The actions a person takes are:

Ring bell, enter, wait, leave.

After obtaining the decision table, ensure that it has no logical errors and that it is in minimal form.

9.9 A University has the following rules for a student to qualify for a degree with Physics as the main subject and Mathematics as the subsidiary subject:

(i) He should get 50% or more marks in Physics and 40% or more marks in Mathematics.

(ii) If he gets < 50% marks in Physics, he should get 50% or more marks in Mathematics. He should, however, get at least 40% marks in Physics.

(iii) If he gets < 40% marks in Mathematics and 60% or more marks in Physics, he is allowed to reappear in Mathematics examination only so that he can qualify.

(a) obtain an EEDT for the rules
(b) obtain an LEDT for the rules
(c) check the logical correctness of the decision table.

9.10 In instalment buying where payments are made on a weekly basis the action taken on "an account goes into arrears" is a crucial aspect of the operation. Table 9.22 illustrates a simplified arrears procedure. Answer the following questions:

(i) Are all elementary rules logically possible?
(ii) Is the table complete?
(iii) Has the table any ambiguities?
(iv) Are there logical errors in the table? If yes, point them out.
(v) Use reasonable assumptions and correct the table.
(vi) Remove any redundancies in the corrected table (Table 9.22)

Table 9.22 A Simplified Arrears Procedure

C1: This week's cash > weekly rate	Y	Y	N	N	–	–	–	–	–
C2: This week's cash > 0	–	–	Y	Y	–	–	N	N	N
C3: Any cash during last month	–	–	–	–	N	N	Y	Y	Y
C4: Arrears > 2 * weekly rate	–	–	Y	N	–	–	N	Y	–
C5: Arrears > 4 * weekly rate	N	Y	N	–	N	Y	–	N	Y
Send arrears letter A	–	X	–	–	–	–	–	–	–
Send arrears letter B	–	–	X	–	–	–	–	–	–
Send arrears letter C	–	–	–	–	X	–	–	–	–
Send arrears letter D	–	–	–	–	–	–	–	X	–
Notify accounts	X	–	–	X	–	–	X	–	–
Take special action						X	–	–	X
	A1	A2	A3	A1	A4	A5	A1	A6	A5

10

Use Case Method

LEARNING GOALS

After reading this chapter a student should be able to:

1. Identify actors participating in a system.
2. Specify requirements of systems with use cases.
3. Document use cases systematically.
4. Draw use case diagrams.
5. Develop test cases to test a system.

We have seen in the previous chapters how system requirements specifications are expanded from a fuzzily stated user requirements to a set of document flow diagrams, data flow diagrams, process specifications and decision tables. This set of methods together is known as *structured systems analysis and design*. Another method called *object oriented systems analysis and design* has also become popular during the last few years. There are some advantages in using object oriented methods. However, the tools used in structured methods, namely, data flow diagrams, documents flow diagrams, decision tables are all useful in designing information systems. Use case method was proposed by Jacobson [33] in the 90s as one of the tools useful in object oriented systems analysis and design primarily to specify users' requirements. He developed a diagramming method to develop use cases. The method was adopted by several designers and modified based on their experience of using it in many large software design projects. A textual method of writing use cases describing the behaviour of a system was proposed by Cockburn [17] in early 2000. It will be described in this chapter along with use case diagrams.

10.1 WHAT IS A USE CASE?

Simply stated a use case specifies in detail the behaviour of a system being developed to meet a set of goals. We will first explain some terms used in writing use cases. This is a commonly accepted terminology used in the literature. Every use case has a *title*. The title states the goal to be met by the system. It should use an action verb such as : Receive items, Bill customer. The system we are concerned with is called the *scope* of a use case. All parties who interact with the system and are interested in its success are called the *stakeholders* in the system. All the persons or entities who actually perform some tasks in order to fulfill the goal of a use case are called *actors*. In a use case scenario there may be *primary* as well as *secondary* actors. Primary actors initiate first action. *Preconditions* specify what holds true before a use case. These are often assumptions made by a systems analyst. *Guarantees* specify what holds true after a use case is completed. *Main success* scenario is a case when all goes according to plan. *Exceptions* are what could go wrong during a use case and what actions are needed in such cases. We will now give an example of a use case description corresponding to the example requirement statement given in Section 5.1 which we reproduce below for ready reference.

User's narrative of requirements. "Our company receives a number of items from many vendors and they are received at the receiving office. As we receive over 1000 items a day it is now virtually impossible for the receiving office to check whether the vendor has supplied items against an order, or sent a wrong item and inform the purchase office. We are also not able to find out if there are excesses or deficiencies in delivery and whether the vendor adhered to the delivery schedule as per the order. The items received at the receiving office are sent for physical inspection. The physical inspection consists of checking whether the quantities stated in the delivery note agree with the physical count, whether the item is the correct one ordered, and a check on the quality of item. We would like to keep a record of rejections due to bad quality, incorrect items, excess or deficient supply, etc., determined during inspection. This will enable us to assess vendors' reliability to guide us in placing orders in the future, besides keeping track of supplies. Items cleared by the inspection office are taken into the inventory by the stores office which keeps a ledger of items stocked and quantity available of each item. Customers send requisitions to the stores. The stores fulfill the requests based on availability and updates the ledger. Currently, we are not able to meet some of our customers' requests. We would like to incorporate automatic reordering by our purchase office if the inventory level of an item is low. We would also like to keep track of unfulfilled requests and meet them when items reach the store. Currently, we are not able to pay our vendors promptly due to delays in payment order reaching our accounts office. We would like to rectify this. We would also like to bill our customers promptly and keep track of customers' payments".

We refer to Figure 10.1 in which there is a block labeled receiving office. We will develop first a use case corresponding to the receiving office.

The scope of this use case is the behaviour of the receiving office software. The stakeholders of the system are the receiving office clerk, purchase office, vendor and inspection office. The primary actor is the receiving office clerk. We will not be concerned with secondary actors in this case.

Use Case 1: Receive items delivered by vendors

Primary actor: Receiving office clerk (ROC).
Scope: Software which records items received.
Stake holders: Vendor, Receiving office, Purchase office, Inspection office.

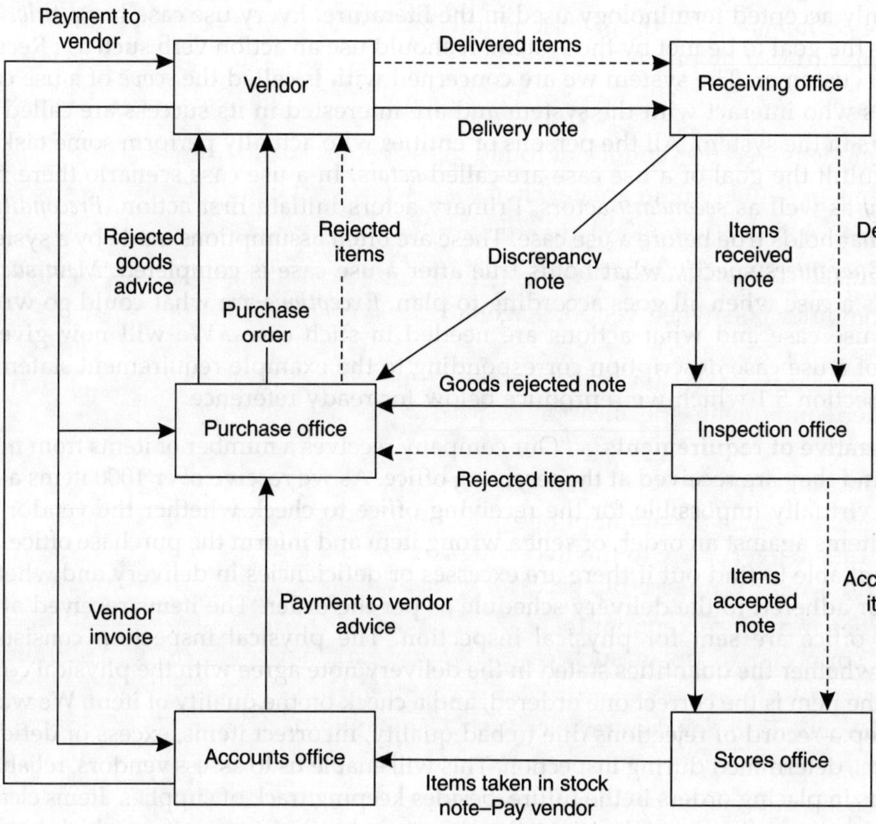

Figure 10.1 Document flow diagram for goods received.

Goal: To store in a database relevant data on items received from vendors.
Preconditions: Items ordered database available. Items delivered to ROC by vendors' agents with delivery note.

Main Success Scenario

1. ROC receives an item with delivery note from vendor. Delivery note has order id, vendor id, item id, quantity delivered, date.
2. ROC retrieves using order id as index a record from items ordered data base.
3. If order id, vendor id, item id match, ROC compares quantity ordered with quantity in delivery note and finds they are equal.
4. ROC compares delivery date specified in order with date in items delivered note and finds it is OK.
5. ROC enters delivery note details in items received database.
6. ROC sends items to inspection office with delivery note.
7. ROC signs delivery note copy and returns it to delivering agent.

Guarantees: Items received database is updated if items received by receiving office match items ordered by purchase office.

Exceptions

1. Item delivered by vendor not found in items ordered database. Action?
2. Items in delivery note greater than that ordered. Action?
3. Items in delivery note less than that ordered. Action?
4. Items in delivery note received on a date outside specified tolerance. Action?
5. Computer fails when vendor's agent arrives. Action?

A number of points should be noted from the above example. They are:

1. The requirements specification is structured in a form which is readable by any user.
2. It has much more detail than users' fuzzily stated requirements.
3. It clearly states the assumptions made, i.e., items are delivered physically along with a delivery note to the Receiving office and that the clerk is to acknowledge with a receipt in case of success.
4. When there are exceptions there are question marks. In other words what the clerk is supposed to do in such cases is not clearly specified in the requirements statement.
5. The actions to be taken on exceptions have to be clearly stated and documented in the use case.
6. The system will be developed based on the use case. Separate documentation is not usually required once actions to be taken in the exception cases are specified.
7. Use case gives in detail primarily what behaviour is expected of the primary actor.

10.2 USE CASE EXAMPLES

We develop below a use case corresponding ordering items from vendor (Figure 10.2).

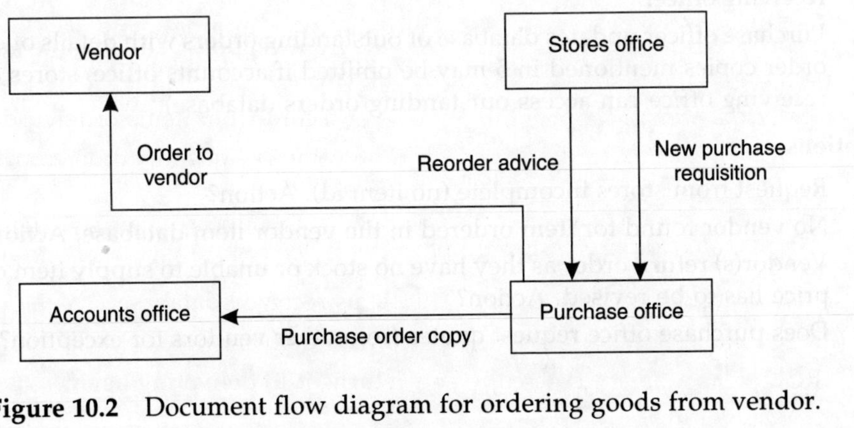

Figure 10.2 Document flow diagram for ordering goods from vendor.

Use Case 2: Order items from vendor

Primary actor: Purchase officer [Actual data entry by clerk(s)].

Scope: Order processing software.

Stake holders: Stores officer, Accounts officer, Vendor.

Goal: Generate purchase order and send to vendor. Store it in outstanding items ordered database.

Preconditions

1. An order request received from stores office which has item id, name and quantity to be ordered and date by which item is needed. The order request should specify tolerance (if any) in delivery date. The order maybe a reorder advice or a new purchase requisition.
2. Purchase office already has negotiated prices for common items. It is stored in a database along with vendor details.

Main Success Scenario (or Success Guarantees)

1. Purchase order with delivery address sent to vendor(s).
2. Vendor(s) instructed on mode of delivery to receiving office (by truck or courier).
3. Vendor acknowledges order and agrees to supply items ordered and sends invoice.
4. When items are shipped an invoice is sent to purchase office with copy of delivery note.

Success Scenario details

1. Request received from stores (with authorization by purchase officer) specifying item name, item id, quantity, date needed with tolerance (if any) in delivery date.
2. Purchase officer finds from vendor—item database vendor capability, price, reliability index.
3. Purchase officer decides quantities to be purchased from each vendor and negotiates price.
4. Purchase officer generates purchase order and enters it in item ordered database.
5. Purchase officer sends purchase order copies to accounts office, stores office and receiving office.
6. Purchase officer updates database of outstanding orders with details of order (The order copies mentioned in 5 may be omitted if accounts office, stores office and receiving office can access outstanding orders database).

Exceptions

1. Request from stores incomplete (no item id). Action?
2. No vendor found for item ordered in the vendor item database. Action?
3. Vendor(s) return order as they have no stock or unable to supply item on time or price has to be revised. Action?
4. Does purchase office request quotes from other vendors for exception?

Observe that the use case gives in reasonable detail the behaviour expected from the purchase office process. There are four exceptions which need clarification before the system can be developed. Use case often specify so called *minimum guarantees*. These specify whether all the stake holders interests are protected. In the above use case we can state minimal guarantee as "If purchase office order process is unsuccessful for some reason the outstanding orders database will remain unchanged. Attempt at ordering will be logged along with reason why it failed. No other database in the system accessible to stake holders will be disturbed".

We will now give four more use cases corresponding to the document flow diagrams of Figure 10.1.

Use Case 3: Inspection of items sent by Receiving Office

Primary actor: Inspection office staff (IOS).

Scope: Inspection Software.

Stake holders: Receiving office, Stores office, Purchase office.

Goal: Inspect actual item received with item received note. Find discrepancies (if any) and report to purchase office. Send accepted items to store.

Pre-conditions: Items received note has all requisite fields filled (order no., item id, item name, quantity, vendor id, vendor name and address, date delivered). Delivered items received from ROC with relevant note.

Main Success Scenario

1. IOS receives items with items received note from receiving office staff, (Physical items arrive with accompanying physical slip and also electronic record of the physical slip).
2. IOS inspects items with note for quantity and quality. Quality OK and quantity also OK.
3. If checks successful sends items to store along with items received and accepted note.
4. If checks not successful sends rejected note with rejected items to Purchase office.
5. The partial accepted note sent to stores office along with accepted items.

Exceptions

1. Items delivered by vendor without corresponding delivery note. Action?
2. If most items pass quality and few rejected. Action? (Is serial no. 5 given in success scenario OK?)
3. If excess items come and pass quality inspection what to do with excess?
4. If items found lesser than that written in delivery note. Action?

We see that there are a number of exceptions in which action taken is not specified. They have to be specified for the use case to be completed.

We will now develop the use case of stores office (Figure 10.1)

Use Case 4: Update inventory (by stores office)

Scope: Inventory update software

Primary actor: Stores officer (SO)

Stake holders: Inspection officer, Accounts officer, Purchase officer.

Goal: Update inventory when inspection office sends accepted items with corresponding note.

Pre-conditions: Items inspected and accepted note received from inspection office. Physical items delivered and taken in stock and the corresponding note correct.

Main Success Scenario

1. Stores inventory database properly updated.
2. Items taken into stock note sent to accounts for its action.

Success Scenario Details

1. Items accepted note received from Inspection office containing (order no., vendor id, vendor name and address, item id, item name, quantity accepted, price per item).
2. Pay vendor note sent to accounts office after updating inventory.
3. Copies of item taken into stock note sent to purchase office and accounts office.

Exceptions

1. Stores inventory database corrupted while updating. Action?
2. Accounts office computer not responding when items taken into stock note sent to it. Action? (does it acknowledge receipt?)
3. Accounts office expected to pay vendor for items taken into stock. How is inventory database linked to corresponding vendor database?
4. Should inspection office send pay to vendor note to accounts office when items accepted or should stores office send it?

In the above case we have taken only the case corresponding to document flow diagram of Figure 10.1. We have assumed that stores primarily receive goods and store it. A store also has to issue goods when requisitioned by a customer. Three situations can arise here. In one case a store is a part of an organization such as a University in which "customers" are various departments and there is no explicit billing and collection of real money. In this case there is an internal debit which debits the budget of the requisitioning department by the value of the goods issued. If the organization is, say, a supermarket it will give bills to customers and collect cash or accept credit/debit card payments. A third case would be one in which a distributor buys in bulk from a manufacturer. In this case the customer will be sent an invoice along with goods and payments collected based on period allowed for payment (may vary from one week to three months). We will consider the case corresponding to a store internal to an organization such as a University or a manufacturer who requisitions items from an internal store to proceed with its own manufacturing operations (Figure 10.3).

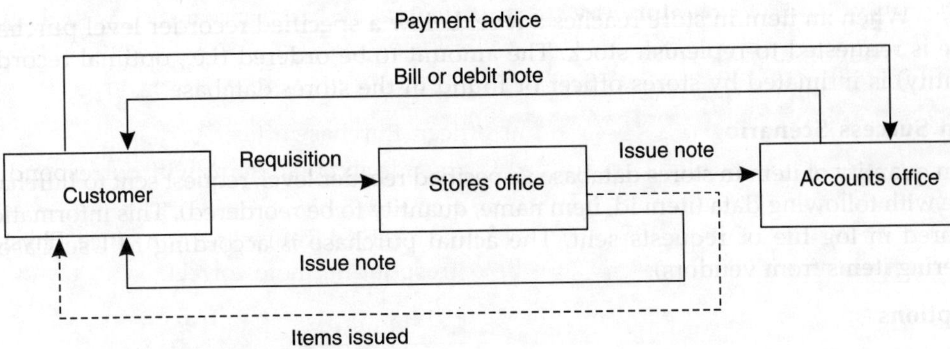

Figure 10.3 Document flow diagram for items issued to customer.

Use Case 5: Stores issues items requisitioned

Primary actor: Stores officer (SO).

Scope: Software to issue items by stores and debit accounts of customers.

Stake holders: Customers, Accounts office, Purchase office.

Goal: Issue items requisitioned to customers. In case items not available notify customer and purchase department. If issued notify accounts to debit customer's account.

Pre-conditions: Items requisitioned note has full details on item (customer id, item code, item name, quantity). Customer has budget allocation for requisition.

Main Success Scenario

1. Stores office receives requisition and finds it complete. Requested item available in store. Issue item.
2. Intimate accounts office to debit customer's account by (item quantity × cost per item).

Exceptions

1. Quantity requisitioned not available in store. Action ? (Partial issue or no issue)
2. Insufficient credit in customer's account. Action?
3. Item requisitioned not normally stored by the store. Should it add it to its normal holding?

In Use Case 5 we have limited the scope to only issuing items to customers. We have not included automatic recorder if the quantity remaining after issue is found below a reorder level. Normally stores would intimate purchase to initiate a purchase order in such a situation. We will now develop a use case for that situation (Figure 10.3).

Use Case 6: Stores requests Purchase office to replenish stock in store

Primary actor: Stores Officer (SO).

Scope: Software to initiate automatic reorder process to purchase items.

Stake holders: Stores officer, Purchase officer.

Goals: When an item in store reaches a level below a specified recorder level purchase office is requested to replenish stock. The amount to be ordered (i.e., optimal recorder quantity) is intimated by stores officer or found in the stores database.

Main Success Scenario

When quantity of item in stores database ≤ specified reorder level, request sent to Purchase office with following data (item id, item name, quantity to be reordered). This information is stored in log file of requests sent. The actual purchase is according to Use Case 2 (ordering items from vendors).

Exceptions

1. A requisition generated is already in the log file of items ordered. Action particularly if the order is for a large quantity or there is a long delay from the date of first requisition?

The use case given above has only two stake holders and this is very simple and is to be linked to Use Case 2. The next Use Case we will consider is payment by accounts for goods purchased and taken into stock. In other words, payments are made only for items which passed inspection. This Use Case corresponds to document flow diagram of Figure 10.1 and data flow diagram of Figure 10.4. We will first consider the case where the accounts office pays vendor for items which passed inspection and taken into stock by the stores office.

Use Case 7: Accounts department pays vendors

Primary actor: Accounts Officer (AO).

Scope: Software to pay vendors for goods taken into stock.

Stake holders: Stores office, Purchase office, Vendor.

Goals: When items supplied by a vendor taken into stock by stores office, accounts office should send pay order (or cheque) to vendor along with vendor invoice reference to reconcile payment.

Pre-conditions: Item taken into stock note received from stores containing (order no., item id, item name, cost per item, quantity taken in stock, vendor id). Vendor invoice database with data (order no., vendor id, vendor name, vendor address, item id, item name, quantity supplied, cost per item, amount payable) accessible to accounts office.

Main Success Scenario

1. Accounts office receives item taken into stock intimation from stores office.
2. Accounts office calculates amount payable and creates pay order with (order no., vendor id, vendor name, vendor address, item id, item name, quantity invoiced, vendor invoice id, quantity accepted, amount payable). Accounts office prints cheque favouring vendor. It mails cheque along with pay order to specified vendor address. Alternately, accounts office electronically transfers the amount to vendor's bank account along with pay order.
3. AO intimates purchase office on payment by sending copy of pay order with invoice reference.
4. Payment data stored in paid database.

Exceptions

1. Stores office intimation corrupted/not received by AO.
2. Stores office intimation has quantity accepted not equal to quantity specified as supplied in vendor invoice. Action?
3. Vendor address or bank account details has changed since supply. How does AO know about this?

We stated earlier that there are two possible situations when items are issued to customer. One is as given in Case 5 in which the "customer" is a part of an organization and there is no real money transaction. We will consider another case in which the organization is a distributor who buys items in bulk from vendors and distributes it to customers who pay for goods received usually by cheque. A variation would be a retail store in which the customer may pay cash or by credit/debit card. We will not be concerned with this case. We will be primarily describing the use case corresponding to document flow diagram of Figure 10.3 and data flow diagram of Figure 10.4.

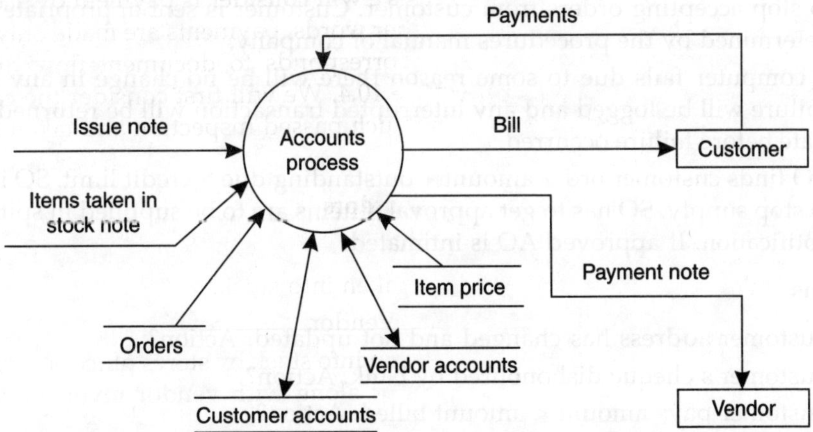

Figure 10.4 Data flow diagram for accounts process.

Use Case 8: Accounts department sends bill to customer and follows up payment

Primary actor: Accounts Officer (AO).

Scope: Software to generate bill and dispatch to customer with a deadline for payment and follows up payment.

Stake holders: Stores Office (SO), Accounts Office (AO), Customer.

Goal: When stores office intimates accounts office on item sold to a customer, the accounts office generates a bill to be sent along with items to be supplied. This bill is sent to stores office to be enclosed with the items. If payment corresponding to the bill is not received from the customer, after a specified time, a reminder is sent. If payment is not received in a specified time after the reminder is sent the customer's credit rating is downgraded so that future supplies are not made without special approval.

Accounts office receives document with the following data (customer id, customer name and address, quantity supplied, cost/item, date) from stores office. Accounts office has a database with (customer id, amount outstanding on date, credit limit).

Main Success Secenario

1. AO receives from SO data (customer id, customer name and address, quantity supplied, cost/item, date).
2. AO creates bill (customer id, name, amount billed, date due) using above data.
3. AO checks in its database amount due from customer. [If (amount due + amount billed) < allowed credit limit] it sends bill to SO for onward transmission. It updates its customer amount due database.
4. Each day AO checks amount received from all customers with the outstanding dues database and subtracts amount received from amount outstanding. AO issues receipt to customer for the amount received. It also lists amounts outstanding with due dates and sends it to the customer.
5. Each day AO checks amount due database and issues reminders to all customers with dues which becomes outstanding on that date. A second reminder is sent on due date + 10 days. If dues not received on due date + 20 days, AO intimates SO to stop accepting orders from customer. Customer is sent appropriate notice as determined by the procedures manual of company.
6. If computer fails due to some reason there will be no change in any database. Failure will be logged and any interrupted transaction will be returned to initial state before failure occurred.
7. AO finds customer order amount + outstanding due > credit limit. SO is notified to stop supply. SO has to get approval if items are to be supplied in spite of AO's notification. If approved AO is intimated.

Exceptions

1. Customer address has changed and not updated. Action?
2. Customer's cheque dishonoured by bank. Action?
3. Customer pays amount < amount billed. Action?
4. Customer pays amount > amount billed. Action?

Minimum Guarantees

Customer has paid for items received and if not appropriate procedure has been followed. Customer's credit rating updated based on payment history. No database other than that related to payment has been disturbed.

10.3 USE CASE DIAGRAMS

We have so far used a style of writing use cases which is textual. Another graphical style has been proposed as a part of Universal Modeling Language (UML) by Jacobson [24, 33]. In his style so called stick diagram is used to represent actors and each use case is written within an ellipse. For Use Cases described so far, such a diagram is shown in Figure 10.5. Observe that the stake holders are shown as actors whose behaviour affects the use case. In my personal view the stick diagram gives an overview of use cases. A full textual description of the use case is essential.

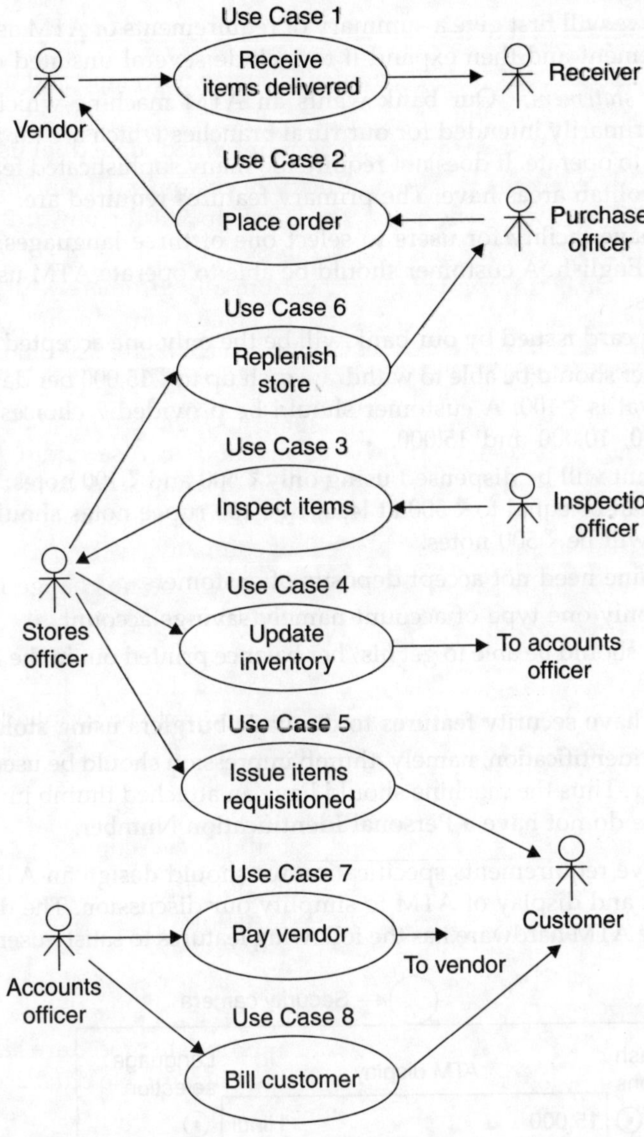

Figure 10.5 Use case diagram for examples of Sec. 10.1, 10.2.

10.4 USE CASE EXAMPLE—AUTOMATED TELLER MACHINE (ATM)

We will now consider an example of Automated Teller Machine (ATM) for which a use case model of requirement is given in detail. It is a favourite example of several authors as ATM is widely used by most of us without appreciating the great amount of attention one must pay to specify the requirements to be met by ATM. Our example is different as the ATM is specialized to meet requirements of rural areas of India.

In what follows we will first give a summary of requirements of ATM as normally stated by bank management and then expand it to include several unstated details.

The requirements statement: Our bank wants an ATM machine which should do the following. It is primarily intended for our rural branches which are far flung. It must be rugged and easy to operate. It does not require too many sophisticated features which our ATMs in metropolitan areas have. The primary features required are:

1. It must have facility for users to select one of three languages; local language, Hindi or English. A customer should be able to operate ATM using any of these languages.
2. The ATM card issued by our bank will be the only one accepted by the machine.
3. A customer should be able to withdraw cash up to ₹ 15,000 per day. The minimum withdrawal is ₹ 100. A customer should be provided 7 choices: 100, 500, 1000, 2000, 5000, 10,000 and 15,000.
4. The amount will be dispensed using only ₹ 500 and ₹ 100 notes. If the amount is greater than or equal to ₹ 500 at least five 100 rupee notes should be dispensed. The rest will be ₹ 500 notes.
5. The machine need not accept deposits of customers.
6. There is only one type of account namely savings account.
7. Customer should be able to get his/her balance printed out in the language of his/her choice.
8. It should have security features to dissuade burglers using stolen cards.
9. Biometric identification, namely, thumb impression should be used for identifying a customer. Thus the machine should have an attached thumb impression reader. Customers do not have a Personal Identification Number.

Based on the above requirements specification we should design an ATM. We will first show the buttons and display of ATM to simplify our discussion. The display is shown in Figure 10.6. The ATM hardware has the following features to satisfy users' requirements.

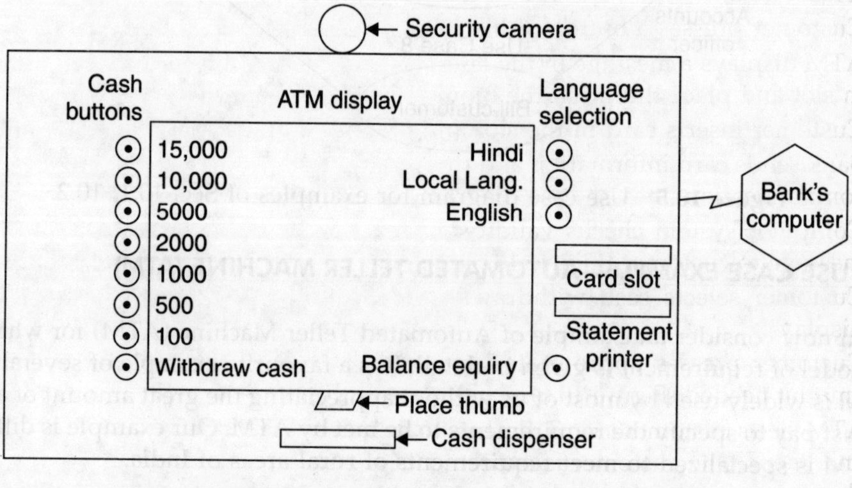

Figure 10.6 ATM with buttons and other features.

Initially the display will show a welcome message in three language: Hindi, local language and English and ask the customer to press button 1 for Hindi, 2 for Local language, and 3 for English by displaying press in the respective languages in windows opposite the button. It then guides the user with appropriate instructions on the display. The requirement is specified in detail as Use Case 9.

Use Case 9: Customer withdraws cash from ATM

Primary actor: Bank Customer.

Scope: The software built in the ATM.

Stake holders: Bank's Computer System, ATM Security officer, ATM maintenance officer.

Goal: To dispense cash requested in predetermined denominations and print balance statement.

Pre-conditions

1. The machine can dispense ₹ 15,000, 10,000, 5000, 2000, 1000, 500 and 100. Five notes dispensed will be ₹ 100. Rest ₹ 500 notes.
2. Maximum withdrawal per day per customer is ₹ 15,000.
3. Customer has a valid ATM card of the bank.
4. Bank has for each card issued a corresponding thumb print of customer in its customer database.
5. There is sufficient cash stored in the ATM to cater to at least a day's estimated withdrawal with sufficient ₹ 100 denomination currency notes.
6. Power OK. Communications line active from ATM to Bank's Computer System.

Main Success Scenario

1. A customer inserts card in slot at the door of ATM cubicle and removes it. The door opens allowing customer inside provided the card is valid.
2. ATM displays welcome banner and tells customer to select language of choice. Three languages displayed one opposite each button on the right panel of ATM (Figure 10.6).
3. Customer presses a button of his choice.
4. ATM displays a message in the language selected by the customer to insert card in slot and place thumb in the biometric thumb reader.
5. Customer inserts card in the slot and places thumb on the reader.
6. Bank sends card information and thumb print to the bank's computer system via communication lines.
7. Computer system checks validity of card and thumb print. If OK, it displays message to customer (Figure 10.6) to select cash withdrawal or balance enquiry.
8. Customer selects cash withdrawal, by pressing the button on the left of the display.
9. Customer presses appropriate button on the left of the display based on the amount needed (Figure 10.6).
10. ATM ejects customer's card, flashes message requesting customer to take the card and beeps. The customer is expected to take the card within one minute. If he/she does not take the card it is an exception condition.

11. If amount requested < minimum balance in the customer's account and the total cash withdrawal for the day by the customer is less than 15,000 then the ATM dispenses money in specified denominations. It beeps and instructs the customer to take money by flashing message on ATM display. Customer is expected to take the cash within a minute. If not it is an exception condition.
12. After the customer takes the cash ATM sends message to the computer to debit his/her account, beeps and flashes message to the customer to collect ATM transaction slip printed by ATM machine. The customer's withdrawal for the day is updated with dispensed amount and his current balance is printed.
13. The ATM reverts to initial state and is ready to take another transaction after 30 seconds.

Exceptions

1. Power failure, customer cannot enter ATM cubicle.
2. Communication failure, ATM displays "ATM not available". Bank's computer system should check communication lines, by sending message once a minute and see if it is received back. In case of failure the Bank's Computer System should inform ATM maintenance officer.
3. Card information and thumb print do not match. Instruct customer to place thumb again. If second time also there is failure one possible action is not to eject card and ask customer to report to Bank. Another action may be to eject card and give one more attempt by a flashing message to the customer to insert card again, wipe thumb and reader, and place thumb on reader. If it fails again do not return card and ask customer to report to bank (Exception action to be clarified by ATM Security officer. ATM may have a video camera to take a picture of customer).
4. Communication line fails during success scenario 6. Eject card. Display ATM failed. Ask customer to try again after 10 minutes. Inform ATM maintenance staff.
5. Power fails during success scenario 6. Ask customer to report to bank (Bank has to decide exception action).
6. Power or communication fails during success scenarios 7, 8 or 9. Do not dispense cash. No change in balance. Inform ATM Maintenance.
7. If customer's balance is less than the amount requested flash message "insufficient balance". Eject card. Print balance statement.
8. If customer's withdrawal during the day including current request greater than ₹15,000 flash message "Exceed maximum, withdrawal for the day". Eject card.
9. ATM does not have sufficient balance to dispense money requested. Flash message ATM has no cash. Abort transaction. Banking system should inform ATM maintenance staff. Print balance statement slip.
10. ATM does not have enough paper to print balance or printer is defective (e.g. no ribbon or printer jammed). Flash message printer defective. Bank's computer system should inform ATM maintenance staff.
11. Customer, after taking cash, reinserts card within 30 seconds. ATM does not accept card and displays "wait for 30 seconds" and beeps.
12. Customer does not take card within one minute after it is ejected in spite of the message to take card and warning beeps. Do not dispense cash. Do not change balance in account. Flash message and beep requesting customer to take card. If he still does not, abort transaction and report to ATM security officer.

13. Customer does not take cash within one minute. Flash message and beep asking customer to take cash within one minute. If he/she does not, inform ATM security officer.

Minimum Guarantee

If a customer has valid credit card and his or her thumb print matches the one in the data base and his credit is greater than the cash requested and if he or she has not drawn more than ₹ 15,000 on that date money would have been dispensed. If an attempt at fraud had been detected ATM security would have been informed. In case of any malfunction of ATM or communication/power lines, the ATM maintenance staff would have been intimated. Data belonging to all other clients of ATM would not be affected. Customer's account would have been debited by the amount dispensed and a correct balance statement would have been printed.

Observe the large number of exception situations which may arise. All these are not specified in the original statement of requirements. Use case methodology has forced the systems analyst to think deeper and raised queries to be answered by the management before designing the system.

In the style used in this section we wrote one use case with a large number of exceptions. Use cases may be written for actions to be taken by the appropriate stake holder for selected exceptions. One such exception is when an ATM needs attention by maintenance staff. Use case 10 describes this.

Use Case 10: ATM maintenance staff attends to fault

Primary actor: ATM's maintenance staff.

Scope: Procedure to fix faults.

Stake holders: Customer, ATM maintenance staff, ATM security staff, Banking Computer System.

Goal: To attend to fault in ATM operation and fix it.

Pre-conditions

1. Power monitor beeps indicating failure.
2. Signal from Banking Computer to ATM not acknowledged.
3. ATM printer sends failed signal.
4. ATM cash dispenser sends failed signal.
5. ATM card reader sends failed signal.
6. ATM thumb print reader sends failed signal.

Main Success Scenario

1. Whenever failure occurs appropriate information received by maintenance staff.
2. Based on the information ATM machine is repaired.

Exceptions

1. Sufficient cash not available to replenish. Action?
2. Spare parts not available to fix printer. Action?
3. Spare parts for card reader, thumb print reader not available. Action?

Minimum Guarantees

If sufficient cash and spare parts available ATM is repaired within stipulated time.

Similar use cases may be written for ATM security staff and ATM banking system. These are left as exercises to the student.

Use case 9 was for cash withdrawal. A separate use case may be written for balance enquiry. Most of it will be found in use case 9 and we may think of this use case as being included in it.

10.5 STICK DIAGRAMS FOR ATM

In addition to the detailed written use cases, stick diagrams are useful to get an overview of the use cases. We give in Figure 10.7 the overview of use case withdraw cash as a stick diagram. In Figure 10.8, 10.9 and 10.10 we give use cases for the other stake holders.

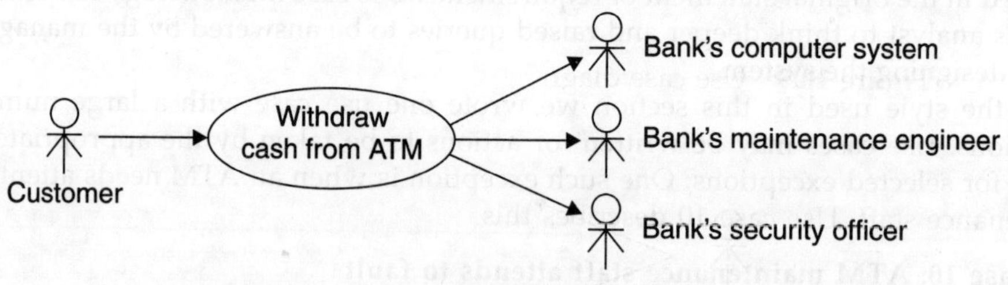

Figure 10.7 Context diagram of ATM use case.

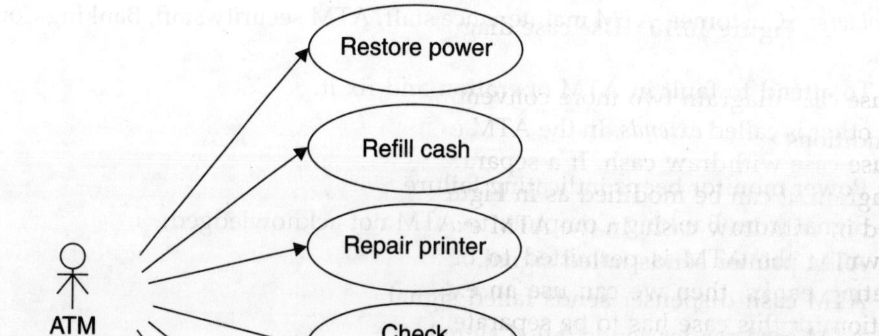

Figure 10.8 Use case diagram for ATM maintenance engineer.

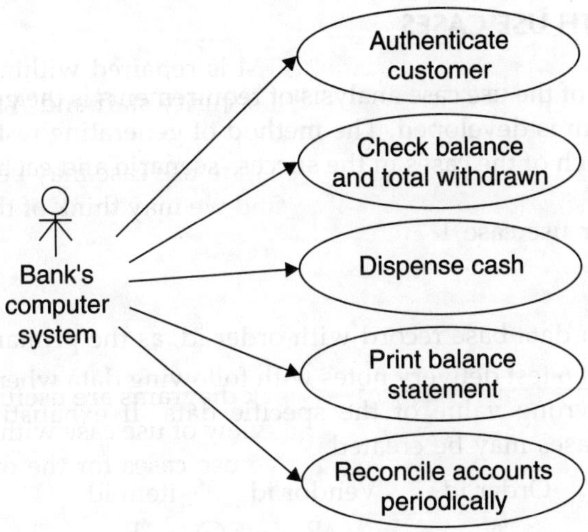

Figure 10.9 Use case diagram of Bank's Computer System.

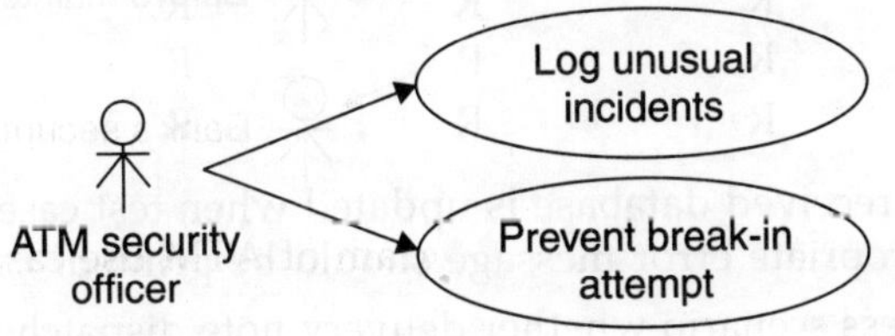

Figure 10.10 Use case diagram for ATM Security Officer.

In use case diagram two more conventions are used. One of them is called *includes* and the other is called *extends.* In the ATM example, balance statement printing was part of the use case withdraw cash. If a separate balance enquiry is to be shown in the use case diagram, it can be modified as in Figure 10.11, where balance enquiry is shown as included in withdraw cash. In the ATM example we assumed that only the card of the bank owning the ATM is permitted to be used. If we want to allow cards of other cooperating banks, then we can use an extension as shown in Figure 10.11. Use case description for this case has to be separately written.

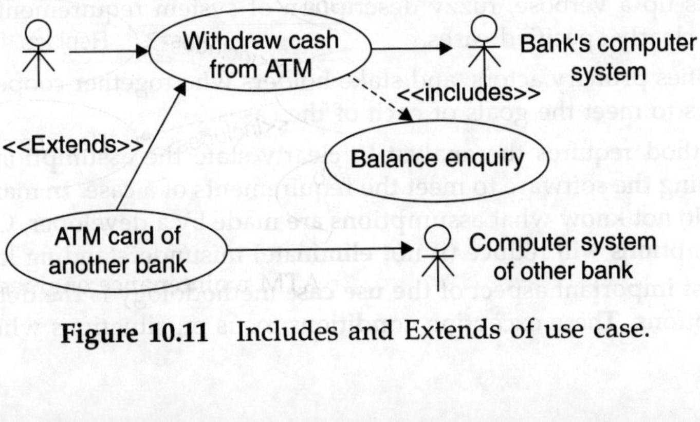

Figure 10.11 Includes and Extends of use case.

10.6 TESTING WITH USE CASES

An important aspect of the use case analysis of requirements is the generation of test cases when the final system is developed. The method of generating test cases is to take data which will invoke each of the cases in the success scenario and each one of the exception conditions.

We illustrate this for use case 1

Test Cases

1. Retrieve from data base record with order id, as the primary index.
2. Create a set of 6 test delivery notes with following data where R is the right value and W the wrong value of the specific data. If exhaustive test needed then $2^5 = 32$ test cases may be created.

	Order id	Vendor id	item id	qty.	date
tc1 :	W	R	R	R	R
tc2 :	R	W	R	R	R
tc3 :	R	R	W	R	R
tc4 :	R	R	R	W	R
tc5 :	R	R	R	R	W
tc6 :	R	R	R	R	R

3. Check if item received database is updated when test case 6 is used. In all other cases, an appropriate error message should be printed.
4. Check in success scenario whether delivery note dispatched to inspection office.

10.7 CONCLUSIONS

One of the most difficult problems in designing information systems is a well stated specifications of the requirements to be met by the proposed system. This is primarily due to the communication gap between clients who state their requirements and system analysts who interpret them and design the system. During the past 50 years several methods of specifying requirements to bridge the communication gap have been proposed. Use case method is one of those which is currently being adopted by many organizations. The primary advantages of use case methodology are:

1. It breaks up a verbose, fuzzy description of system requirements into a set of smaller clearly specified cases.
2. It identifies primary actors, and stake holders who together cooperate and carry out tasks to meet the goals of each of the cases.
3. The method requires the analyst to clearly state the assumptions made while developing the software to meet the requirements of a case. In many projects the clients do not know what assumptions are made by a developer. Clear statement of assumptions will reduce (if not eliminate) misunderstanding later.
4. The most important aspect of the use case methodology is the detailed write up of exceptions. These exception conditions focus on situations where there is no

clarity in the client's statement of requirements. If actions to be carried out for each of these exceptions are clarified by clients before beginning the design, changes in the system design later can be avoided.

5. Specification of behavior in success scenario as well as the listing of exceptions aid in designing test cases which are to be used to test the software developed for the system. The creation of test data is systematized when use case analysis is employed.

The primary disadvantages of use case methodology are:

1. Use case description could become very long, particularly for complex systems. One must thus modularize and break up a system appropriately.
2. Use cases provide a *behavioural description*. The stick diagrams used in UML is too concise and does not provide much information.
3. How documents and data flow in a system are not part of the specifications.

The advantages are numerous as we have seen and this has led to the use case methodology being adopted widely by industry.

There are several styles available to describe use cases. This style we have adopted consists of the following parts:

1. *Use case title*—specifies action performed by use case to meet a specified goal.
2. *Primary actor*—initiates the action performed by a use case.
3. *Scope* specifies the boundaries of the subsystem being described by the use case.
4. *Stake holders* are the other actors who cooperate with the primary actor to meet the goals of the use case.
5. *Goal* specifies the primary objective or guarantee of delivery promised by the use case.
6. *Pre-conditions* are the assumptions made by the primary actor while designing the system to meet the specified goal(s).
7. *Main success scenario* lists the behaviour of the use case when everything goes according to plan.
8. *Minimum guarantees* specifies what holds true after the use case is completed. It specifies whether all stake holders interests have been protected.
9. *Exceptions* list what could go wrong during a use case and what actions are needed in such cases. Usually these contain situations which have not been foreseen by the client and a system designer requires clarifications.

In addition to this a stick diagram would be useful to get a bird's eye view of a system.

SUMMARY

1. A use case specifies a goal to be met by a system being developed.
2. A system requirement specification consists of a set of use cases.
3. Each use case has a primary actor whose responsibility is to meet the specified goal by appropriate actions.

4. A use case is a behavioural description of the primary actor and the stake holders who cooperate to meet the specified goal.
5. There are several formats suggested in the literature to describe use cases. The one we have used in our description has the following main parts:
   - Use case title which specifies the action to be performed.
   - A primary actor who initiates actions to meet the specified goal(s).
   - Scope specifies the software system being developed.
   - Stake holders are other actors who cooperate with the primary actor to carry out the specified actions.
   - A statement of goal to be met.
   - A set of pre-conditions which are assumed to be true by the software developer. In other words, these are assumptions made by the developer.
   - A set of guarantees specifying what holds after a use case is completed.
   - Exceptions list situations when things can go wrong and appropriate actions need to be clarified by the requirements specified.
6. Stick diagrams showing the actors and use cases they are involved in. They give only a bird's eye view of use cases. A detailed documentation of use cases is essential.
7. Use cases are effective to develop test cases.

EXERCISES

10.1 What is use case? What are its advantages and disadvantages in evolving requirement specifications.

10.2 Does use case analysis concern itself with data and document flow? In what way is it different from data and document flow methods used in a systems requirement specifications.

10.3 Give a template to specify use case description given in this chapter.

10.4 Consider Exercise 5.3. Give use case description for that exercise.

10.5 Consider the hostel information system case example of Section 4.7. Identify the actors in this case.

10.6 Develop use case corresponding to mess billing described in Section 4.7.

10.7 Develop use case corresponding to store issue of reorder example described in Section 4.7.

10.8 Develop use case corresponding to the purchase of items for the hostel mess described in Section 4.7.

10.9 In the ATM example considered in this chapter we assumed that a customer's identity is established using his or her thumb print. Most ATMs require a customer to use a 4-digit PIN code. Modify use case 9 given in this chapter when PIN code is used. If a wrong PIN is used more than thrice the ATM should not return the ATM card and inform security.

10.10 Modify use case 9 to allow a customer to use ATMs of a group of cooperating banks.

10.11 Develop a use case where ATM security officer is the primary actor (See Figure 10.10).

10.12 Develop a use case where the Banking System is the primary actor (See Figure 10.9).

10.13 Develop a stick diagram corresponding to Exercise 10.6.

10.14 Develop a stick diagram corresponding to Exercise 10.7.

10.15 Develop a stick diagram corresponding to Exercise 10.8.

10.16 Develop test cases for the mess billing system of Section 4.7 using the use case of Exercise 10.6.

10.17 Develop test cases for Exercise 10.7.

10.18 Develop test cases for Exercise 10.8.

11

Logical Database Design

LEARNING GOALS

After reading this chapter a student should be able to:

1. Develop a conceptual model of data using entity-relationship modeling.
2. Organize the data required in an application as relations.
3. Explain why relations should be normalized.
4. Normalize relations from first to fifth normal form.

In Chapters 5 and 7 we discussed what data items are needed in a system and how they flow; we also discussed how a data dictionary is developed for a system. The dictionary only lists data used and does not say anything about the relationships among individual data items. It is necessary to develop a *conceptual model* of data which specifies the relationships between data. This is used to group data into a number of tables. The tables are organized to:

1. reduce duplication of data,
2. simplify functions like adding, deleting and changing data, in other words, updating data, and
3. select data required for processing, in other words, retrieving data.

The method of organizing or structuring data into tables is known as *normalization*. The collection of tables of data is called a *database*. The normalized database is later converted to a physical database. In this chapter we will use a method called *entity-relationship analysis* to obtain a conceptual model of data used in applications. This analysis is then used to organize data as relations, normalizing relations, and obtaining a relational database.

11.1 ENTITY RELATIONSHIP MODEL

The entity-relationship model for data uses three features to describe data. These are:

1. **Entities** which specify distinct real-world items in an application.

2. **Relationships** which connect entities and represent meaningful dependencies between them.
3. **Attributes** which specify properties of entities and relationships.

We will illustrate these terms with an example. A vendor supplying items to a company, for example, is an *entity*. The items he supplies is another entity. A vendor and an item are related in the sense that a vendor *supplies* an item. The act of supplying defines a *relationship* between a vendor and an item. An *entity set* is a collection of similar *entities*. We can thus define a *vendor set* and an *item set*. Each member of an entity set is described by some *attributes*. For example, a VENDOR may be described by the attributes.

(vendor code, vendor name, address)

An ITEM may be described by the attributes

(item code, item name)

We saw that two entities may be related by a relationship. A *relationship set* relates two entity sets. Thus a relationship set will have several members each connecting a member of an entity set with a member of another entity set.

In the entity-relationship (E-R) diagram (Figure 11.1), an entity set is represented by a rectangle and relationship set by a diamond-shaped box. We may define another entity set called ORDER and a relationship set "PLACED WITH". The statement "Orders Placed with Vendors" is shown in the E-R diagram of Figure 11.2. (The notation for E-R diagram is not standardized. We have used an abbreviated Chen's notation. In Appendix A we describe two other notations). In the E-R diagram entities and relationships are named using uppercase letters in singular form. Thus in the E-R diagram we have named entities as VENDOR, ITEM, ORDER and relationship as SUPPLIES and PLACED WITH. The relationships in these examples relate two entities and are called *binary* relationship.

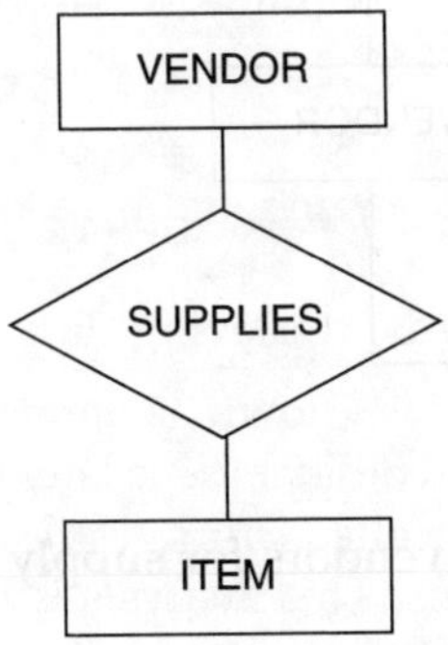

Figure 11.1 An E-R diagram for vendors, items and their relation.

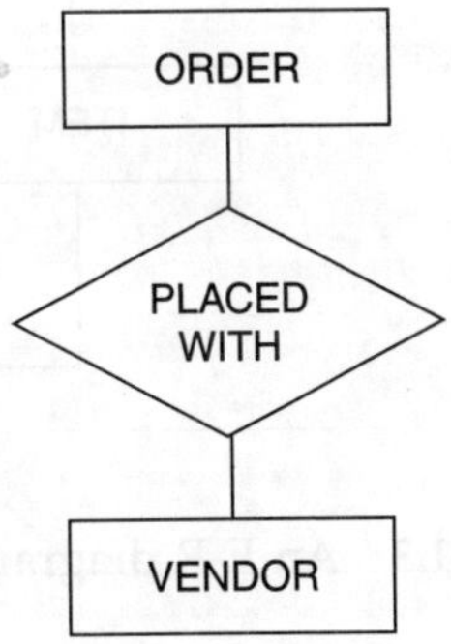

Figure 11.2 An E-R diagram for orders placed with vendors.

The interesting design problem is identifying *entities* which can be in a set. Only items with the same property can be in any entity set. For example, students form an entity set and courses another entity set. The relationship between these entities is "attend". Students attend courses. Courses are held in class rooms. An entity set is thus class rooms whereas the relationship is "held in". Once entities in an application are identified, the relationships are reasonably simple to identify. A thumb rule is that entities are normally described by nouns whereas relationships are normally described by verbs. There is no `automatic' or algorithmic method for identifying entities and relationships. One primarily has to use his commonsense and information provided by users. The best way to learn is thus through examples and exercises. In what follows we use the word entity as synonymous to entities belonging to an entity set. This should not cause any confusion.

An entity may be related to more than one entity. For example, orders are placed for items with vendors. Vendors supply items which are ordered. This is shown in Figure 11.3. Observe that the relationship SUPPLIES connects three entities. It is thus called a *ternary* relationship. A ternary relationship is used to represent the statement "Vendor supplies items which are ordered". Having identified entities and relationships, the next step is to specify their attributes. Attributes specify the properties of entities in an entity set and relationships in a relationship set. For example, the attributes of the entity VENDOR, which describes a vendor fully would be: (vendor identification or code, vendor name and vendor address). The attributes of the entity ITEM would be (item code, item name). The attributes of the relationship SUPPLIES would normally include the unique identification (or code) of all the entities it connects, namely, vendor code, item code, and order no. In addition it has attributes: quantity supplied, date of supply and price/unit.

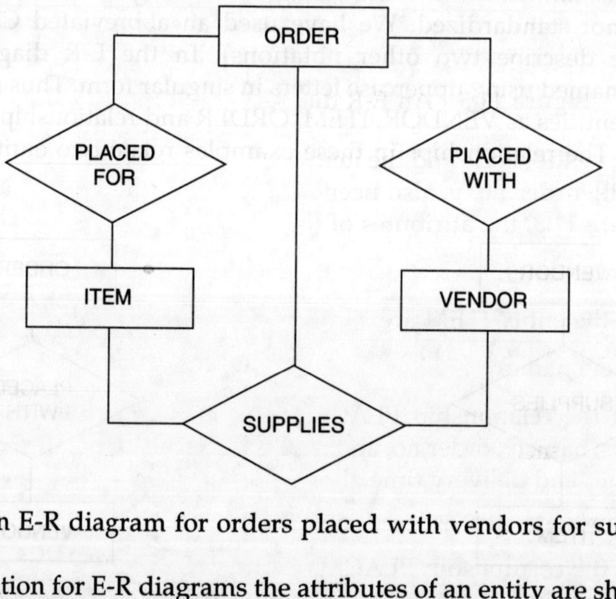

Figure 11.3 An E-R diagram for orders placed with vendors for supply of items.

In Chen's notation for E-R diagrams the attributes of an entity are shown inside ovals connected to the rectangle repressenting an entity (Figure 11.4). As this notation takes up a lot of space without adding clarity, in this book the attributes of entities and relationships are shown next to the entity box enclosed in square brackets (Figure 11.5). The attribute used as the identifier or key of the entity is underlined. A vendor may supply many items and an item may be supplied by many vendors. Therefore, we need both vendor identification and item code to uniquely specify the quantity of items supplied and date of supply. Thus the key for the relationship SUPPLIES is a composite key which

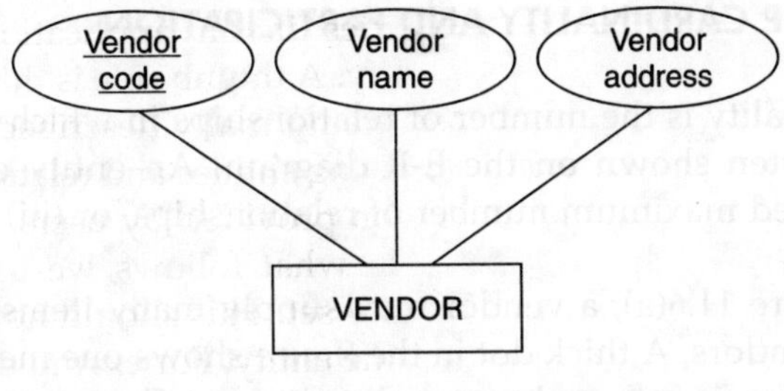

Figure 11.4 Chen's notation used to represent attributes of an entity.

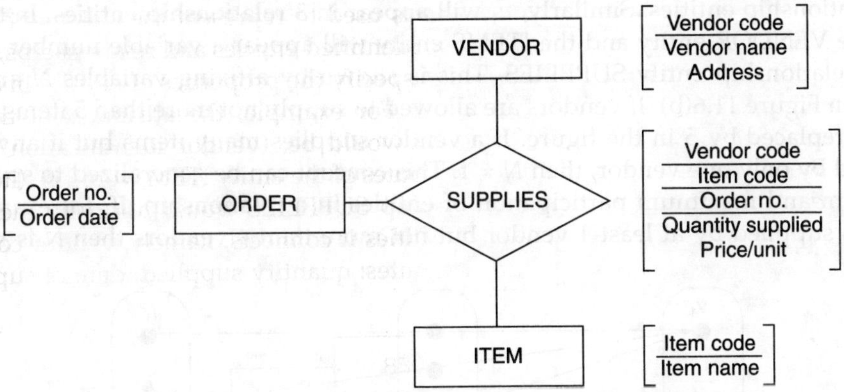

Figure 11.5 An E-R diagram showing attributes.

includes both the vendor code and the item code. As a vendor may supply the same item in different orders, order no. is also needed as a composite key attribute in SUPPLIES. Referring to Figure 11.3, the attributes of the entity ORDER are:

(order no., order date)

The attributes of the entity ITEM are:

(item code, item name)

The attributes of the relationship PLACED FOR should include the keys of both the entities it connects, namely, order no. and item code. Beside these, it has qty. ordered with vendor, price/unit, and delivery time allowed.

(order no., item code, qty. ordered, delivery time allowed)

The attributes of the relationship PLACED WITH are:

(order no., vendor code, item code)

The attributes of the entity VENDOR are:

(vendor code, vendor name, vendor address)

The attributes of the relationship SUPPLIES are:

(vendor code, item code, order no., qty. supplied, date of supply, price/unit).

11.2 RELATIONSHIP CARDINALITY AND PARTICIPATION

A relationship cardinality is the number of relationships in which an entity can appear. This cardinality is often shown on the E-R diagram. An entity can appear in (i) one relationship, (ii) a fixed maximum number of relationships, or (iii) a variable number of relationships.

Referring to Figure 11.6(a), a vendor may supply many items and an item may be supplied by many vendors. A thick dot in the figure shows one member in an entity set. We see that v_1 supplies i_1, i_2, i_3 and v_2 supplies i_2, i_3, i_4. Corresponding to i_2, there are two thick dots in the relationship entity SUPPLIES which means that i_2 will appear in two relationship entities. Similarly, v_1 will appear in 3 relationship entities. In this case both the VENDOR entity and the ITEMS entity will appear a variable number of times in the relationship entity SUPPLIES. This is shown by affixing variables N and M as shown in Figure 11.6(b). If vendors are allowed to supply not more than 5 items, then M will be replaced by 5 in the figure. If a vendor supplies many items but if any item is supplied by only one vendor, then $N = 1$. This concept can be generalized to specify the minimum and maximum participation of entities in a relationship. If, for example, an ITEM is supplied by at least 1 vendor but not more than 4 vendors then N is replaced

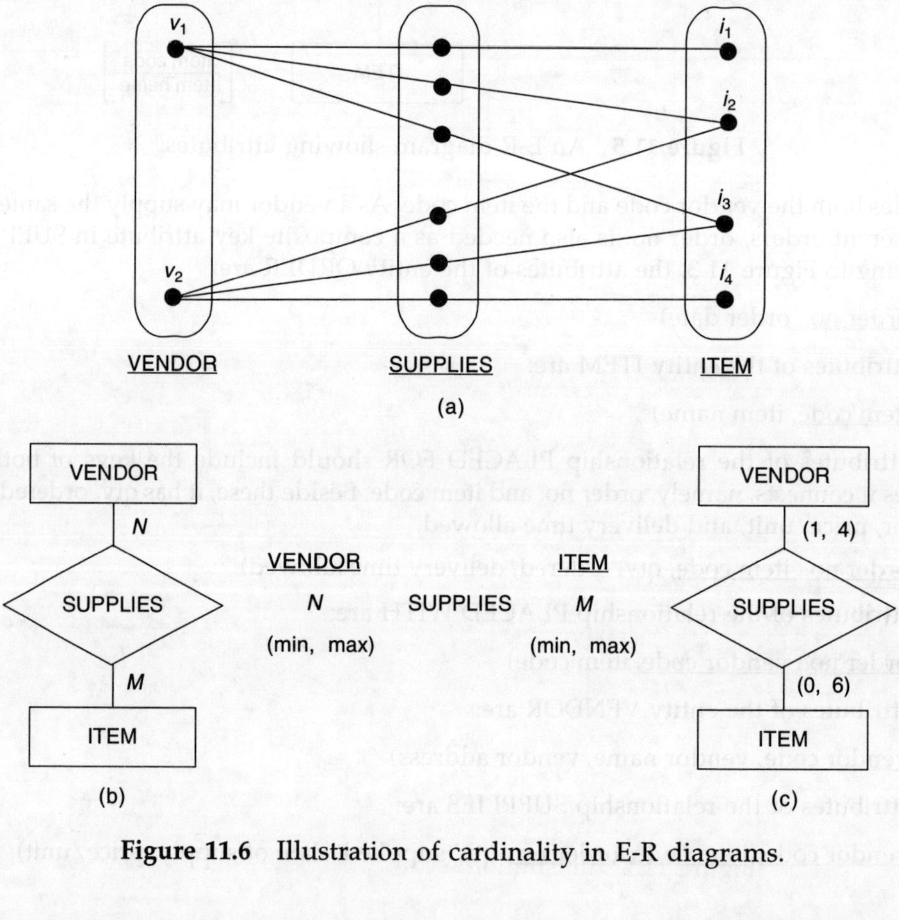

Figure 11.6 Illustration of cardinality in E-R diagrams.

by (1, 4) where 1 is the minimum and 4 the maximum participation of VENDOR in the relation SUPPLIES. A similar participation constraint for ITEM may also exist. For example, if a VENDOR supplies maximum 6 ITEMS and a minimum of 0 ITEMS (i.e., if a vendor is not required to supply any item) then the participation M is replaced by (0, 6). This is shown in Figure 11.6(c). One may read Figure 11.6(c) as "1 to 4 VENDOR may supply ITEM. 0 to 6 ITEMS may be supplied by a VENDOR".

This idea of cardinality of relationship can be clarified with another example. Consider a college where students have *advisers* assigned to them to help students and monitor their progress. A teacher will normally *advise* many students. A student, however, will have only one adviser assigned to him. This is shown in Figure 11.7. The number (i, 1) below TEACHER entity implies that only 1 teacher advises a STUDENT and N above STUDENT entity means that he advises N students. This is a 1:N relationship and for clarity this is shown on the left-hand side of the figure. If the maximum number of students allocated to a teacher is 10, then N will be replaced by 10 in the E-R diagram. If we use the generalized (min, max) notation then 1 is replaced by the tuple (1, 1) below TEACHER entity. This implies that only one teacher advises a student. N is replaced by the tuple (1, N) below student entity which means that a least 1 upto N students may be advised by a teacher. If the maximum value of N is 10 then (1, N) will be replaced by (1, 10) in the E-R diagram.

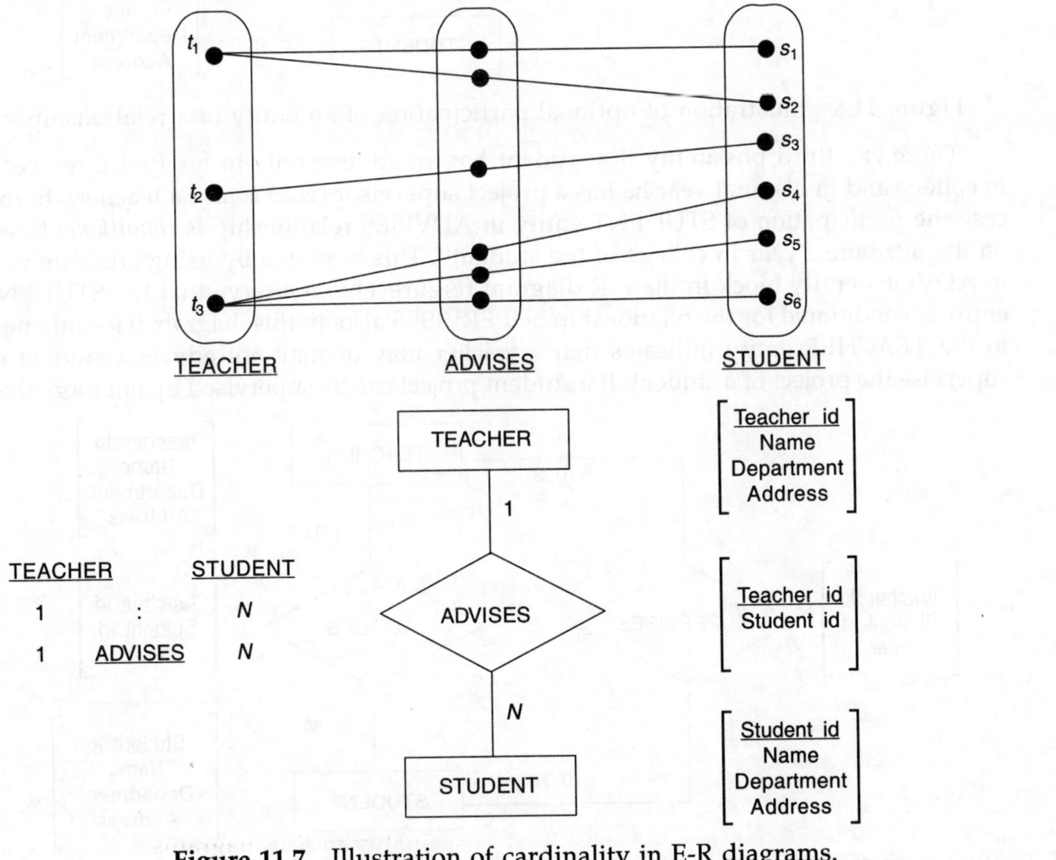

Figure 11.7 Illustration of cardinality in E-R diagrams.

So far we have assumed that every member in the entity set will definitely appear in the relationship set. In the teacher-student E-R diagram (Figure 11.7) we have assumed that every teacher has to advise one or more students and every student must have an adviser. In other words, the participation of TEACHER entity and STUDENT entity in ADVISES relationship is compulsory (or mandatory). If we relax this condition and say that some teachers may not advise students it is shown by replacing 10 by (0, 10) (see Figure 11.8). This means that some teachers may not be found in the ADVISES relationship. In other words, the participation of TEACHER entity in ADVISES relationship is *optional*.

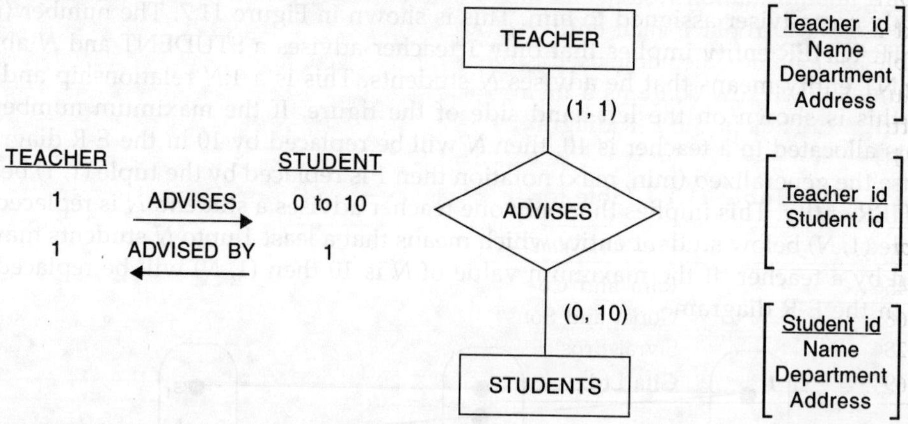

Figure 11.8 Illustration of optional participation of an entity in a relationship.

There is a third possibility if a student has an adviser only in his first three years in college and in his final year he has a project supervisor (who is also a teacher). In this case the participation of STUDENT entity in ADVISES relationship is *conditional* based on the attribute, "year in college of the student". This is shown by using attribute year in ADVISES entity block in the E-R diagram (Figure 11.9). Observe that the STUDENT entity is conditional for the relationship SUPERVISES also. In this diagram 0 for min next to the TEACHER entity indicates that a teacher may or may not advise a student or supervise the project of a student. If a student project can be supervised by not more than

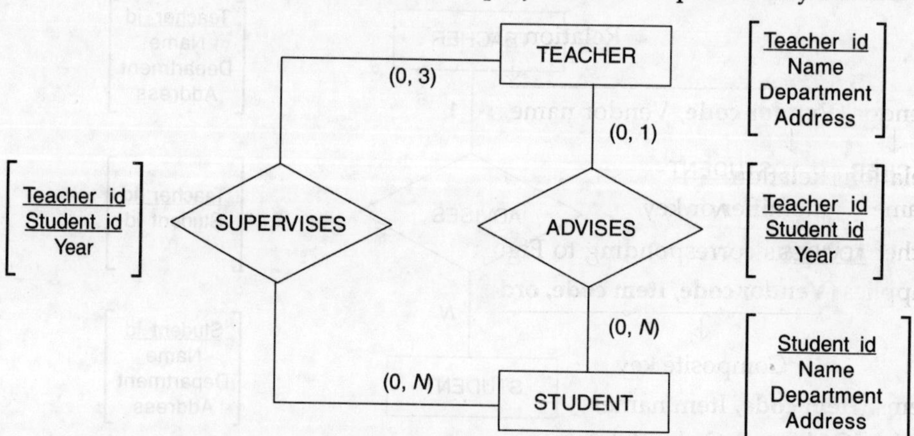

Figure 11.9 Illustration of conditional participation of an entity in a relationship.

three teachers, then number 3 will be placed as max in the line connecting the TEACHER entity to SUPERVISES relationship. The entry (0, N) next to STUDENT states that a TEACHER may advise or supervise 0 to a variable number of students.

11.3 RELATIONS

Entity-relationship diagrams are a means to an end, the end being the logical design of a *database*. E-R diagrams allow us to have an overview of the important entities for developing an information system and their relationships. Having obtained E-R diagrams, the next step is to replace each entity set and relationship set by a *table* or a *relation*. Each table has a name. The name used is the *entity name*. Each table has a number of rows and columns. Each row contains a member of the *entity set*. Each column corresponds to an attribute. Thus in the E-R diagram (Figure 11.5), the vendor entity is replaced by Table 11.1.

Table 11.1 Table for the Entity *Vendor* (Figure 11.5)

Vendor Code	*Vendor Name*	*Address*
1456	Ram and Co.	112 First Cross, Bangalore 12
1685	Gopal and Sons	452 Fourth main, Delhi 8
1284	Sivraj Bros.	368 M.G. Road, Pune 8
1694	Gita Ltd.	495 N.S.C. Road, Calicut 9

Table 11.1 is also known as a *relation*. Vendor is the relation name. Each row of a relation is called a *tuple*. The titles used for the columns of a relation are known as *relation attributes*. Each tuple in the above example describes one vendor. Each element of a tuple gives a specific property of that vendor. Each property is identified by the title used for an attribute column. In a relation the rows may be in any order. The columns may also be depicted in any order. No two rows can be identical.

Since it is inconvenient to show the whole table corresponding to a relation, a more concise notation is used to depict a relation. It consists of the relation name and its attributes. The identifier or key of the relation is shown in bold face (We will formally define the term *key* later in this section). The relation of Table 11.1 is depicted as:

Relation attributes
↓
Vendor (**Vendor code**, Vendor name, Address)
↓ ↓
Relation Name — Relation identifier or key

The other relations corresponding to Figure 11.5 are:

Supplies (**Vendor code, Item code, order no.**, qty. supplied, date of supply, price/unit)
↓
Composite key

Items (**Item code,** Item name)

Order (**Order no.,** Order date)

A specified value of a relation key uniquely identifies the row of a relation.

If a relationship is $N:M$, then the key of the relationship entity is a composite key which includes the keys of the entity sets which are related. On the other hand, if the relationship is $1:N$, then the key of the relationship entity is the key of one of the entity sets in the relationship. For example, the relations and keys corresponding to the E-R diagram of Figure 11.7 are:

Teacher (**Teacher id.**, name, department, address)

Advises (**Teacher id.**, Student id.)

Students (**Student id.**, name, department, address)

One may ask why an entity set and relationship set are represented as relations. The main reasons are ease of storing relations as flat files in a computer and, more importantly, the existence of a sound theory on relations, which ensures good database design. The raw relations obtained as a first step in the above examples are transformed into *normalized relations*. The rules of transformation called *normalization* are based on sound theoretical principles and ensure that final normalized relations obtained reduce duplication of data, ensure that no mistakes occur when data are added or deleted and simplify retrieval of required data. We will not discuss in detail the theory of normalization. We will, however, describe what these normalization steps are, why they are needed, and how they are done. The end product of all these steps is a good database design for an application.

11.4 NORMALIZING RELATIONS

Consider the relation shown in Table 11.2. In this relation, an order no. includes many items. The attribute "item lines" is not a single attribute but is composed of many attributes.

Table 11.2 An Unnormalized Relation

Order No.	*Order Date*	*Item Lines*		
		Item code	Quantity	Price/unit
1456	26021999	3687	52	50.40
		4627	38	60.20
		3214	20	17.50
		Item code	Quantity	Price/unit
1886	04031999	4629	45	20.25
		4627	30	60.20
		Item code	Quantity	Price/unit
1788	04041999	4627	40	60.20

Besides this, the number of item lines is variable. This form is not suitable for storage as a file in a computer. Further, retrieval of data based on a component of a composite attribute is difficult. For example, to find out how many items with a specified item code are ordered, one must break up the composite attribute first before attempting a search. Thus a relation with a format such as the one in Table 11.2 is not allowed. It is said to be *unnormalized.* To normalize this relation, a composite attribute is converted to individual attributes. The normalization step consists of first identifying fields within a composite attribute as individual attributes. After doing this the common attributes for a composite attribute are duplicated as many times as there are lines in the composite attribute. The normalized relation corresponding to the relation given in Table 11.2 is shown in Table 11.3.

Table 11.3 Normalized Form of the Relation given in Table 11.2

Order No.	*Order Date*	*Item Code*	*Quantity*	*Price/Unit*
1456	26021999	3687	52	50.40
1456	26021999	4627	38	60.20
1456	26021999	3214	20	17.50
1886	04031999	4629	45	20.25
1886	04031999	4627	30	60.20
1788	04041999	4627	40	60.20

The relation shown in Table 11.3 is said to be in *First Normal Form*, abbreviated as 1NF. There are no composite attributes, and every attribute is atomic and describes one property. Converting a relation to the 1NF form is the first essential step in normalization. There are successive higher normal forms known as 2 NF, 3NF, BCNF, 4NF and 5NF. Each form is an improvement (from the point of view of reducing the repetitive storage of data) over the earlier form. In other words, 2NF is an improvement on 1NF, 3NF is an improvement on 2NF, and so on. A higher normal form relation is a subset of lower normal form as shown in Figure 11.10. The higher normalization steps are based on three important concepts:

1. Dependence among attributes in a relation.
2. Identification of an attribute or a set of attributes as the *key* of a relation.
3. Multivalued dependency between attributes.

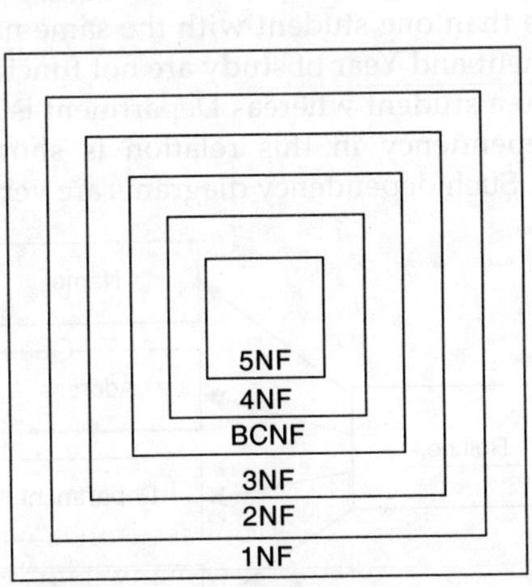

Figure 11.10 Illustration of successive normal forms of a relation.

(i) Functional dependency. As the concept of dependency is very important, it is essential that we first understand it well and then proceed to the idea of normalization. There is no fool-proof algorithmic method of identifying dependency. We have to use our commonsense and judgement to specify dependencies. Dependencies are also obtained using problem specifications given by users of the system. Let X and Y be two attributes of a relation. Given the value of X, if there is only one value of Y corresponding to it, then Y is said to be functionally dependent on X. This is indicated by the notation:

$$X \rightarrow Y$$

For example, given the value of item code, there is only one value of item name for it. Thus item name is functionally dependent on item code. This is shown as:

Item code → Item name

Similarly in Table 11.3, given an order number, the date of the order is known. Thus:

Order no. → Order date

Functional dependency may also be based on a composite attribute. For example, if we write

$$X, Z \rightarrow Y$$

it means that there is only one value of Y corresponding to given values of X, Z. In other words, Y is functionally dependent on the composite X, Z. In Table 11.3, for example, Order no., and item code together determine Qty. and Price. Thus:

Order no., Item code → Qty., Price

As another example, consider the relation

Student (**Roll no.,** Name, Address, Dept., Year of study)

In this relation, Name is functionally dependent on Roll no. In fact, given the value of Roll no., the values of all the other attributes can be uniquely determined. Name and Department are not functionally dependent because given the name of a student, one cannot find his department uniquely. This is due to the fact that there may be more than one student with the same name. Name in this case is not a key. Department and Year of study are not functionally dependent as Year of study pertains to a student whereas Department is an independent attribute. The functional dependency in this relation is shown in Figure 11.11 as a *dependency diagram.* Such dependency diagrams are very useful in normalization.

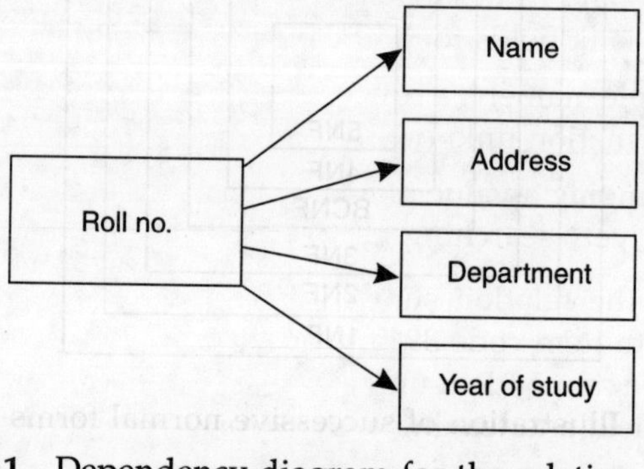

Figure 11.11 Dependency diagram for the relation "Student".

(ii) Relation key. Consider the relation of Table 11.1. Given the Vendor code, the vendor name and address are uniquely determined. Thus Vendor code is the *relation key*. Given a relation, if the value of an attribute X uniquely determines the values of all other attributes in a row, then X is said to be the *key* of that relation. Sometimes more than one attribute is needed to uniquely determine other attributes in a relation row. In that case such a set of attributes is the key. In Table 11.3, Order no. and Item code together determine Order date, Qty. and Price. Thus *Order no.* and *Item code* together form the key. In the relation Supplies (Vendor code, Item code, order no., Qty. supplied, Date of supply, Price/unit), *Vendor code, Item code* and *order no.* together form the key. All three are needed as a given vendor may supply the same item against several orders and each combination will determine Qty. supplied, Date of supply and Price. This dependency is shown in the dependency diagram of Figure 11.12. Observe that in the figure the fact that Vendor code, Item code and order no. together form a composite key is clearly shown by enclosing all of them in a rectangle.

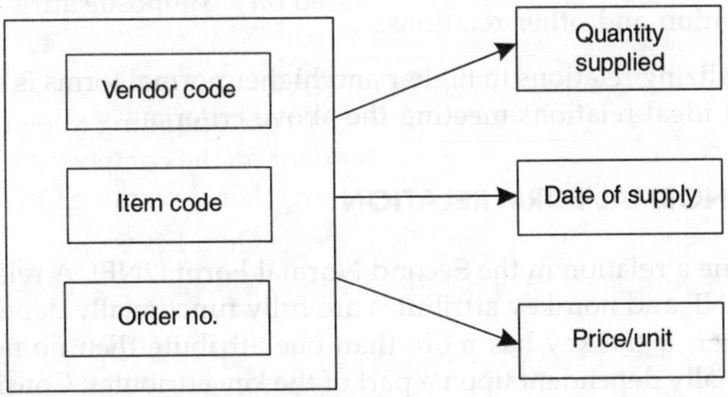

Figure 11.12 Dependency diagram for the relation "Supplies".

11.5 WHY DO WE NORMALIZE A RELATION?

Relations are normalized so that when relations in a database are to be altered during the lifetime of the database, we do not lose information or introduce inconsistencies. The type of alterations normally needed for relations are:

1. **Insertion** of new data values to a relation. This should be possible without being forced to leave blank fields for some attributes.
2. **Deletion** of a tuple, namely, a row of a relation. This should be possible without losing vital information unknowingly.
3. **Updating** or changing a value of an attribute in a tuple. This should be possible without exhaustively searching all the tuples in the relation.

Consider, for example, the relation shown in Table 11.3. If we wish to enter in our database a new item with item code 3945, whose price/unit is ₹40.50 but for which no order has been placed, we cannot do it unless we leave blank fields for order no. and order date. *Order no.* is a key field and leaving a blank for it would make retrieval impossible.

If Order no. 1886 in Table 11.3 is deleted then we lose the information that Item code 4629 costs ₹20.25. Such an accidental loss of information should not occur. If the price of item code 4627 is changed from ₹60.20 to ₹62.00, then in the relation of Table 11.3, it is necessary to find out all the tuples (rows) where Item code 4627 occurs and then change the Price/unit in all these places. In the table, three rows should be changed. If by mistake one row is missed there will be inconsistency in the database.

Ideal relations after normalization should have the following properties so that the problems mentioned above do not occur for relations in the (ideal) normalized form:

1. No data value should be duplicated in different rows unnecessarily.
2. A value must be specified (and required) for every attribute in a row.
3. Each relation should be self-contained. In other words, if a row from a relation is deleted, important information should not be accidentally lost.
4. When a row is added to a relation, other relations in the database should not be affected.
5. A value of an attribute in a tuple may be changed independent of other tuples in the relation and other relations.

The idea of normalizing relations to higher and higher normal forms is to attain the goals of having a set of ideal relations meeting the above criteria.

11.6 SECOND NORMAL FORM RELATION

We will now define a relation in the Second Normal Form (2NF). A relation is said to be in 2NF if it is in 1NF and non-key attributes are fully functionally dependent on the key attribute(s). Further, if the key has more than one attribute then no non-key attributes should be functionally dependent upon a part of the key attributes. Consider, for example, the relation given in Table 11.3. This relation is in 1NF. The key is (Order no., Item code). The dependency diagram for attributes of this relation is shown in Figure 11.13. The non-key attribute price/unit is functionally dependent on Item code which is part of the relation key. Also, the non-key attribute Order date is functionally dependent on Order no., which is a part of the relation key. Thus the relation is not in 2NF. It can be transformed to 2NF by splitting it into three relations as shown in Table 11.4.

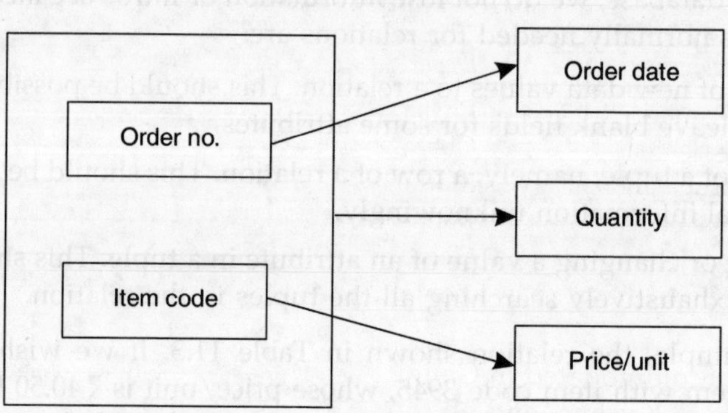

Figure 11.13 Dependency diagram for the relation given in Table 11.3.

In Table 11.4 the relation "Orders" has *Order no.* as the key. The relation "Order details" has the composite key *Order no.* and *Item code.* In both relations the non-key attributes are functionally dependent on the whole key. Observe that by transforming to 2NF relations the repetition of Order date (Table 11.3) has been removed. Further, if an order for an item is cancelled, the price of an item is not lost. For example, if Order no. "1886" for item code "4629" is cancelled in Table 11.3, then the fourth row will be removed and the price of the item is lost. In Table 11.4 only the fourth row of the Table 11.4(b) is omitted. The item price is not lost as it is available in Table 11.4(c). The date of the order is also not lost as it is in Table 11.4(a).

Table 11.4 Splitting of Relation given in Table 11.3 into 2 NF Relations

(a) Orders

Order no.	*Order date*
1456	26021999
1886	04031999
1788	04041999

(b) Order Details

Order no.	*Item code*	*Qty.*
1456	3687	52
1456	4627	38
1456	3214	20
1886	4629	45
1886	4627	30
1788	4627	40

(c) Prices

Item code	*Price/unit*
3687	50.40
4627	60.20
3214	17.50
4629	20.25

These relations in 2NF form meet all the "ideal" conditions specified. Observe that the three relations obtained are self-contained. There is no duplication of data within a relation.

11.7 THIRD NORMAL FORM

A Third Normal Form normalization will be needed when all attributes in a relation tuple are not functionally dependent only on the key attribute. If two *non-key* attributes are functionally dependent, then there will be unnecessary duplication of data. Consider the relation given in Table 11.5. Here, *Roll no.* is the key and all other attributes are fully functionally dependent on it. Thus it is in 2NF. If it is known that in the college all first

Table 11.5

Roll no.	*Name*	*Department*	*Year*	*Hostel name*
1784	Raman	Physics	1	Ganga
1648	Krishnan	Chemistry	1	Ganga
1768	Gopalan	Mathematics	2	Kaveri
1848	Raja	Botany	2	Kaveri
1682	Maya	Geology	3	Krishna
1485	Singh	Zoology	4	Godavari

year students are accommodated in Ganga hostel, all second year students in Kaveri, all third year students in Krishna, and all fourth year students in Godavari, then the non-key attribute Hostel name is dependent on the non-key attribute Year. This dependency is shown in Figure 11.14. Observe that given the year of student, his hostel is known and

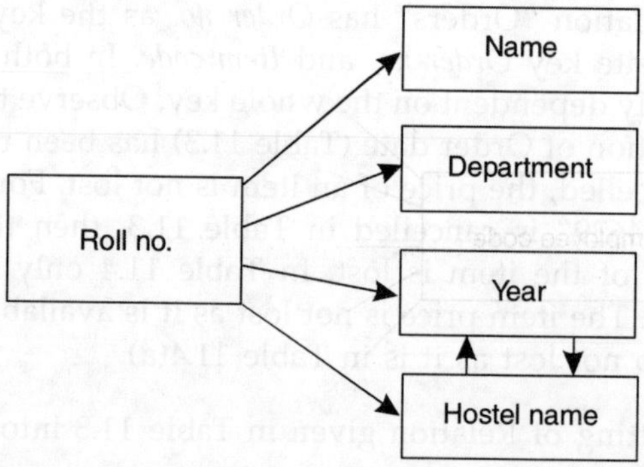

Figure 11.14 Dependency diagram for the relation given in Table 11.5.

vice versa. The dependency of hostel on year leads to duplication of data as is evident from Table 11.5. If it is decided to ask all first year students to move to Kaveri hostel, and all second year students to Ganga hostel, this change should be made in many places in Table 11.5. Also, when a student's year of study changes, his hostel change should also be noted in Table 11.5. This is undesirable. A table is said to be in 3NF if it is in 2NF and no non-key attribute is functionally dependent on any other non-key attribute. Table 11.5 is thus not in 3NF. To transform it to 3NF, we should introduce another relation which includes the functionally related non-key attributes. This is shown in Table 11.6. It should be stressed again that dependency between attributes is a semantic property and has to be stated in the problem specification.

Table 11.6 Conversion of Table 11.5 into Two 3NF Relations

Roll no.	*Name*	*Department*	*Year*
1784	Raman	Physics	1
1648	Krishnan	Chemistry	1
1768	Gopalan	Mathematics	2
1848	Raja	Botany	2
1682	Maya	Geology	3
1485	Singh	Zoology	4

Year	*Hostel name*
1	Ganga
2	Kaveri
3	Krishna
4	Godavari

In this example the dependency between Year and hostel is clearly stated. In case hostel allocated to students do not depend on their year in college, then Table 11.5 is already in 3NF. Let us consider another example of a relation. The relation "Employee" is given below and its dependency diagram in Figure 11.15.

Employee (**Employee code**, Employee name, Dept., Salary, Project no., Termination date of project)

As can be seen from the figure, the termination date of a project is dependent on the Project no. Thus this relation is not in 3NF. The 3NF relations are:

Employee (**Employee code**, Employee name, Dept., Salary, Project no.)
Project (**Project no.**, Termination date)

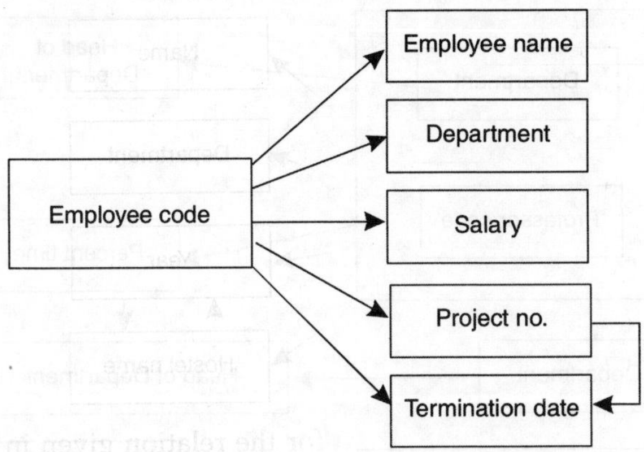

Figure 11.15 Dependency diagram of employee relation.

11.8 BOYCE-CODD NORMAL FORM (BCNF)

Assume that a relation has more than one possible composite key. Assume further that the composite keys have a common attribute. If an attribute of a composite key is dependent on an attribute of the other composite key, a normalization called BCNF is needed. Consider, as an example, the relation "Professor".

Professor (**Professor code, Dept.**, Head of Dept., Percent time)

It is assumed that

1. A Professor can work in more than one department
2. The percentage of the time he spends in each department is given.
3. Each department has only one Head of Department

The relationship diagram for the above relation is given in Figure 11.16. Table 11.7 gives the relation attributes. The two possible composite keys are *Professor code* and *Dept.* or *Professor code* and *Head of Dept.* Observe that department as well as Head of Dept., are both part of a composite key. It is assumed that Heads of Dept. have unique ids.

Table 11.7 Normalization of Relation "Professor"

Professor Code	*Department*	*Head of Dept.*	*Percent Time*
P1	Physics	Ghosh	50
P1	Mathematics	Krishnan	50
P2	Chemistry	Rao	25
P2	Physics	Ghosh	75
P3	Mathematics	Krishnan	100
P4	Mathematics	Krishnan	30
P4	Physics	Ghosh	70

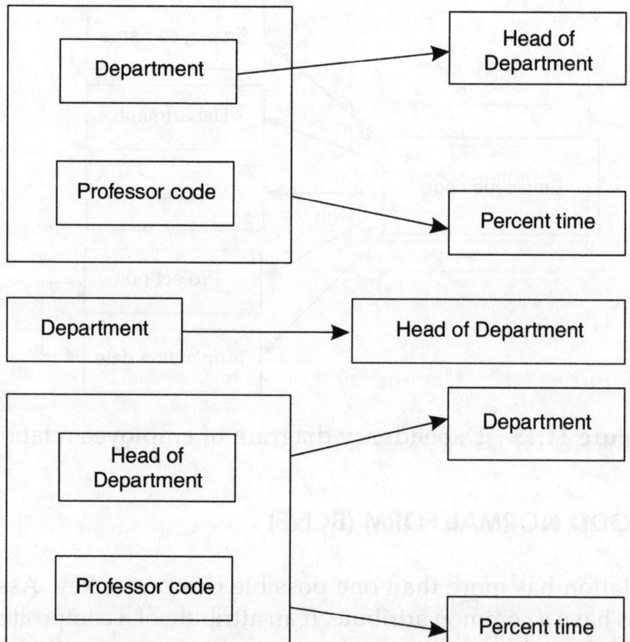

Figure 11.16 Dependency diagram of Professor relation.

The relation given in Table 11.7 is in 3NF. Observe, however, that the names of Dept. and Head of Dept. are duplicated. Further, if Professor P2 resigns, rows 3 and 4 are deleted. We lose the information that Rao is the Head of Department of Chemistry.

The normalization of the relation is done by creating a new relation for Dept. and Head of Dept. and deleting Head of Dept. from Professor relation. The normalized relations are shown in Table 11.8 and the dependency diagrams for these new relations in Figure 11.17. The dependency diagram gives the important clue to this normalization step, as is clear from Figure 11.16 and Figure 11.17.

Table 11.8 Normalized Professor Relation in BCNF

(a)

Professor Code	*Department*	*Percent Time*
P1	Physics	50
P1	Mathematics	50
P2	Chemistry	25
P2	Physics	75
P3	Mathematics	100
P4	Mathematics	30
P4	Physics	70

(b)

Department	*Head of Dept.*
Physics	Ghosh
Mathematics	Krishnan
Chemistry	Rao

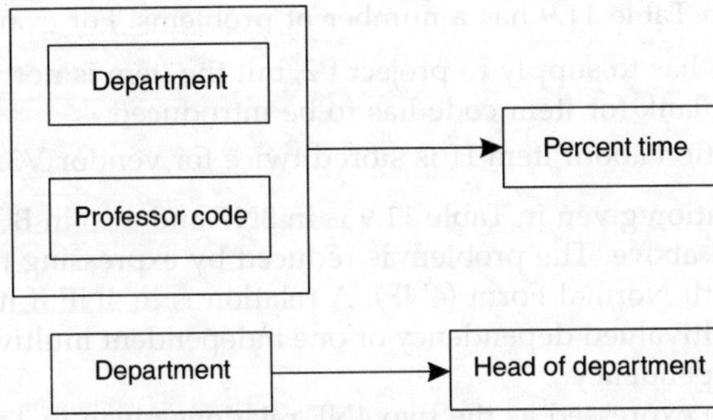

Figure 11.17 Dependency diagram of Professor relation.

11.9 FOURTH AND FIFTH NORMAL FORMS

When attributes in a relation have multivalued dependency, further normalization to 4NF is required. We will illustrate this with an example. Consider a vendor supplying many items to many projects in an organization. The following are the assumptions:

1. A vendor is capable of supplying many items.
2. A project uses many items.
3. A vendor supplies to many projects
4. An item may be supplied by many vendors.

Table 11.9 gives a relation for the problem and Figure 11.18 the dependency diagram(s).

Table 11.9 Vendor-Supply-Projects Relation

Vendor Code	*Item Code*	*Project No.*
V1	I1	P1
V1	I2	P1
V1	I1	P3
V1	I2	P3
V2	I2	P1
V2	I3	P1
V3	I1	P1
V3	I1	P2

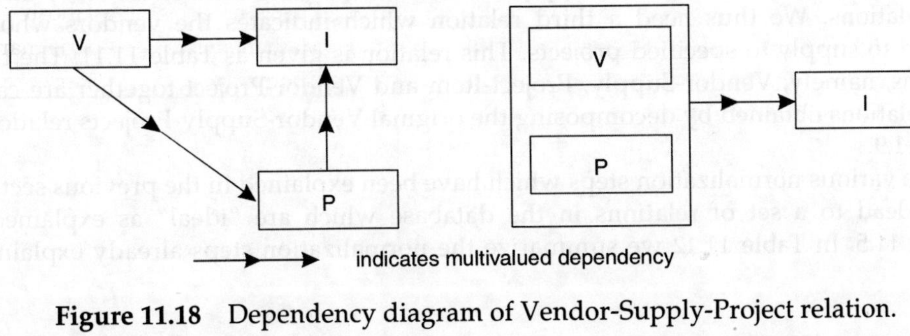

Figure 11.18 Dependency diagram of Vendor-Supply-Project relation.

The relation given in Table 11.9 has a number of problems. For example

- If vendor V1 has to supply to project P2, but the item is not yet decided, then a row with a blank for item code has to be introduced.
- The information about item I1 is stored twice for vendor V3.

Observe that the relation given in Table 11.9 is in 3NF and also in BCNF. It still has the problems mentioned above. The problem is reduced by expressing this relation as two relations in the Fourth Normal Form (4NF). A relation is in 4NF if it has no more than one independent multivalued dependency or one independent multivalued dependency with a functional dependency.

Table 11.9 can be expressed as the two 4NF relations given in Table 11.10. The fact that vendors are capable of supplying certain items and the requirements of items by Projects are independently specified in the 4NF relation.

Table 11.10 Vendor-Supply-Project Relations in 4NF

(a) Vendor Supply

Vendor code	*Item code*
V1	I1
V1	I2
V2	I2
V2	I3
V3	I1

(b) Project-Item

Project no.	*Item code*
P1	I1
P1	I2
P1	I3
P2	I1
P3	I1
P3	I2

These relations still have a problem. From Project-item relation it is seen that project P2 requires I1. The vendor supply relation indicates that I1 is supplied by V1 and V3. We may thus infer that (V1, P2, I1) and (V3, P2, I1) should be tuples in the original Vendor-Supply-Project relation. We see, however, that (V1, P2, I1) is not a tuple in it whereas (V3, P2, I1) is a valid tuple. This spurious tuple has occurred because Vendor V1 may not be allowed to supply item I1 to project P2. Similarly, we see from the Project-item relation that P3 uses item I1. Vendor Supply relation states that I1 is supplied by vendors V1 and V3. Thus we get two tuples (V1, P3, I1) and (V3, P3, I1). The tuple (V1, P3, I1) is a valid tuple whereas (V1, P3, I1) is a spurious tuple. Such spurious tuples occur when Vendor-Supply and Project-Item relations are combined because the constraint that some vendors are allowed to supply to only some projects is not reflected by the two 4NF relations. We thus need a third relation which indicates the vendors who are allowed to supply to specified projects. This relation is given as Table 11.11. The three relations, namely, Vendor-Supply, Project-Item and Vendor-Project together are called 5 NF relations obtained by decomposing the original Vendor-Supply-Projects relation of Table 11.9.

The various normalization steps which have been explained in the previous sections would lead to a set of relations in the database which are "ideal" as explained in Section 11.5. In Table 11.12 we summarize the normalization steps already explained.

Table 11.11 5NF Additional Relation

Vendor Supply	
Vendor Code	*Project No.*
V1	P1
V1	P3
V2	P1
V3	P1
V3	P2

Table 11.12 Summary of Normalization Steps

Input Relation	*Transformation*	*Output Relation*
All relations	Remove multiattribute lines in table with varying number of records.	INF
INF relation	Remove dependency of non-key attribute on part of a multiattribute key	2NF
2NF	Remove dependency of non-key attributes on other non-key attributes	3NF
3NF	Remove dependency of an attribute of a multiattribute key on any attribute that does not form a complete key	BCNF
BCNF	Remove more than one independent multivalued dependency from relation by splitting relation	4NF
4NF	Add one relation relating attributes with multivalued dependency to the two relations with multivalued dependency	5NF

11.10 SOME EXAMPLES OF DATABASE DESIGN

Example 1

In Chapter 5 we considered a problem where items are supplied by a vendor and checked by a receiving process against orders for detecting any discrepancy. An order file was used to check whether the deliveries are against orders and whether there was any discrepancy. We will now see how these data can be organized as relations. The E-R diagram of Figure 11.3 applies for this case and the relations are:

ORDER (**order no.**, order date)

ORDER PLACED FOR (**order no., item code**, qty. ordered, delivery time allowed)

ORDER PLACED WITH (**order no., vendor code,** item code)

VENDOR (**vendor code**, vendor name, vendor address)

ITEM (**item code**, item name)

SUPPLIES (**vendor code, item code, order no**., qty. supplied, date of supply, price/unit).

The key attribute(s) are in bold letter(s) in each relation.

Let us examine whether the relations are in appropriate normal form. ORDER and ORDER PLACED FOR are simple relations. In the relation ORDER PLACED WITH, the key is the composite attributes **order no.**, and **vendor code**. However, the entity item code is not needed in the ORDER PLACED WITH relation. Given an **order no.** all the items supplied against this order can be found from ORDER PLACED FOR relation. The vendor with whom the order has been placed can be found from the ORDER PLACED WITH relation given below as each order no. has only one vendor.

ORDER PLACED WITH (**order no., vendor code**)

The two relations ORDER PLACED WITH and ORDER PLACED FOR have composite keys. The non-key fields are not related to one another. In a key with more than one attribute, the individual attributes are not functionally dependent. Thus these two relations are in 3NF form and do not need any further change. The relations VENDOR and ITEM are simple and are in 3NF form. The relation SUPPLIES is, however, not properly normalized. **vendor code** and **order no**. are functionally dependent. There is a multivalued dependency between vendor code and item code as a vendor can supply many items. We thus split the relations into two relations:

ACTUAL SUPPLY (**order no.**, **item code**, qty. supplied, date of supply, price/unit)
VENDOR SUPPLY CAPABILITY (**vendor code, item code**)

Observe that the relation VENDOR SUPPLY CAPABILITY will have a (**vendor code, item code**) table without a vendor having supplied any item. The relation ORDER PLACED WITH will have a tuple only when a vendor actually supplies an item.

Example 2

We now consider another problem. Let a database contain the following: Teacher code, Teacher's name, Teacher's address, rank, department, courses taught by the teacher, course name, credits for course, no. of students in the class, course taught in semester no., student no., name, dept., year, and courses taken in semester no. The following information is given on dependencies.

- A teacher may teach more than one course in a semester.
- A teacher is affiliated to only one department.
- A student may take many courses in a semester.
- The same course may have more than one section and different sections will be taught by different teachers.
- A course may be taught in more than one semester

An entity-relationship diagram for this problem is given in Figure 11.19. The relations corresponding to the E-R diagram are:

TEACHER (**Teacher code**, teacher's name, teacher's address, rank, dept.)
TEACHES COURSES (**Teacher code, course no.**, semester no., no. of students, section no.)
COURSE (**course no., semester no.,** course name, credits)
STUDENT (**student no.**, student's name, dept., year)
STUDENT-COURSES (**student's no., course no.**, semester no.)

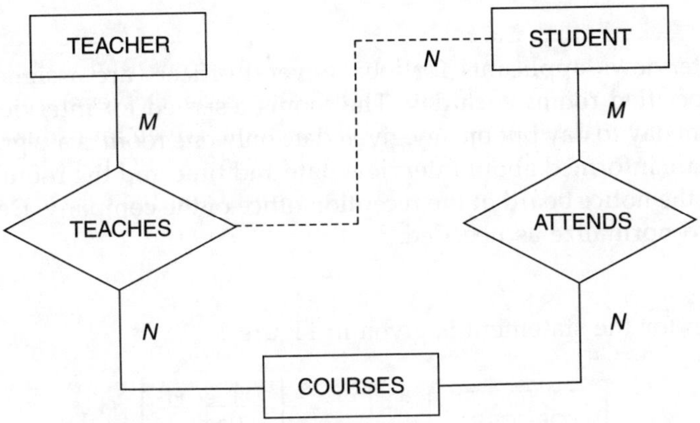

Figure 11.19 An E-R diagram for teacher database.

TEACHER relation has only one key. All non-key attributes are functionally dependent only on the key. There is no functional dependency among non-key attributes. Thus the relation is normalized in 3NF.

STUDENT relation is also, similarly, in 3NF. In the COURSE relation, course number and semester no. have multivalued dependency. Thus it is better to split it into two relations SEMESTER-COURSE (**semester no.**, course no.)

and COURSE (**course no.**, course name, credits)

Thus we can answer a query: in semester 1 what courses are taught? The relations TEACHES COURSES and STUDENT-COURSES have multiattribute keys. The relations STUDENT-COURSES has no problem. However in TEACHES COURSES a teacher may not teach a course in a given semester but he may have capability to teach it. Thus we add another relation

TEACHER-COURSE CAPABILITY (**Teacher code**, course no.)

The only point which is not clear, from these relations, is the relation between teacher and student. This has been missed in the E-R diagram of Figure 11.19. If we connect the relationship teacher with student (dotted line in Figure 11.19) it will model the relationship between the teacher, courses taught and students. In other words, we should be able to answer the question "Which teacher is teaching course no. X to student no. Y in Semester 2?" Let us add a relation.

TEACHER-STUDENT (**Teacher code, student no., course no.**).

In this relation **Teacher code** and **course no.** have a multivalued dependency. Similarly, **Teacher code** and **student no.** as well as **student no.** and **course no.** have multivalued dependency. However, TEACHES COURSES (**Teacher code, course no.**, semester no., no. of students, section no.) and STUDENT COURSES (**student no., course no.**, semester no.) relations are already in the database. Thus the relation TEACHER-STUDENT as it is specified above is sufficient to give the idea that student Y takes course X from Teacher Z in semester 1.

Example 3

A company interviews applicants for jobs. Several officers are assigned to interview applicants in specified rooms each day. The rooms reserved for interview by an officer may change from day to day but on any given date only one room is assigned to an officer. The applicants are informed about interview date and time and the room assigned is put up each day on the notice board at the reception office of the company. Design a database for this task and normalize as neeeded.

Solution:

An E-R diagram for the statement is given in Figure 11.20.

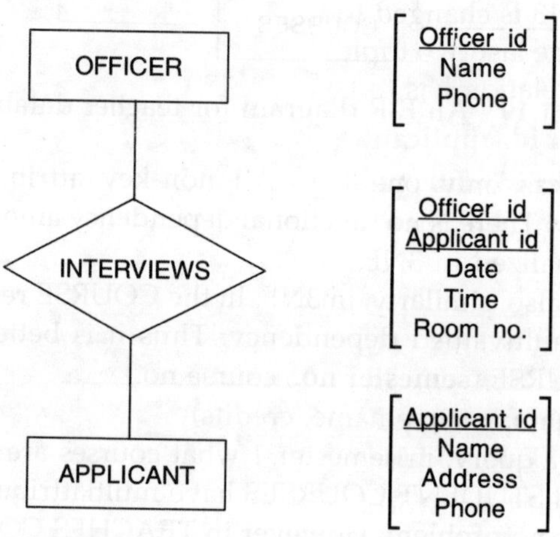

Figure 11.20 E-R diagram for Example 3.

APPLICANT (**applicant id**, name, address, phone no.)
OFFICER (**officer id**, name, phone no.)
INTERVIEW (**officer id, applicant id**, date, time, room no.)
Let us now normalize these relations:
APPLICANT relation has only one key and non key attributes are fully dependent on key attribute. Thus it is in 3NF and no more normalization is required. OFFICER is also in 3NF.

The relation INTERVIEW given below needs examination:
INTERVIEW (**officer id, applicant id**, date, time, room no.)
The relation in 3NF as non key attributes are not related. However due to the fact that room no. assigned for interview may change from day to day there is additional dependencies.
Officer id, date → room no.
This may cause a problem as may be seen from Table 11.13.

Table 11.13 INTERVIEW relation

Officer id	*Applicant id*	*Date*	*Time*	*Room No.*
512	1452	11052010	11.00	R01
512	1453	11052010	12.00	R01
513	1456	11052010	11.00	R02
512	1457	12052010	11.00	R02
513	1458	12052010	12.00	R04
513	1459	12052010	13.00	R04

If the room no. of officer 513 is changed to R03 from R04 on 12052010 then whole table needs to be searched and the last two tuples have to be changed. An error in update can occur. Thus INTERVIEW relation has to be split as shown below:

INTERVIEW-DATE (**officer id, applicant id,** date, time)

INTERVIEW-ROOM (**officer id, date,** room no.)

These relations are in 3NF and BCNF.

SUMMARY

1. A conceptual model of data used in an application is obtained by using an entity-relationship model (E-R model).
2. **Entities** specify distinct real world items in an application, **relationships** represent dependencies between entities, and **attributes** specify properties of entities and relationships.
3. Entity set describes items which are similar. Students, courses, teachers are examples of entity sets.
4. Examples of relationships are students **attend** courses, Teachers **teach** courses. "attend" and "teach" are relationships.
5. E-R models have the facility to describe 1 to 1, 1 to many, many to many, conditional and optional relationships.
6. E-R models assist in designing relational databases.
7. A relation is a table which depicts an entity set. Each column in the relation corresponds to an attribute of the entity. Each row contains a member of the entity set.
8. A relational database consists of a collection of relations relevant for a specified application.
9. Relations in a database are normalized. Normalization is a procedure used to transform a set of relations into another set which has some desirable properties.
10. Normalization ensures that data in the database are not unnecessarily duplicated. It also ensures that addition and deletion of entity rows (or tuples) or change of individual attribute values do not lead to accidental loss of data or errors in database.

11. Normalization is performed by first converting a relation so that all entities are simple. This results in a relation said to be in First Normal Form (1NF).
12. Successive normalization steps are carried out to meet the objectives stated in point 10.
13. These successive normal forms are called 2NF, 3NF, BCNF, 4NF and 5NF.
14. In practice in most problems normalization upto 3NF is sufficient.
15. A summary of how normalization steps are carried out is shown in Table 11.12.

EXERCISES

11.1 What is an entity? Give examples of entities.

11.2 What is a relationship? In what way is it different from an entity?

11.3 In what way is an attribute different from an entity?

11.4 What do you understand by a 1 to 1, 1 to many, many to many relationships? Give one example for each.

11.5 What is the difference between a relation and a relationship?

11.6 What is normalization of a relation?

11.7 Why should relations be normalized?

11.8 What is a functional dependency?

11.9 What is a key attribute in a relation?

11.10 What is the difference between a 2NF and a 3NF relation?

11.11 When is BCNF required?

11.12 When are 4NF and 5NF required?

11.13 What is the difference between 4NF and 5NF?

11.14 Develop E-R diagram for the following:

(i) Customer withdraws money from his account
(ii) Students write examinations.
(iii) Students attend classes
(iv) Professors write books
(v) Driver drives a car

11.15 Draw an E-R diagram showing the cardinality for the following:

(i) A bill is sent to a customer. A customer can receive many bills.
(ii) A clerk works in a bank. The bank has many clerks
(iii) A part is used in many products and a product uses many parts.
(iv) Students apply for seats in colleges. Each student can atmost get one seat. A college has many seats. A student can send many applications.
(v) A car is owned by a person. The person can own many cars.

11.16 For Exercise 11.14, obtain relations for each entity. Normalize the relations.

11.17 For the following word statement, obtain E-R diagram and relations. Use any reasonable assumptions. "A machine shop produces many parts which it takes on contract. It employs many machinists who operate any of the machines. A part needs working on only one machine. A record is kept on the quantity of material needed for producing each part. The production of each part is tracked by giving a job number, start time and end time and machinist identification".

11.18 For the problem on library, periodical management stated in Exercise 6.23, obtain E-R diagrams and relations. Make any reasonable assumptions and state the assumptions.

11.19 In Section 6.2 Solution B, a subsystem for mess bill calculation is defined. Obtain the E-R diagram and relations for the problem.

11.20 For the problem statement of Exercise 7.16, obtain the E-R diagram and a set of relations by using the diagram.

11.21 For the problem statement of Exercise 7.17, obtain the E-R diagrams and relations.

11.22 For the problem statement of Exercise 7.19, obtain E-R diagrams and relations.

11.23 Refer to Figure 5.8. Obtain the E-R diagram and relations for the data relevant to the stores process.

11.24 Refer to the Figure 5.9. Obtain the E-R diagram and relations for the data relevant to the purchase process.

12 Database Management Systems (DBMS)

LEARNING GOALS

After reading this chapter a student should be able to:

1. Explain the need for an integrated database in organizations.
2. State the objectives of Database Management Systems.
3. Describe the organization of Database Management Systems.

We have already discussed various ideas regarding conceptual model of data and logical model of data as relations. In this chapter we will see how these ideas are integrated in a database management system.

The concept of *Database Management Systems (DBMS)* is as important as that of high level languages and compilers which allow application programs to be written independent of the architecture of a computer. Similarly, DBMS aims to allow writing application programs independent of the physical organization of files in secondary storage. Application programs may be written using a logical data model. The logical to physical mapping is achieved by a DBMS.

12.1 PROBLEM WITH FILE-BASED SYSTEMS

We have already seen how to model data needed in applications. In the conventional data processing method, the program is developed for each application. The programmer also designs the file appropriate for the application. For example, data processing in a bank may have many applications such as maintaining savings accounts, current accounts, fixed deposit accounts, and miscellaneous loan accounts. Each application may have its own file of data. For example, a savings account customer file will have customer account number, address, balance, etc. A similar file will be maintained for current account. There are a number of disadvantages when this approach is used. These will be discussed now.

(i) **Data redundancy.** The same data may be duplicated at many places. For example, a customer may have a savings account, a current account and some fixed deposits. In such a case, the name and address will be found in three different files.

(ii) **Data integrity.** The problem of data integrity arises due to redundancy. The same data may be found in different forms in different files. For example, if a customer changed his address and notified it to the bank, the address may have been changed in the savings account file, but not in the fixed deposit file. Thus two different addresses may be found for the same person. Further, checking the validity of data may not be uniformly implemented with the result that data in one file may be correct and in another file wrong.

(iii) **Data availability.** Data is scattered in many files. If a customer's credit worthiness in the bank has to be checked before he is given a loan, it would be necessary to look at many files. Due to non-uniformity in file design, the same customer may have different identification numbers in different files and hence obtaining the necessary data will be difficult.

(iv) **Management control.** As data is scattered in different files and relating them is not easy, it is difficult to implement uniform policies in an organization. For instance, if a bank decides that loan accounts will be given only to current account holders, it will be difficult to implement this decision due to non-uniform coding of customers.

Databases and systems to manage databases have been developed primarily to reduce these problems.

12.2 DATABASE AND DATABASE MANAGEMENT SYSTEMS

A *database* is a collection of related data necessary to manage an organization. It excludes transient data such as input documents, reports and intermediate results obtained during processing. It has a long life; often several years.

A database models the data resource of an organization using the relationships between different data items. The model is independent of any application program.

Parts of one database may be used by different application programs for different applications. Application programs may require different "views" of portions of data from a database, and it should be possible to provide them.

A Database Management System is a set of procedures that manage the database and provide access to the database in the form required by any application program. It effectively ensures that necessary data in the desired form is available for diverse applications of different departments in an organization. Figure 12.1 illustrates the distinction between a database and a Database Management System.

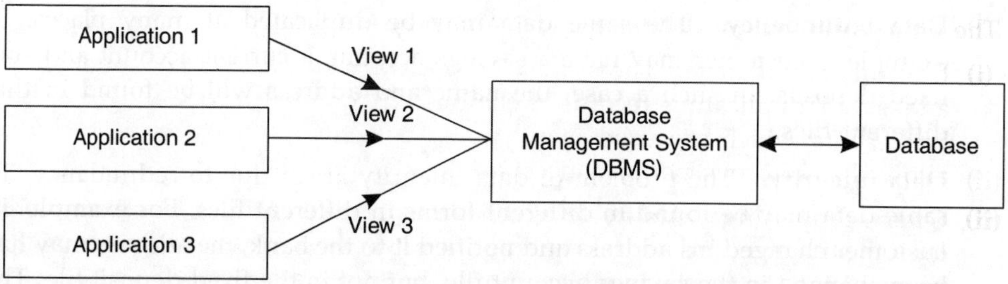

Figure 12.1 Illustration of the distinction between a DBMS and a database.

12.3 OBJECTIVES OF DATABASE MANAGEMENT

The main motivation for designing a system to manage a database is the ability to *share data.* Data is considered as an important resource of an organization to be shared by a number of applications. Data is collected, validated, protected, logically organized and stored with controlled redundancy. It should be always kept in mind that the data is not only to be used for a number of current applications but it has to serve also future applications. This flexibility of accommodating new requirements enables one to meet the cardinal principle of information systems design: *The system should be built to accommodate changing needs.*

Data independence is an essential requirement to enable the design of application programs independent of the physical format, storage and access of data. If an application program needs information on where data is stored and how it is accessed, then any changes in data location and access method will require changes in the application program too. The application program must then be recompiled and maintained. As the needs of an organization are bound to change, the format of data and storage systems will also change to accommodate new data, thereby affecting all application programs.

With improvements in computer systems and operating systems, the method of accessing and storing data is likely to change. If the access and storage methods are built into the application logic, then programmers must spend considerable amount of their time in program maintenance and change. This is a source of error; it is also an unnecessary expense. Thus application programs should be oriented towards information content of data and should not be concerned with details of data representation, location or access. The ability to write an application program without knowing the representation details of database is called *data independence.*

To summarize, data independence allows:

1. the contents and organization of a database to be changed without affecting application programs;
2. changing data processing equipment and software without requiring rewriting of application programs;
3. sharing of data by different application programs by allowing data to appear to be organized properly for each application;
4. simplifying program development by removing from the purview of a programmer low-level data organization; and
5. central control of data which ensures the security and integrity of data.

The other objectives of database management are as follows:

(i) Control redundancy. The system should identify existence of common data and avoid duplicating records. Relationships or pointers should be used to locate data which are used many times. Selective redundancy is sometimes allowed to improve performance or for better reliability.

(ii) Query processing. Ability to allow users to create new databases and their logical structure and to ask queries from these databases. Queries may be grouped into transactions which should be correctly processed.

(iii) Data access by many users. Allow several users to access database concurrently without one user accidentally corrupting the data.

(iv) Failure recovery. Ability to recover from hardware/software failures which may occur during query processing.

(v) Relating data items. Relationship between data items should be specified.

(vi) Data integrity. Consistency of data values and relationship must be preserved. In order to achieve this, the system must ensure validity of data by using good data editing, updating of data, and propagating changes to other related data elements. It also involves maintaining audit trails to enable recovery if errors are detected.

(vii) Data security. This is concerned with protecting access to data. Protection is needed at many levels for access, modification, deletion or display. Access restriction may be for an individual data item or groups of items. For example, access to the salary of employees may be allowed to the financial controller but access to medical records would be barred.

(viii) Database performance. The system should be able to provide timely information as required. The cost of storing and retrieving data should be commensurate with the value of information provided.

(ix) Management controls. As the dependence of an organization on a database increases, positive management controls should be exercised over addition, deletion, change and disposition of data. Data must be protected to satisfy legal, accounting, and auditing requirements.

12.4 OVERVIEW OF DATABASE MANAGEMENT SYSTEMS

The design of a DBMS begins by determining the conceptual requirements of current applications and applications expected to be implemented in the near future. The data needs for all these requirements are determined. Using *entity-relationship* (E-R) *data modelling*, a *conceptual data model* is derived.

The conceptual model derived by using the entity relationship modelling is next translated to a data model compatible with the DBMS chosen. If a relational DBMS is chosen, the conceptual model is translated to relations. Relations constitute the *logical data model*. Application programmers are presented with a subset of the logical data model (A subset of relations in the case of relational DBMS). Such subsets are called *external data models*. (They are also known as subschemas in the literature).

The logical model is mapped onto a disk storage. The data in a physical storage device is known as the *physical model*. The physical model design involves selecting access methods, organizing indices or hashing functions, and ensuring optimal use of disk space. This model is also known as the *internal model*.

The external model should not be affected by changes in access methods or indices or change of disk drives. In other words, it should not be affected by changes in the internal model. Similarly, if the conceptual model has foreseen new applications, then such changes should not affect the external model.

If the DBMS is changed then the logical model (external model) too will change. Otherwise, the external model is invariant, which implies that applications are data independent. Figure 12.2 illustrates these ideas.

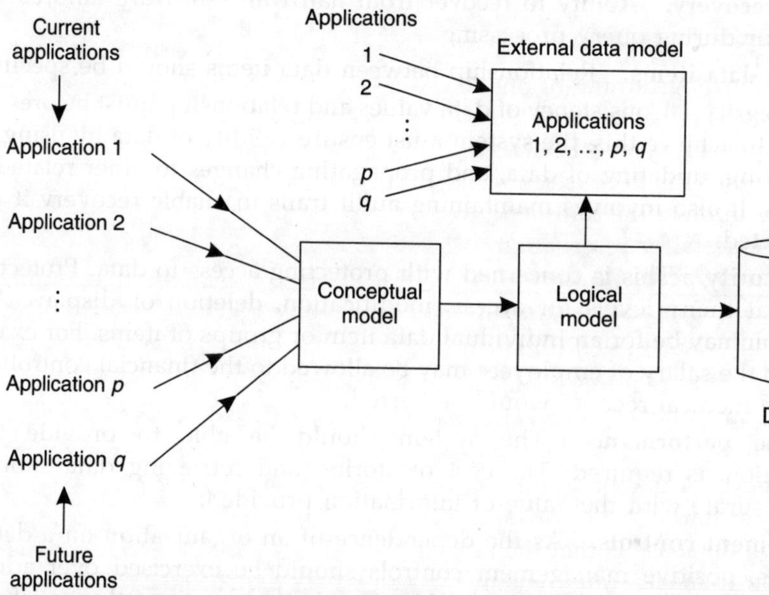

Figure 12.2 Application independence of DBMS.

The structure of a typical DBMS is given in Figure 12.3

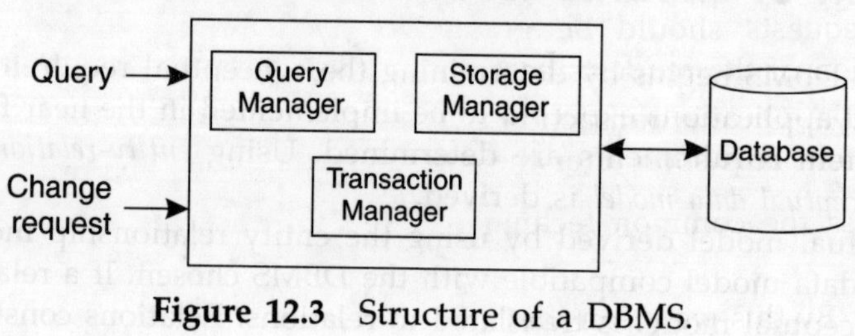

Figure 12.3 Structure of a DBMS.

1. *Database* normally resides on a disk. It contains data as well as *metadata,* i.e. information about the structure of the data. Besides this *indexes* for the data to enable fast retrieval is stored in a specified part of the disk.

2. *Storage Manager* is a software which resides in the main memory of the computer. Its main purpose is to obtain blocks of data from the disk as and when required to satisfy queries.
3. *Query Manager* is a software whose primary task is to convert queries which may be in a high level query language to a sequence of requests for data stored in the disk. These requests are handled by the storage manager. Besides managing queries, it also manages requests for change of data or metadata.
4. *Transaction Manager* ensures that when several queries are asked concurrently they do not interfere with one another. It also ensures that no data is lost even when there is a system failure. It interacts with both the query manager and the storage manager.

12.4.1 Transaction Processing

A typical DBMS allows grouping of one or more queries or requests for modifications into what is known as a *transaction*. One may loosely define a transaction as a group of operations that appear to have been executed sequentially as a unit. Often a DBMS allows several transactions to be executed concurrently and a transaction manager must ensure correct execution of the transactions. For example, in a railway ticket reservation system, several requests for reservation/cancellation may be initiated concurrently by different clerks having access to the system. The transaction manager must ensure that the same seat is not assigned to two different persons. To achieve this, execution of transactions should satisfy four properties often called *ACID properties*. These are:

- **Atomcity.** This requires that an *entire* transaction is executed or nothing is executed. In a ticket reservation system a ticket is issued only after the reservation is made and the seat is marked as sold.
- **Consistency.** A database must satisfy a consistency condition as defined by a user. For example, an appropriate consistent condition for a railway ticket reservation database is that the same seat is not assigned to two passengers. A transaction manager must ensure that this condition is met at the end of a transaction.
- **Isolation.** When two or more transactions run concurrently the tasks they perform must be isolated from one another. For example, if two clerks in a railway reservation system request for a seat and only one seat is available one of the requests should be granted and the other denied. This requires serialization of the requests based on some criteria.
- **Durability.** Once a transaction is completed its effect should not be lost even if the system fails soon after the transaction completes.

We briefly discuss the common techniques used for ensuring the ACID properties. These are:

1. **Locking** By this we mean that a transaction can lock the items accessed by it. When one transaction locks items in the database no other transaction can access those items. For example, when a seat is reserved, that seat is locked by a transaction disallowing any read or write in that location till the transaction is complete.

2. **Logging.** Transaction manager records in a non-volutile storage a log of all transactions initiated, the changes in the database caused by each transaction and end of each transaction. Logging is essential to ensure durability.
3. **Commitment.** Normally transactions are carried out tentatively. In other words all changes required by the transaction are computed and kept ready. They are then logged. Only when the transaction is complete the changes are made in the database. This is called commitment. The main point is that the log keeps record of all actions and only when all actions are complete the database is altered. Thus if there is a failure in the middle of a transaction log can be used to restart or abort it as appropriate.

12.5 QUERY LANGUAGE

Currently the Relational Database Model is the defacto standard and the corresponding DBMS is known as RDBMS (Relational DBMS). One of the primary purposes of a DBMS is to enable retrieval of data required for applications by formulating simple queries to the database. In order to formulate queries there is a need for a precise language. Currently a language called Structured Query Language (SQL) is a popular language. We will briefly describe the features of SQL. There are four parts to SQL. These are:

1. **Data Definition Statements.** Some common statements are: CREATE TABLE, DROP TABLE, ALTER TABLE and CREATE VIEW.

   In SQL a relation is named TABLE. CREATE TABLE is used to create a relation with appropriate attributes and rows. DROP and ALTER are to delete or change a relation. Relations defined using CREATE TABLE are stored in the database in a physical form whereas those defined using CREATE VIEW do not exist physically but are virtual. Queries can be formulated on such "views".
2. **Database Modification Statements.** Some common statements are: INSERT INTO <relation> VALUES <values>, DELETE FROM <relation> WHERE <condition>, UPDATE <relation> SET <new values> WHERE <condition>.
3. **Query Statements.** These are mainly used to retrieve from relations data meeting specified requirements. For example, given a student relation one may want to retrieve the phone numbers of guardians of students studying in the first year. A query in SQL will resemble the following:

   ```
   SELECT    phone–number, guardian-name
   FROM      student
   WHERE     year = 1
   ```

   In general a query is of the type

   ```
   SELECT    <column names>
   FROM      <relation table names>
   WHERE     <conditions>
   ORDER     By <sequence of columns>
   ```

4. **Data Control Statements.** These are primarily used to control access to the data. The most common statements of this type are:

   GRANT (to assign privilege to access), REVOKE (to remove privilege)

   To implement ACID properties of transactions statements of the type COMMIT (to end a transaction) and ROLLBACK (to abort a transaction) are available.

12.6 DATABASE ADMINISTRATOR

We have seen that with a centralized database, data is recognized as an important resource of the organization. A Database Administrator (DBA) is a controller of data, whose responsibility is to ensure the integrity, security and privacy of data. DBA coordinates the function of collecting data, obtaining the organization's conceptual view of data, including future needs, designing the logical database, implementing the database, maintaining it, and ensuring its integrity, security and privacy.

The Database Administrator maintains a *data dictionary* which is a central repository of information about entities, relationships, data elements corresponding to entities and relationships. It also includes information on the origin of data elements, their meaning, uses, representation and format.

12.7 DATABASE DESIGN

Database design begins with designing the conceptual model of a database. Conceptual modelling uses both current and potential applications and ends with an E-R model and a data dictionary.

The next step is to design a logical model of the database. The logical model depends on the DBMS software. If it is a relational DBMS then relations form the logical data model.

There are two other DBMS, called the *network DBMS* and *hierarchical DBMS.* Logical data model for these are different from relations. We will not discuss these in this text.

The physical data model is designed to ensure good performance. In this step data on frequency of use of data elements, access time needs, etc., are taken into account. The physical data model is implemented and evaluated. The system structure is shown in Figure 12.4.

A good database system should

- Satisfy current and future application needs of an organization,
- Cater to unanticipated user requirements in the best possible way,
- Be expandable with the growth and changes in an organization,
- Be easy to modify with changes in hardware and software environment,
- Validate data before storage (validated data should be correct and remain correct with changes in the system), and
- Allow only authorized persons to have access to the data stored in the database.

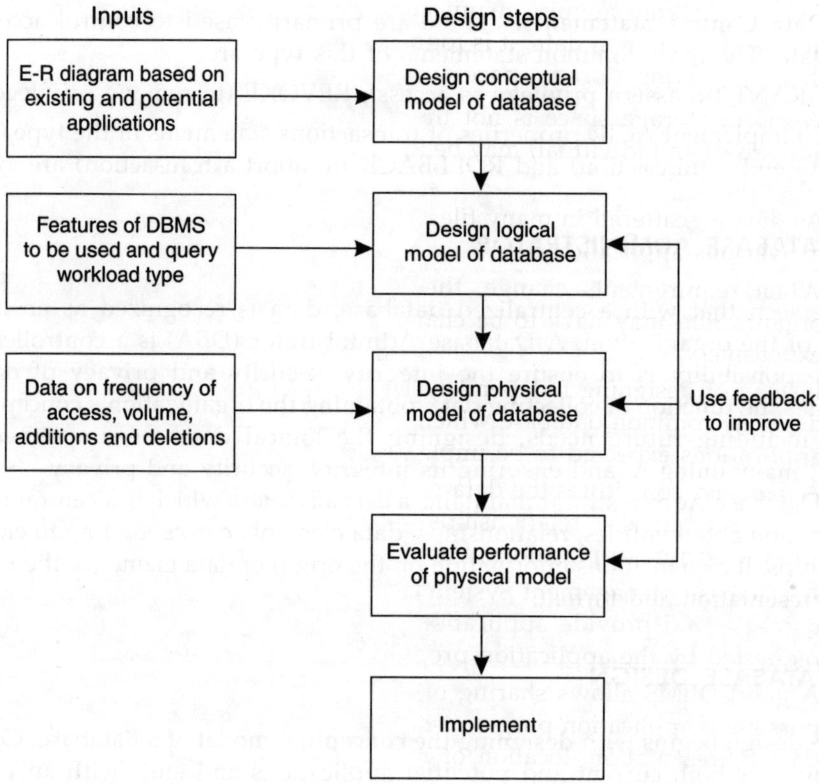

Figure 12.4 Steps in the design of a database system.

12.8 CONCLUSIONS

In this chapter we have briefly presented the concepts of database and Database Management Systems. In Chapter 11 we looked at data modelling using E-R models and logical modelling using relations. Once a set of normalized relations are obtained for a set of applications and potential applications, these constitute the logical model of data which can be used as inputs to a Relational Database Management System (RDBMS). The relations have to be represented by using the syntax rules of the specific RDBMS being employed. Besides relations, an RDBMS also supports languages to extract rows, columns, parts of relations, combine relations and, in general, manipulate relations. Query languages are also supported to retrieve required data from relations. MS Access, for example, is an RDBMS for personal computers and incorporates its own data editing, manipulating and query languages. Most currently available commercial RDBMS are easy to use and quite versatile.

SUMMARY

1. Traditionally, application programs were written after designing data files appropriate to the application.

2. As organizations require many application programs and these applications need some common data, it is found that the same data is stored in many files, thereby wasting storage space.
3. Waste of storage space is not the only problem. The same data (such as an address of an individual) may be in different forms in different files, leading to confusion.
4. As data is scattered in many files, it is difficult to correlate them and use them for various applications.
5. When requirements change, the application programs and the entire data organization may have to be changed. Such a change is time consuming and expensive.
6. Instead of designing independent files for different applications, it is better to design a common database which can be used for all current applications and applications expected to be implemented in the future.
7. A database constitutes the data resource of an organization. It incorporates not only data but also relationships between data items. It is modelled not for specific current applications, but all current and potential applications.
8. A Database Management System (DBMS) is a set of procedures that manage a database and provide application programs access to the database in a form requested by the application programs.
9. A good DBMS allows sharing of data among applications. Sharing of data is possible if application programs can be written without having a knowledge of data representation, location or access. In other words, application programs should be free to have their own view of data by DBMS.
10. Other major objectives of DBMS are to minimize storage of redundant data, maintain consistency of data values, ensure validity of stored data, protect data, prevent unauthorized access, control insertion, deletion and alteration of data and ensure fast access to data required by application programs.
11. To use a DBMS, a conceptual model of data needed for current and future applications is designed. E-R modelling presented in Chapter 10 is useful for this. The conceptual model is next converted to a logical model appropriate for the DBMS. For RDBMS, normalized relations constitute the logical model.
12. The physical data model is then designed to ensure good performance. User's view of data for application is primarily the logical view.
13. Query languages are available to conveniently retrieve required data from RDBMS.

EXERCISES

12.1 What are the advantages and disadvantages of systems using separate data files?

12.2 What do you understand by the term *data integrity?*

12.3 If redundant data is stored can it lead to data integrity problem?

12.4 Students' records in a University are kept by various sections: Hostel, Health Centre, Academic Office, major departments, Accounts Section and Library. If each of these sections maintains its own file-based system for processing, what problems do you foresee? Give examples.

12.5 Define a database of an organization.

12.6 What is the difference between a database and a Database Management System (DBMS)?

12.7 What are the basic objectives in evolving a database for an organization?

12.8 What do you understand by the term *data independence?*

12.9 What advantages are available to a programmer from data independence?

12.10 Draw an analogue between the advantages of high level language programming and data independence of application programs.

12.11 How can data integrity be maintained in a database?

12.12 Distinguish between issues of privacy and security in a database.

12.13 What is the role of E-R diagrams in database design?

12.14 What is the difference between a conceptual model and a logical model of a database?

12.15 What is an internal model of a DBMS?

12.16 What data models are used by application programs in a database oriented system?

12.17 Why is a DBMS divided into three layers, namely, conceptual model, logical model, and internal model?

12.18 How is data independence of application programs ensured in a DBMS?

12.19 What is RDBMS?

12.20 Define a transaction. What properties must be satisfied by a transaction processing system?

12.21 What are the common techniques used to ensure ACID properties of a transaction?

12.22 What is a structured Query Language? Give sample of a query language statement.

12.23 What are the responsibilities of a Database Administrator (DBA) in an organization?

12.24 What are the ideal features of a database system?

12.25 Are DBMS relevant to Personal Computers (PCs)?

13

Object-Oriented System Modelling

LEARNING GOALS

After reading this chapter a student should be able to:

1. Define objects.
2. Identify objects in an application given requirement specifications.
3. Model information systems using objects.

In the earlier chapters of this book we described how the requirement specifications are captured and understood using data flow diagrams. We saw how the processing of data in a system can be described using structured English and decision tables. Finally we also described how the data which is to be processed is organized logically in a database. This methodology of design has been used extensively and is often known as "Structured Analysis and Design". With the improvement of hardware systems and large scale use of computers in organizations, the systems being developed have been continuously becoming more complex. As the complexity of systems grow it has been felt that there is a need for better methods of analysis and design which can assist in developing such complex systems. Complexity is usually handled by breaking up a large system into smaller subsystems or blocks which can be designed independently by small teams and "assembling" these subsystems into the complete system. The major question that arises in implementing this idea is: "How do we break up a complex system into subsystems in such a way that each subsystem can be *reasonably independent and self-contained* and how these subsystems can *cooperate* to get the job done"? This idea of breaking up a large system into smaller subsystems is not new. Functions and procedures have been used extensively in programming. The major difficulty with decomposing large programs as procedures and functions has been organizing their cooperation. Procedures and functions cooperate by exchanging data. Inadvertent modification of data takes place unless one is very careful. Further, it has been found difficult and time consuming to track such errors. This problem is compounded when systems become large. Thus there has been a search for other methods of breaking up complex systems. A good method of breaking up a system into cooperating subsystems should have the following properties:

1. Each subsystem should have clearly specified responsibility of performing a part of the overall task.
2. The subsystem should be able to perform this task when it gets an "order" or a "stimulus" to do the task.
3. Other parts of the system need not know *how* a subsystem performs the task assigned to it. They should know only *what* task a subsystem does.
4. When a subsystem does a task it should not affect or alter other subsystems. In other words each subsystem should be self-contained and independent.
5. A subsystem should know *what* other subsystems do and how to send requests to them for assistance so that it can cooperate with them and get its own job done.
6. A subsystem should be like a black box. It should hide the data it uses. Other subsystems should not be able to access data stored in a subsystem.
7. The subsystems should be designed in such a way that they can be *reused* as components of other systems. In other words they should not be too specific with restricted capabilities—they should be reasonably general.
8. In order not to have too many varieties of subsystems one method of generalization would be to let some subsystems *inherit* properties of another subsystem which is similar so that it need not be designed from scratch.
9. Another aid to generalization is to design subsystems which can accept generic commands which can be interpreted by it based on its own capability and responsibility. For example, a generic command "Admit" may be interpreted as an order to admit students if the subsystem's responsibility is student admission processing and as an order to admit a patient to a hospital if it is a hospital management system. This property of a subsystem is known as *polymorphism*.

A methodology of analyzing requirements with the aim of identifying subsystems with the desirable properties stated above is now being used in many projects. This is known as *object-oriented modelling*. The subsystems which are identified using this method are known as *objects*. The main motivation for object-oriented modelling are:

- It facilitates changing systems at low cost. Changes may be either extending functionality of the system or repairing faults in a system.
- It promotes *reuse* of code in designing *large* systems.
- It reduces problems encountered while integrating subsystems into a larger system.
- It is simpler to implement in distributed systems.

In this chapter we will briefly describe objects and their properties, how to identify objects given the requirements specifications of an information system and finally model an information system using co-operating objects.

13.1 OBJECTS AND THEIR PROPERTIES

An object is usually a *tangible* entity which is *uniquely identifiable* and is *permanent* relative to the life time scale of an application. Some examples of common tangible objects are:

a student, a passenger, a vendor, a customer and an item in an inventory. An object can also be *intangible*. Examples of intangible objects are: a bank account, an insurance policy, a tree data structure, a push down stack, etc.

Objects having similar meaning and purpose may be grouped together to form what is known as a *class*. For example, each student in a college is an instance of an object belonging to the class "student". In other words one may consider a class as a *template* for objects of a particular type. Other example of a class is vendor. Yet another example of a class is train tickets. A particular vendor is an instance of the class vendors.

The term "object" is often used to refer to a group of objects. Strictly speaking one should use the term object class (or class) when we refer to group of objects and object instance when referring to an individual object.

All objects have *attributes*. For example, a student attending a college may be modelled as an object. The attributes of this object may be: Roll-no., name, address, department, courses taken, grades and year of study. The term attribute is a property of a class. The values of attributes are those of an object instance. For example the values of the attributes of a particular student may be:

Roll-no : 56789
name : A.B. Gupta
address : 25 A Block, IISc Hostel, Bangalore 560 012
department : Physics
courses taken : Physics 101, Chemistry 201, Math 205
grades : A, B, B
year of study : Second

Within a class the names of attributes should be unique. The attributes of an object are picked based on the application. The attributes of a student which are relevant when he/she is admitted to a hospital may be: Roll no., name, address, age, blood type, height, weight and guardian's address.

All objects have *a state*. A state is one of many possible positions in which an object may exist. A state is described by the values of one or more attributes at a given time and which may change. For example, the state of a student object may be: Passed, failed, suspended. One may also think of a state as information retained by an object.

All objects have a set of *operations* which they can perform to fulfill specified responsibilities. The operations performed by an object determine its *behaviour*. The set of operations an object can perform determines how an object reacts to requests from other objects. Operations can also change the state of an object. For example, when a student reappears in an examination the state may change from failed to pass. The operations which are relevant for an object would depend on the application. From the point of view of admissions section of a college the relevant operations would be: add new student, remove a student, and change department. From the point of view of the examination section they would be: find average grade, promote student, detain student, suspend, etc.

Every object has a unique *identity* which uniquely identifies an object in a class. We will now look at some more objects and their specifications.

A Bank Account

Attributes:	Account no., Type of account, Name, Address, Balance in account, Account status.
Operations:	Open account, Close account, Find balance, Withdraw from account, Pay into account and Calculate interest.
Identity:	Account no. attribute

Other examples of objects are:

A Vendor

Attributes:	Vendor code, Vendor name, Vendor type, Vendor address
Operations:	Add a vendor, Delete a vendor, Find vendor address, Change vendor address, Find vendor type.
Identity:	Vendor code attribute

An inventory item

Attributes:	Item code, item name, qty. in stock, price per unit, reorder level, reorder qty.
Operations:	Find stock of item, Issue item (s), Add item(s), Reorder item, Change price, Find price, Find total inventory.
Identity:	Item code attribute

A train ticket

Attributes:	Ticket no., PNR no., Berth nos. reserved, Sex and Age of holder(s), Departure station, Destination station, Date of travel, Train no., Date of issue of ticket, Cost of ticket, Ticket status.
Operations:	Issue, Cancel, Compute cost of ticket, Compute cancellation charges.
Identity:	PNR no. attribute

A notation called UML (Unified Modelling Language) has been proposed by three leading experts (Booch, Jacobson and Rumbaugh) in object-oriented systems analysis and design. The UML notation is useful in graphically depicting an object-oriented model of a system. There are several diagrams defined in UML. We will however use only a part of it called *class diagrams* which represent classes and their interactions in a system. Those interested in details of UML are referred to the book "The Unified Modelling Language—User Guide" [26]. It is expected that UML will become industry standard. We will thus use this notation to describe classes and the association between classes.

Classes are depicted by *class diagrams* in UML. The class diagrams for bank account class and vendor class are shown in Figure 13.1.

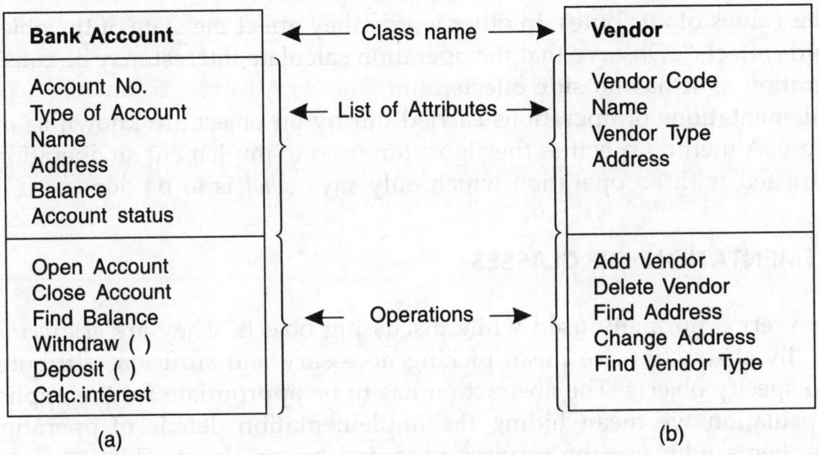

Figure 13.1 UML class diagrams for Bank Account and Vendor.

A class diagram shows the static structure of a class in an object-oriented model. A class is represented by a rectangle divided into three parts. The class name is given in the top part, the list of attributes in the middle part and the operations in the bottom part.

In addition to a class diagram UML defines an object diagram also called an instance diagram. In Figure 13.2(a) and (b) we have shown object diagrams corresponding to the class diagram of 13.1(a) and (b) respectively. An object diagram is a rectangle with two parts. The names of the object and its class are underlined and shown in the top part using the syntax:

Object name : class name

The object's attributes and their values are shown in the bottom part.

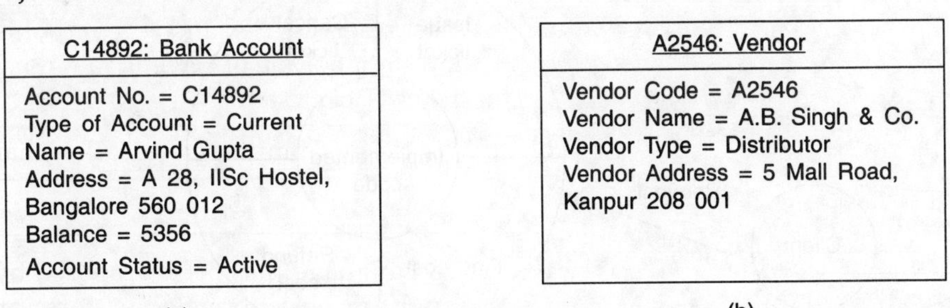

Figure 13.2 Object diagrams for Bank Account object and Vendor object.

We have given operations for various classes. The operations may be classified as (i) constructor (ii) query and (iii) update. In the example of Bank Account class constructor is open account which creates a new instance of an object belonging to a class. Closing an account is also analogous. Find balance is a query operation. It only accesses the state of an object in the class without altering its value. In other words it has no "side effects". The operations of withdrawal and deposit are update operations.

They alter the values of attributes. In other words they affect the state of the object and thus have "side effects". Observe that the operation calculate interest may be considered a query operation as it has no side effects.

The implementations of operations carried out by an object are known as *methods* in the literature. A method specifies the algorithm used to implement an operation. This is to be contrasted with an operation which only says *what* is to be done.

13.2 IMPLEMENTATION OF CLASSES

There are two terms normally used while discussing objects. They are *abstraction* and *encapsulation*. By abstraction we mean picking necessary and sufficient attributes and operations to specify objects. The abstraction has to be appropriate for the application.

By encapsulation we mean hiding the implementation details of operations on objects from clients who use the services provided by an object. This is ensured by specifying the operations which an object can perform and publicizing them while hiding how these operations are implemented by the object. This will allow an object to modify the code used by it (e.g., improving it by using a better algorithm or correcting errors in code) without affecting the way in which clients use the services provided by the object. In other words, the client's code does not change when the object's code changes. This is also known as *information hiding* and it greatly assists in modularizing applications. Consider for example the train ticket object. The interface and code are depicted in Figure 13.3. For example, if a client asks for refund on cancellation, the request is passed on to the object. A client need not know *how* the refund calculation is implemented. If the railways change the refund rules the code for this is changed. The client's code is not changed and the client also need not know of the code change by the object as long as the interface presented to the client remains the same.

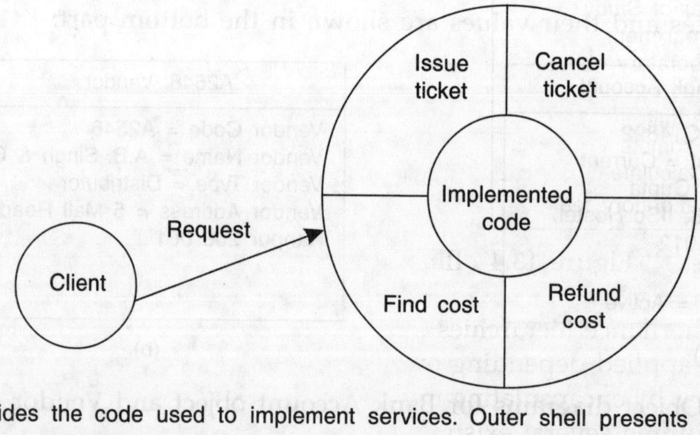

(Inner circle hides the code used to implement services. Outer shell presents interface to clients i.e., operations used by clients).

Figure 13.3 Interfaces and implementation.

13.2.1 Inheritance

As we explained earlier, in object-oriented approach, we identify objects and classes relevant to an application. We may realize while modelling that some classes need further refinement. In other words some new classes are required which are very similar to existing classes but with additional attributes, behaviour and operations. For example while modelling students in a college it may be realized that some of the operations required for students in science departments may be different from those in arts departments. For instance additional rules may be imposed for admission, payment of laboratory fee, etc. We may thus model students in two subclasses; a science student class and an arts student class. Both these classes will *inherit* the attributes and operations of a student class. There may be additional attributes and operations.

In Figure 13.4 we have illustrated inheritance. Observe that science student and arts student are subclasses of the parent class college student. These two classes inherit all the attributes and operations of the parent class. Besides these, the science student class has an additional attribute, laboratory name to which the student is assigned. One additional operation each have also been specified for the two subclasses.

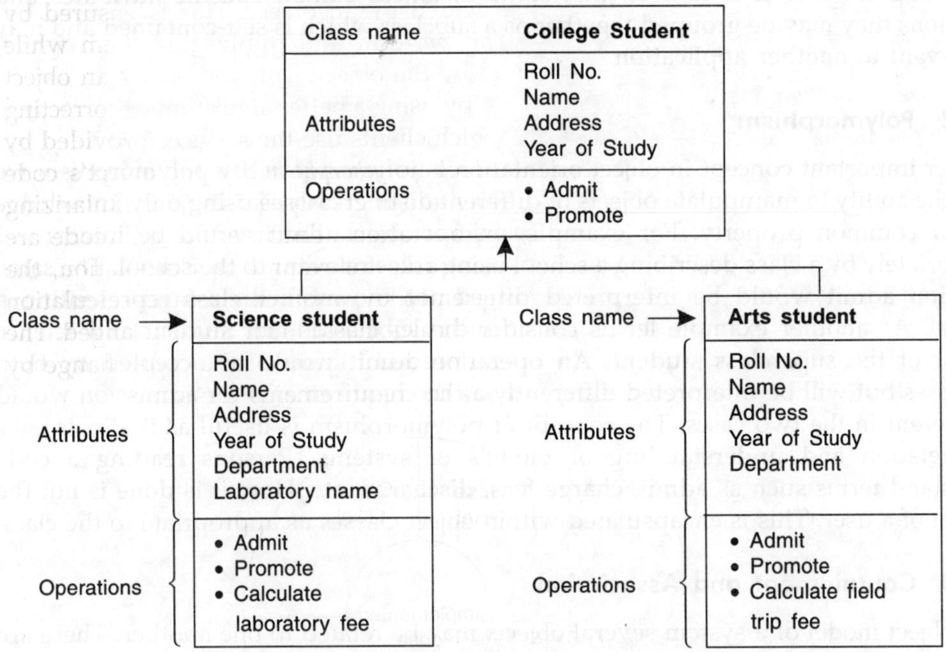

Figure 13.4 Illustrates inheritance.

In developing inheritance hierarchies in object-oriented systems both *generalization* and *specialization* are applied depending on the situation. Generalization is the process of creating a superclass by encapsulating the attributes and operations of several classes. This is done by examining several existing classes and grouping together those which have similar attributes and operations. These are all used to create a *superclass* which encapsulates the common attributes and operations. The original classes used in this process are made into *subclasses* in the inheritance hierarchy. Why do we generalize? The main reason is to promote reuse of existing classes which may become part of a superclass. Generalization opportunities are kept in view while analyzing the requirements specification of a problem. During analysis inheritance hierarchies among key classes are established. Classes available in a library are also examined to see if their attributes and operations can be inherited. This is used during design to develop classes which allow reuse.

For example, while creating an application for use by health care professionals there may be objects describing physicians, general surgeons, specialists such as eye surgeon, ear, nose and throat surgeon, orthopaedic surgeon, etc. Each of these are subclasses of a superclass of doctors. The common attribute of all doctors would be name, address, qualifications, hospital attached, appointment times, consulting fee. Common operations would be make appointment, charge fee, order tests. This is a process of generalization.

Specialization is the opposite of generalization. In this subclasses are created from a class to reflect the special requirements of a new subclass. For example if a class hospital exists and specialization may be a dispensary which inherits all the properties of the out-patient department of the hospital. The process of specialization begins with an existing class. If it is noticed that some instances exhibit specific attributes and operations they may be grouped together as a subclass which is self-contained and may be relevant to another application.

13.2.2 Polymorphism

Another important concept in object orientation is *polymorphism*. By polymorphism we mean the ability to manipulate objects of different distinct classes using only knowledge of their common property. For example, an operation admit would be interpreted appropriately by a class describing a school using rules relevant to the school. The same operation admit would be interpreted differently by another class representing a hospital. As another example let us consider the subclasses arts student and science student of the superclass student. An operation admit would be accepted by both subclasses but will be interpreted differently as the requirements for admission would be different in the two cases. The concept of polymorphism is useful as it allows easy interpretation and understanding of models of systems. Persons reading a code understand terms such as admit, charge fees, discharge, etc. How it is done is not the concern of a user. This is encapsulated within object classes as appropriate to the class.

13.2.3 Containment and Association

In an object model of a system several objects may be related to one another. There are two major types of relationships. One of them is "is-part-of" which is used to represent what is called *containment*. For example, a division would normally contain many departments. In this case the department object would be "is-part-of" the object division. Departments are *sub classes* of the class division. Another example is an object "cycle" which contains objects "wheels", "pedals", "chain", "handle", "bell", etc. Each of these objects have their own attributes and are sub-classes of the class cycle. *Associations* are somewhat similar to relationships in the E-R model. For example an object "student" and the object "courses" are associated. One possible representation of this relationship is shown in Figure 13.5. Observe Ref: course in the attribute list. In general it is Ref: <class name>. The relationship is usually a verb phrase. These abstractions are useful in modelling. Models of systems are not unique. Modelling is an iterative process. Further there is no agreement on methods of modelling. Containment is somewhat easier to understand and model and we prefer it in our modelling examples.

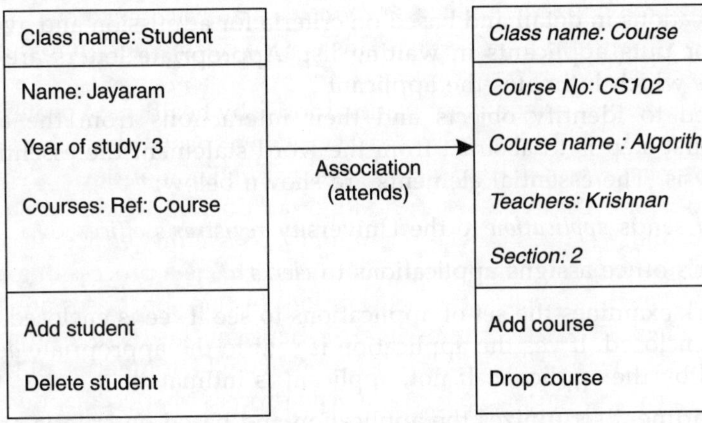

Figure 13.5 Association of Classes.

13.3 IDENTIFYING OBJECTS IN AN APPLICATION

The starting point of analysis and design of an information system is the requirements specification. The requirements specification is a detailed word statement. A simple method of identifying objects from a word statement is to examine nouns as potential objects. Verbs which specify actions to be performed by nouns are potential operations. The number of nouns would be large. We pick objects based on the following criteria adapted from the suggestions given by Coad and Yourdon [27].

1. An object must perform some services required in the application. In other words we should be able to assign specific responsibilities to objects.
2. An object must have attributes whose values are examined and used in performing the service assigned to it. There must be several relevant attributes.
3. An object must be essential for the functioning of the system. This is judged by examining whether it is essential to remember information about it.
4. There must be a common set of attributes and operations which are necessary for all occurrences of the object.

We will consider an example to illustrate how objects are selected.

EXAMPLE 13.1

A word statement giving the admission procedure of a University is given below:

A University releases an advertisement for admission of students giving essential qualifications for each course available in the University. Courses are offered by various departments of the University. The last date for receipt of application and the fee to be enclosed with the application are specified. Several clerks in the Registrar's office receive the applications and check whether the marksheets and fees have been enclosed. Valid applications are sent by the clerks to concerned academic departments. Each department checks the applications in detail and based on criteria for admission and available seats admits, rejects or puts applicants in waiting list. Appropriate letters are sent to the Registrar's office which intimates the applicant.

It is required to identify objects and their interactions from the above word statement. The first step is to abstract from the word statement the essential elements for further analysis. The essential elements are shown below:

- *Applicant* sends *application* to the University *registrar's office.*
- Registrar's office assigns applications to *clerks.*
- Each clerk examines the set of applications to see if fee is enclosed and marks-sheet is enclosed. If yes the application is sent to the appropriate *department* as specified by the applicant. If not, applicant is intimated.
- The department scrutinizes the application and based on criteria and available seats admits, rejects or places an applicant in waiting list. This is intimated to registrar's office.

The words in italics in the above set of statement are potential objects. Let us examine each one in turn and decide whether to accept or reject it as an object in the final system to be designed.

Let us first consider an applicant. The attributes of an applicant are his/her name, address, age, etc. No operations are performed by an applicant. He or she merely sends an application and awaits decision of the University. In our design we will thus not specify it as an object. It is an entity external to the system.

The next potential object is application. The attributes of an application are: application no., applicant name, applicant address, marks obtained in subjects, fees paid, department applied and status of application (which could be admit/reject or wait list). The operations performed by a clerk in registrar's office on an application are:

(i) Scrutinize if marks obtained list is enclosed and (ii) Check if fee is paid.

The next action is admit/not admit decision by the department. Application is an object. Its attributes are clearly specified and operations are to be performed on the application which uses its attributes.

The next potential object is registrar's office. Its attributes are irrelevant to the application. We thus reject it. Clerks in registrar's office have a role to play. No operations are performed on the attributes of a clerk. Thus clerk as an entity is not a relevant object in this application. However, the identification of a clerk scrutinizing an application may be incorporated as an entity in Application object. This would allow relating who scrutinized an application if a need arises later.

The last candidate object is department. The attributes of a department are dept. code, dept. name, number of seats available, no. of seats for wait list and minimum qualification for admission. The operations are performed using these attributes on request from the application object. Thus department qualifies as an object. We give in Figure 13.6 the objects appropriate to the example. In this example the application object will send a request message to the department objects for a service, namely, determine the value of the status attribute which can be in one of three states, namely, admit, place in wait list or reject.

Class name *Application*
Attributes
Application no. Clerk no. Applicant name Applicant address Marks sheet Degree obtained Fees paid Status of the application.
Operations
Scrutinize Send application to dept Send rejection Send admit letter Send wait list letter

Application class

Class name *Department*
Attributes
Department code Dept. name Course No. of students to be admitted No. on wait list Min. qualification Status of application
Operations
Examine application and determine its status.

Department class

Figure 13.6 Objects for Example 13.1.

The department object performs the requested service using the marks sheet sent by the application object with its request and responds. Application class has the responsibility to inform the external entity (client), namely, the applicant the outcome of the application.

Picking objects to model a system is dependent on the requirements specification. It is also not unique. In this example we did not consider Applicant as an object in our model as the requirements did not specify the need to gather information on applicants. If the requirements had added the requirement "The University requires statistical data on rejected and admitted students" then we may add an object Applicant with the attributes shown in Figure 13.7.

Class name *Applicant*
Attributes
Application no. Applicant name Applicant address Applicant age Applicant sex Degree obtained Marks obtained
Operations
Find Avge Age Count applicants Find % female applicants

Figure 13.7 Applicant object.

EXAMPLE 13.2

We give below a portion of the requirement specifications of a stores office which was discussed in Section 5.1.

User's narrative of requirements. "Our company receives a number of items from many vendors and they are received at the receiving office. As we receive over 1000 items a day it is now virtually impossible for the receiving office to check whether the vendor has supplied items against an order, or sent a wrong item and inform the purchase office. We are also not able to find out if there are excesses or deficiencies in delivery and whether the vendor adhered to the delivery schedule as per the order. The items received at the receiving office are sent for physical inspection. The physical inspection consists of checking whether the quantities stated in the delivery note agree with the physical count, whether the item is the correct one ordered, and a check on the quality of item. Any discrepancy found during inspection is intimated to the purchase office. Items cleared by the inspection office are taken into the inventory by the stores office which keeps a ledger of items stocked and quantity available of each item. Currently we are not able to pay our vendors promptly due to delays in payment order reaching our accounts office. We would like to rectify this by requesting the stores office to send a goods taken in stock report to accounts office to enable them to make payment immediately".

We abstract below the relevant aspects from the word statement.

- *Receiving office* receives *items* from several *vendors*
- Receiving office checks *delivery note* against *order* and finds out excess deficiency, late delivery, etc.
- *Discrepancy note* (if any) sent to *Purchase Office*
- Receiving office sends *items received note* to *inspection office*
- Inspection office physically inspects items received and accepts good items. Bad items are returned to the vendor
- Discrepancy note (if any) is sent to Purchase office by inspection office
- A note giving *accepted items* from vendor is sent to *stores office* by inspection office
- Stores updates *inventory*. Sends *goods taken in stock* report to *accounts office* for *payment*
- Accounts office sends payment to vendor.

The potential objects are noun phrases in italics in the requirements specifications.

- Receiving office
- Items
- Vendors
- Delivery note
- Order to vendor

- Discrepancy note
- Purchase office
- Items received note
- Inspection office
- Accepted items
- Stores office
- Inventory
- Goods taken in stock report
- Account office
- Payment voucher to vendor

We will now select the objects from this list. Receiving office merely receives items and checks it against order. The attributes of receiving office have no role to play. However the identification code of a receiving clerk may be relevant for an object whose responsibility is to check delivery against order. Thus receiving office is not an object. However, the transactions of receiving goods is a possible object as this transaction requires the attributes of both delivery note and order. These attributes are required for finding out if the delivery is against an order and also to detect discrepancies in the delivery. We thus identify "receiving office transaction" as a possible object.

The next candidate is items. Items are physical items which are delivered accompanied by a delivery note. Thus items are not objects but delivery note could be. Vendors are objects in this application as vendor address is needed to send payment. Order to vendor is a possible object. In this requirements specification no operations are specified on the order. However order is compared with delivery note to create discrepancies note. We thus create a class called "receiving office transaction" as discussed earlier whose subclass is delivery note and it refers to the class order to vendor. Receiving office transaction will inherit the attributes of delivery note and will refer to the attributes of order. This is shown in Figure 13.10. The attributes of delivery note are given as Figure 13.8 and that of order as Figure 13.9. The responsibility of this class is to compare delivery note against order and send a discrepancy note to the vendor and to purchase department.

Class: Delivery note
Attributes Receiving clerk id Vendor code Order No. Delivery date Qty. supplied Units
Send item received note Send discrepancy note

Figure 13.8 Class delivery note.

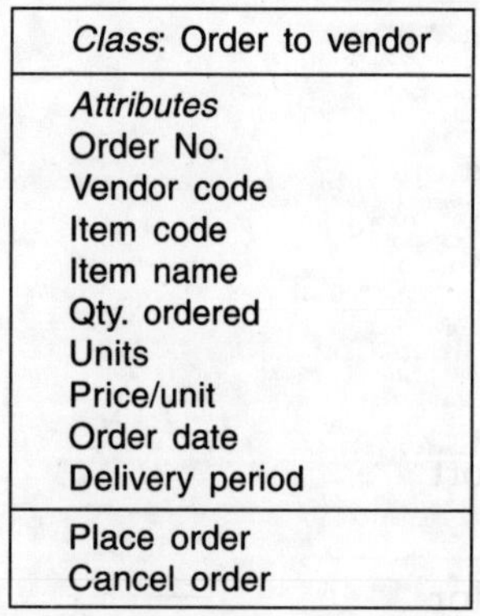

Figure 13.9 Class order to vendor.

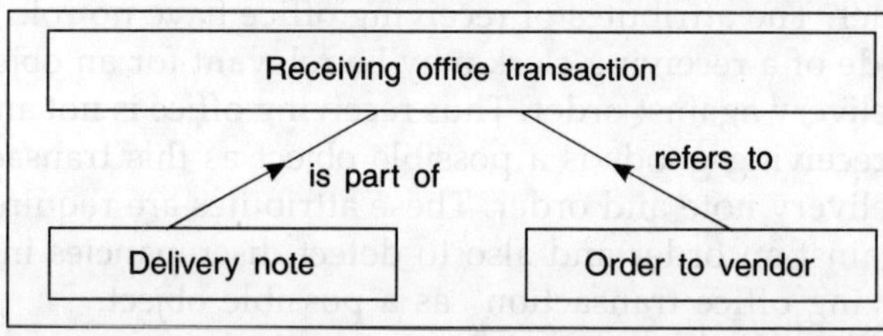

Figure 13.10 Receiving office transaction class.

The attributes and operations of Receiving office transaction class are shown below:

Class:	Receiving office transaction
Attribute:	Attributes of delivery note (Figure 13.8) Refer to attributes of order to vendor (Figure 13.9)
Operations:	• Compare qty. price, delivery date, date of order given in the order to vendor with that in the delivery note • Send discrepancy note (if any) to purchase office and vendor. If no discrepancy send delivery note copy to purchase. • Send delivery note to inspection office transaction.

Discrepancy note is not an object as it has no operations performed on it. Purchase office in this application is an external entity and its attributes are not relevant. Similarly inspection office is not an object but inspection office transaction is a possible class. It inherits the attributes of delivery note and has as its own attribute: number of items accepted. Observe that the attributes of delivery note is inherited by two classes, namely, receiving office transaction and inspection office transaction. This is allowed.

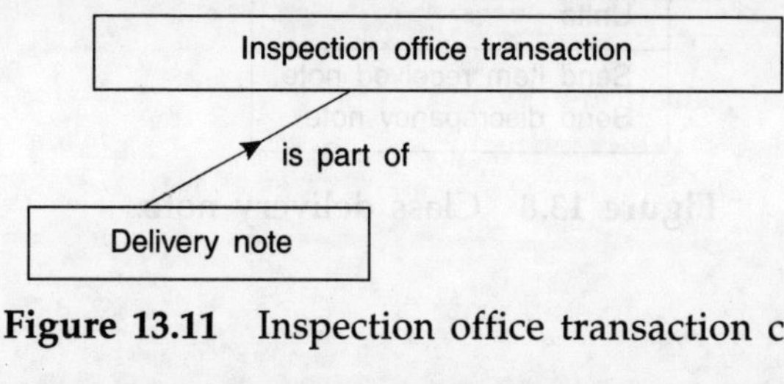

Figure 13.11 Inspection office transaction class.

Class:	Inspection office transaction
Attribute:	Attribute of delivery note + No. of items accepted
Operations:	• Send information on accepted items to stores office transaction. • Send discrepancy note (if any) to purchase office and vendor

We consider stores office next. The stores office is not an object but a stores office transaction is an object. The stores office transaction inherits the attributes of inventory and acceptance notes. The attributes and operations of stores office transaction object are given below:

Class:	Stores office transaction
Attribute:	Item code Item name Qty. in stock Order No. Vendor code Qty. accepted
Operations:	• Update inventory by adding number of items accepted to the quantity in stock. • Send advice to accounts to make payment for items accepted.

Finally we consider accounts office. The attributes of the office are not relevant. However accounts office transaction is an object whose responsibilities are to send payment to vendor for items accepted as intimated by accounts. The attributes of the accounts office transaction are given below:

Class:	Accounts office transaction
Attribute:	Order No. Vendor No. Items accepted Price/unit
Operations:	• Calculate amount to be paid • Print cheque • Request vendor object for vendor address • Print vendor address label • Despatch payment • Intimate purchase office on payment

Observe that one of the operations of Accounts office transaction is "Request vendor object for vendor address". We thus need a vendor object in our design. This is given in Figure 13.12.

Class: Vendor
Attributes Vendor code Vendor name Vendor address
Operations
• Add vendor • Delete vendor • Change address • Send address

Figure 13.12 Class vendor.

We conclude this example by consolidating all the objects identified from the requirements specification. They are:

- Delivery note
- Order to vendor
- Vendor
- Receiving office transaction
- Inspection office transaction
- Stores office transaction
- Accounts office transaction

The **external entities** are

Purchase office which in this example has no responsibilities but merely receives messages from the relevant objects. Purchase office transaction would, however, be an object if the word statement included order processing responsibility of purchase office.

13.4 MODELLING SYSTEMS WITH OBJECTS

Having identified objects from the requirement specifications the next step is to use them to model the system. Many methods have been described in the literature for this purpose. Describing all these methods is outside the scope of this book. We will take one of the suggested modelling methods which models a system as a set of collaborating objects.

This method suggested by Wirfs-Brock, Wilkerson and Wiener [28] is called Classes-Responsibilities and Collaborator method (abbreviated CRC method). It primarily suggests the following steps to model a system from the given requirements specification. The steps are:

Step 1: Find objects and their classes.

Step 2: Determine the responsibilities of each object.

Step 3: Determine the objects with whom they collaborate.

Step 4: State the responsibility of each object, that is, actions it can carry out on its own using its knowledge.

Step 5: State the contracts each object assigns to its collaborators. A collaborator is another object that is used to perform an action requested by an object or which provides information to it.

Step 6: Prepare a document giving each class, its collaborators, its own responsibilities and the responsibilities it assigned to its collaborators.

Step 7: Develop a graph which shows pictorially the classes and their mutual collaborations.

The modelling is carried out by what the authors call a CRC team which consists of a team of three to six persons. The team normally consists of user's representative, systems analyst(s) and the project coordinator. Their responsibility is to identify objects and play the role of classes to understand requirements modelling. They suggest *class index cards* to document the system. The format of a class index card is shown in Table 13.1.

Table 13.1 CRC Class Index Card Format

Class Name:
Superclasses and subclasses:
Short description of class:
Collaborators:
Private responsibilities of class:
Contract(s) with collaborators:

In addition to these, for each class in the system, a class interaction graph which depicts the interaction between classes would be useful. We now develop this documentation for the examples considered in the last section.

The CRC model for Example 13.1 is developed by preparing CRC Cards for Application Class (Table 13.2) and Department Class (Table 13.3).

Table 13.2 Application Class

Class:	Application
Superclasses:	None
Subclasses:	None
Collaborators:	Department
Description:	This class represents applications received for admission to a University
Private responsibilities *Scrutinize*: Applications are scrutinized to see if fee is paid and marks sheet is enclosed. If yes, application is sent to department. Else a reject letter is sent to applicant. **Contract(s) with collaborator(s)** *Forward application to department*: When application passes scrutiny, forward it to department. *Send letter to applicant*: When Department notifies decision (admit, reject or wait list), send appropriate letter to applicant.	

Table 13.3 Department Class

Class:	Department
Superclass:	None
Subclass:	None
Collaborators:	Application
Description:	This class represents departments whose responsibility it is to admit, reject or place on waiting list an application.
Private responsibility: *Rank order applications:* Rank order applications based on selection criteria. Mark in application: admitted, rejected or on waiting list depending on available seats	
Contract(s) with collaborator(s) *Send reply to application:* Send status of each applicant, namely, admitted, rejected or waiting list to application class.	

The collaboration graph is given in Figure 13.13.

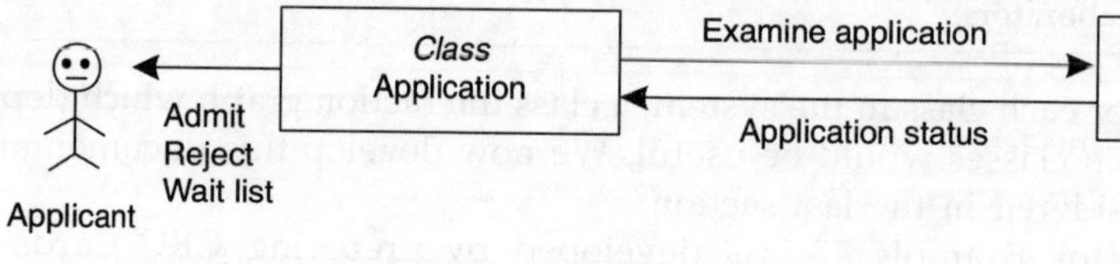

Figure 13.13 Collaboration graph for Example 13.1.

We will now develop CRC documentation for Example 13.2. The documentation for Receiving office transaction is given below:

Class:	Receiving office transaction
Superclass:	None
Subclass:	Delivery note, order to vendor
Collaborators:	Inspection office transaction
Description:	This class represents actions performed by the receiving office of a company
Private responsibilities: Compare delivery note with order to vendor. Report discrepancy if any to purchase office.	
Contract(s) with collaborator(s) Send delivery note to inspection office transaction.	

The documentation of the other classes, namely, inspection office transaction, stores office, transaction and accounts office transaction are given next.

Class:	Inspection office transaction
Superclass:	None
Subclass:	Delivery note
Collaborators:	Stores office transaction
Description:	This class represents actions performed by inspection office when it receives delivered items.
Private responsibilities: Compare actual items received with delivery note and accept items based on acceptance criteria. Create items accepted note. Send discrepancy note to purchase office.	
Contract(s) with collaborator(s) Send accepted items note to stores office transaction	

Class:	Stores office transaction
Superclass:	None
Subclass:	None
Collaborators:	Inspection office, accounts office transactions
Description:	This class represents actions performed by stores office when it receives items acceptance note from inspection office.
Private responsibilities: Compare item code of acceptance note with that in inventory. Add items accepted to inventory. Create payment note for accounts office.	
Contract(s) with collaborator(s) Send payment note containing vendor no., order no., qty., accepted to accounts office transaction.	

Class:	Accounts office transaction
Superclass:	None
Subclass:	None
Collaborators:	Stores office transaction, vendor
Description:	This class calculates payment to be made to vendors.
Private responsibilities: Using note sent by stores office transaction calculate amount to be paid.	
Contract(s) with collaborator(s) Print cheque and send payment to vendor. Intimate purchase office about payment.	

The collaboration graph for this example is given in Figure 13.14.

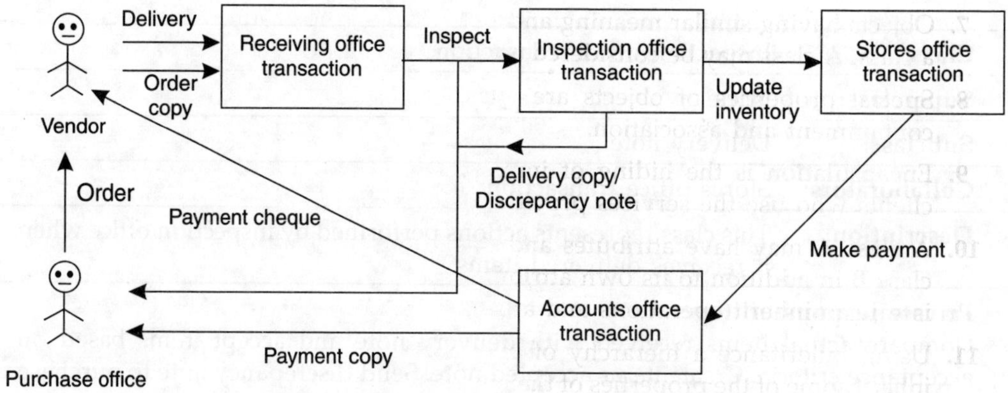

Figure 13.14 Collaboration graph for Example 13.2.

13.5 CONCLUSIONS

We saw in this chapter that there are a number of advantages in modelling systems using objects. The major advantage of object-oriented modelling is that it allows easy maintenance of systems. As objects are self-contained, changes in the algorithms used by an object does not affect other objects. Further, objects can be reused. The major challenge for a systems analyst is to find objects from requirements specifications. Use case analysis is effective in obtaining requirements specifications. There is no unique method of finding objects. However one may recognize objects by realizing that they are like contractors who carry out specific jobs on request. Even though object-oriented modelling may be used to model small systems their main merit is perceived only in modelling large systems. The trend is towards modelling using object-oriented techniques and it is thus important for a student to understand the basic ideas of modelling with object which we presented in this chapter.

SUMMARY

1. Complex systems are designed by breaking down such systems into a number of co-operating subsystems which are reasonably independent and self-contained.
2. One method of design is to use what are known as objects as subsystems.
3. The main motivation for using object-oriented modelling is the realization that individual objects may be modified without affecting the other parts of a system.
4. Objects also promote their reuse in many systems.
5. An object is an entity which is uniquely identifiable and is permanent relative to the life time scale of an application. All objects have attributes and state. A set of operations are defined for each object which determines its behaviour.

6. Objects collaborate by responding to requests from other objects to carry out specified operations.
7. Objects having similar meaning and purpose may be grouped together to form a class. A class may be considered as a template for objects of a particular type.
8. Special properties of objects are encapsulation, inheritance, polymorphism, containment and association.
9. Encapsulation is the hiding of implementation details of an object from the clients who use the services provided by an object.
10. A class A may have attributes and operations which are identical to another class B in addition to its own attributes and operations. In such a case class A is said to inherit operations and attributes of class B.
11. Using inheritance a hierarchy of classes may be constructed where children inherit some of the properties of their parents in addition to their own properties.
12. An important concept in modelling using objects is polymorphism. Polymorphism is the ability to manipulate objects of different distinct classes using only knowledge of their common property.
13. Containment is the property of a class being composed of several subclasses. The term "is-part-of" is used to specify this.
14. Association is relationship between classes through some common attributes.
15. To model an information system using objects the requirements specification is examined and all nouns in it are picked as potential objects.
16. From among potential objects the relevant ones appropriate for modelling are the ones which have specific responsibilities and are essential for the functioning of the system.
17. Having identified objects and their responsibilities the next step is to find the objects with whom they collaborate.
18. A document is prepared for each class giving its own responsibilities, its collaborators and their responsibilities.
19. A graph showing pictorially the classes and their mutual collaboration completes object-oriented modelling.

EXERCISES

13.1 What is object-oriented modelling?

13.2 Why is object-oriented modelling used in practice?

13.3 When is object-oriented modelling particularly useful?

13.4 Define an object.

13.5 What is the difference between a class and an object?

13.6 What are subclasses and superclasses? Give examples of each of these.

13.7 What do you understand by information hiding? Why is it resorted to in designing information systems? How is it achieved?

13.8 What do you mean by inheritance in object-oriented systems? Why is it useful? Give an example of inheritance.

13.9 What do you understand by the term polymorphism in an object-oriented system? Why is it useful? Give an example of polymorphism.

13.10 How do you select objects from a requirement specification? Given the following requirement statement, select potential objects.

A list of employers with their basic pay is sent to a clerk. He calculates the gross pay using standard allowances which are known for each pay slab. Deduction statements such as loan repayment, subscription to association, etc., are also sent to another clerk who matches these slips with the slips of gross pay and calculates net pay. This step is used by another clerk to write out pay cheques for each employee which are sent to respective employees. The total pay bills is also computed.

13.11 Pick objects and model the following requirements statement using objects. "A magazine is printed monthly and posted to its subscribers. Two months before the expiry of subscription, a reminder is sent to the subscribers. If subscription is not received within a month, another reminder is sent . If renewal subscription is not received up to two weeks before the expiry of the subscription, the subscriber's name is removed from the mailing list and the subscriber informed".

13.12 Give a brief requirements specification for a bus ticket reservation system. Model it using objects.

13.13 For the full requirements specifications given in Section 5.1 (See also use case diagram 10.5) obtain CRC model.

13.14 For the ATM example given in Section 10.4 obtain CRC model.

14

Data Input Methods

LEARNING GOALS

After reading this chapter a student should be able to:

1. Design forms used for manual entry of data.
2. Devise methods of coding key data elements.
3. Compute modulus 11 check digit for error detection of important data elements.
4. Identify types of errors to be checked by an input data validation program.
5. Design screens for interactive data input.

The data which is input to a computer-based information system must be correct. If data is carelessly input and errors enter the system, it will lead to incorrect results whose consequences will be expensive and embarrassing to the designer. Hence, one of the important tasks of a system designer is to ensure that the design prevents such errors. An important point to be remembered by the designer is that the *volume* of input data in information systems is very large. This implies that, when an error occurs, it will take a long time to find out the source of the error unless the design includes controls to isolate a subset of data records as the source of the error. In this chapter we will see how data entry can be systematized by designing proper data input forms, designing codes for key fields, designing controls for data inputs, and designing screens for interactive data input.

14.1 DATA INPUT

The input of data to a computer can be either *on-line* or *off-line.* In on-line data entry the user directly enters data into the computer's secondary storage, namely, a disk file. This entry is through a terminal connected to the computer. In this method the computer prompts a person entering data with a menu, or commands or a form displayed on the video display. Any errors in data entry are checked immediately and the person who entered the data is sent a message within one or two seconds to correct the input data. With the advent of e-commerce and e-filing of various forms such as income tax returns, applications to colleges etc., on-line entry of forms is far more common nowadays by untrained persons. On-line form design thus has to be simple and fool-proof to minimize errors.

In off-line data entry, transactions are entered on forms manually at the point where the transaction originates. Examples are sales reports filled by salesmen, application form to open a bank account, application form for a job etc. Many forms are collected to form a *batch.* The batch of transactions are then entered from the forms by a data entry operator into a computer's disk. In this case considerable time elapses between the entry of data on forms and its actual entry into a computer. Errors in data entry are thus found much later than when they occur. Thus there should be a well-designed method of correcting the error in data entry which is independent of elapsed time.

In Figure 14.1 we depict how data is entered in a batch mode and stored in a file. As illustrated in the figure, data entered by operators are checked by a data validation program and data records with errors are printed in an error report. The incorrect transactions in the input are corrected and the data is entered again. The correct transactions are used by a file update program to update the database. To reduce errors

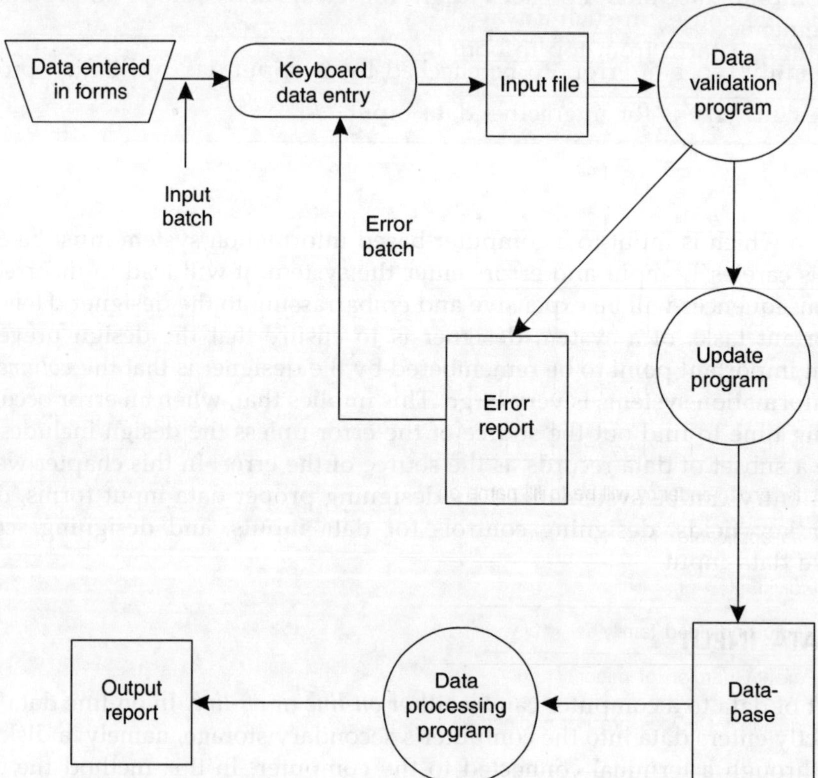

Figure 14.1 Batch data entry, correction and processing.

in data entry, it is essential to design carefully the form in which the data is manually entered. The main principles used in designing forms are to reduce the human error in filling the forms and minimize the chance of mistakes in data entry. The following points should be kept in mind while designing forms:

1. Leave enough space for writing legibly the information
2. Give clear instructions to fill the form. Each instruction is in front of the line where the data is to be entered. Examples of good and bad design are given in Figures 14.2 and 14.3 respectively.
3. Use boxes to enter data. Each box must have enough space to write. The number of boxes should correspond to the size of the data, e.g., PIN code has 6 digits, mobile numbers are 10 digits long.
4. Use standard format and give clear instructions to fill data such as date and time. Use 24 hour clock for time. For example, a standard format for entering date is

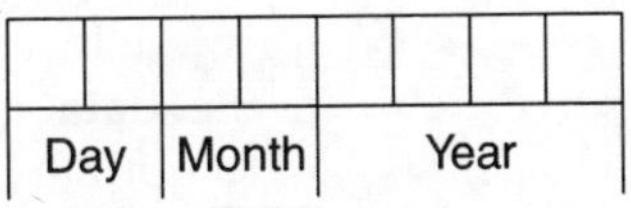

5. Arrange entries in such a way that the data entry operator can enter data by reading a form from top to bottom and left to right.

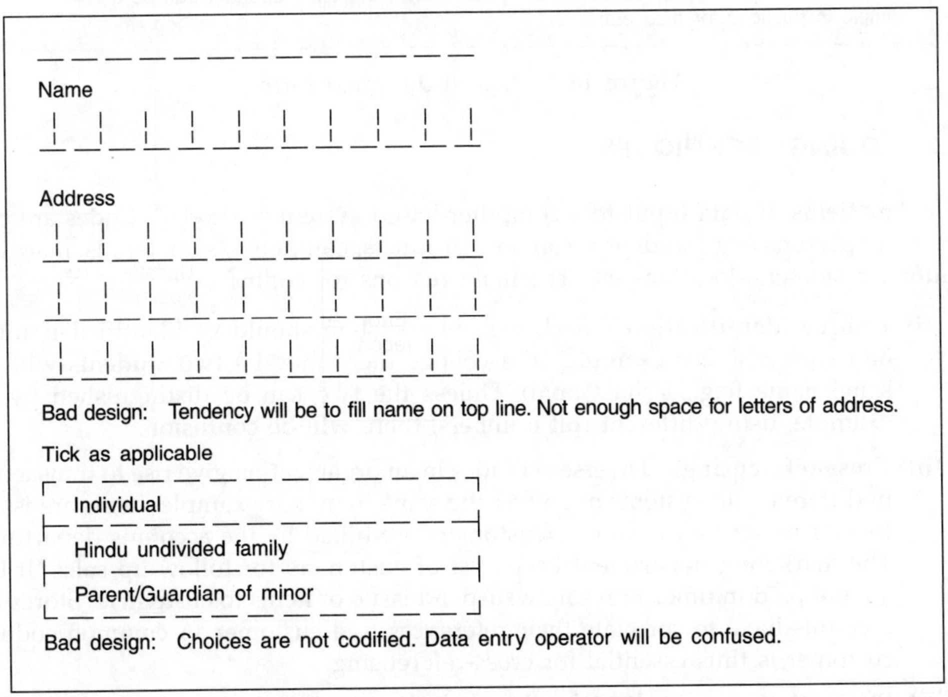

Figure 14.2 A poorly designed form.

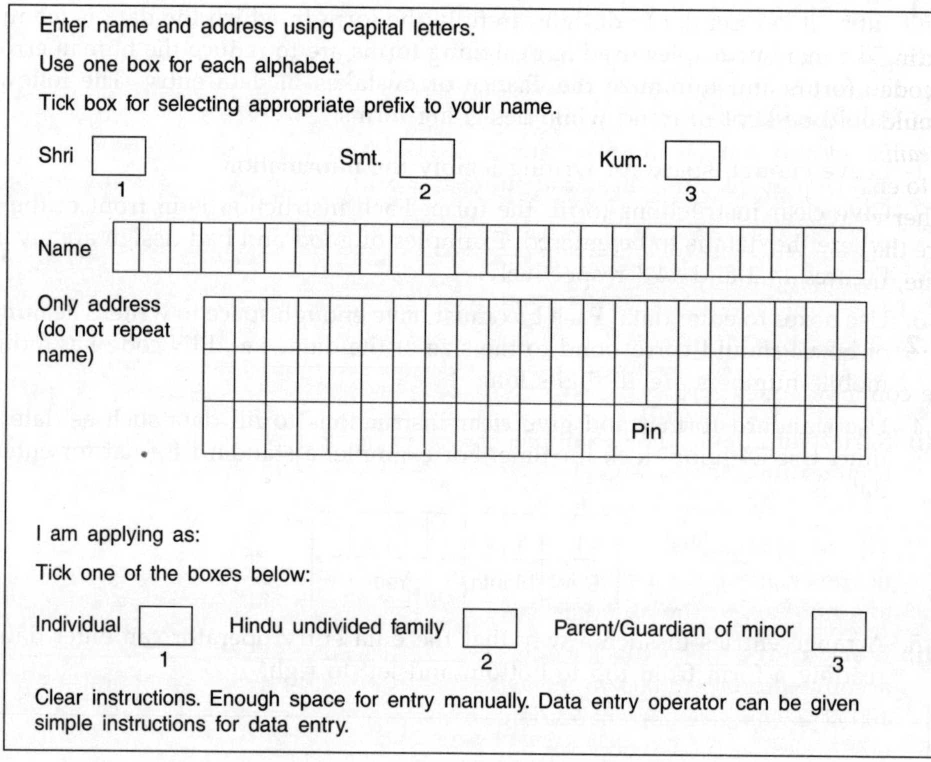

Figure 14.3 A well designed form.

14.2 CODING TECHNIQUES

Many key fields in data input to a computer-based system are coded. Codes are used to identify persons (e.g., student's Roll no.), products, components, materials, machines, vendors, customers, locations etc. The main reasons for coding are:

(i) **Unique identification.** Each item in a system should be identified uniquely and correctly. For example, in a college there may be two students with the same name (e.g., Usha Gopal). Unless the two can be distinguished by, for example, using different roll numbers, there will be confusion.

(ii) **Cross-referencing.** Diverse activities in an organization give rise to transactions in different sub-systems but affect the same item. For example, items are issued to a customer by the stores. Customers are billed by the accounts department. The marketing department keeps list of customers for follow-up sales. If bills are not paid in time, stores may suspend issue of items to customers. Stores and accounts have to correlate their references to a customer. A common code for customer is thus essential for cross-referencing.

(iii) **Efficient storage and retrieval.** Code is a concise representation. It reduces data entry time and improves reliability. Code as a key reduces storage space required for the data. Retrieval based on key search is faster in a computer.

14.2.1 Requirements of a Coding Scheme

A coding scheme must be *concise*. In other words, the number of digits/characters used in a code must be minimal to reduce storage space of the code and retrieval efficiency. It should be *expandable*, that is, it must allow new items to be added easily. It should be *meaningful* and convey to a user some information about the characteristics of the item to enable quick recognition and identification of the item. It should be *comprehensive*. In other words, it should include the characteristics relevant to all the related activities where the item will be involved. It should also be *precise*, i.e., the scheme should produce unique, unambiguous code.

14.2.2 Types of Codes

Some common codes are:

(i) **Serial numbers.** For example, for bank accounts one may assign a serial number for each account holder. As and when a new account is opened, the person is given the next serial number.

The advantage of this method is that it is concise, precise and expandable. It is, however, not meaningful. From the serial number it is not possible to find out anything about the account holder. It is also not comprehensive.

(ii) **Block codes.** The block codes use blocks of serial numbers. For example, account numbers 0000 to 9999 may be used for savings accounts in a bank; 10000 to 99999 for current account; 100000 to 999999 for special deposit accounts, etc. A similar block code can be used in other areas such as coding items in a store, and assigning roll numbers to students. This code is expandable and more meaningful than the serial number coding. It is precise but not concise or comprehensive.

(iii) **Group classification code.** This is an improvement on the block code and is more meaningful. For example, a method of assigning roll numbers to students may use the following system:

87	1	05	2	465
↓	↓	↓	↑	↓
Year of admission	Term admitted	Department	Status (UG/PG)	Serial no. in the department.

Often mnemonics instead of numbers make the code more meaningful. For instance, in the above representation we can write the code as

87	1	CS	UG	465
		↓	↓	
		Computer Science	Undergraduate	

This code is meaningful, comprehensive, expandable and precise. It is, however, not concise. The code can become very long if an attempt is made to make it more comprehensive. This is a good coding method and is widely used. For example, the PIN code used by the Postal Department is based on this idea.

(iv) Significant codes. These codes use some or all of the digits to describe a value of the product being coded. For example, the code

BA	M	95	C	B	R → Style
↓	↓	↓	↓	↓	
Banian	Male	Chest size	Cotton	Colour	

describes a banian for males of chest size 95 cm., made of cotton of blue colour and round neck . The value may be directly used in some calculation. This code is meaningful, precise, comprehensive and expandable. It is not concise.

There are often special coding schemes which are standardized. Simple examples of international standardization are Dewey decimal classification codes used for books for simplicity of locating books in libraries and ISBN (International Standard Book Numbers) used by publishers all over the world to assign a number to a book when it is published.

14.3 DETECTION OF ERROR IN CODES

In Section 14.2 we discussed various methods of coding. We did not specifically describe codes with capability of error detection. When data are entered using a keyboard, typing errors can occur. It is desirable to detect errors in typing at data entry stage to avoid delays and difficulties which may arise if they are not detected early. For instance, an incorrect entry of a customer's account number when he cashes a cheque may, unless detected during data entry, leads to debiting a wrong account. It is thus necessary to design a code such that if there is an error in the code it can be detected during data entry by a simple program. It is practically impossible to design a code so that any arbitrary error(s) is detected. A code can be designed if the types of errors normally committed in data entry are known. The code will then be able to detect all such errors.

By experimental study it has been found that the common types of errors committed during data entry are as shown in Table 14.1. This table also gives the probability of occurrence of each of these errors.

Table 14.1 Common Errors Made during Data Entry

Type of Error	*Example*	*Occurrence (percent)*
Single transcription (one digit incorrectly typed)	45687 → 42687	86
Transposition error (Two digits are interchanged)	96845 → 96485 96845 → 94865	9
All other errors		5

If a code is designed which is able to detect the two types of common errors, namely, single transcription and transposition errors, it will be reasonably good. Such a code has been designed and is called *modulus-11 code*. We will first describe how such a code is constructed, then see how it is able to detect errors, and finally show the theoretical basis of this design.

14.3.1 Constructing Modulus-11 Codes

Given a set of codes, they are transformed to another set of codes with error detecting property as follows:

Let the given code be : 48793

Multiply the least significant digit by 2, the digit to its left by 3, the digit to its left by 4 and so on. Add the products. (The digits 2, 3, 4, etc., with which we multiply the digits of the code are called weights). For the given code, 48793, by applying this operation we get

Weighted sum of digits

$$4 * 6 + 8 * 5 + 7 * 4 + 9 * 3 + 3 * 2 = 125$$

Weights

Divide the weighted sum of digits by 11. Append (11-remainder) to the right of the code. This is the new code to be used. In this case, as the remainder is 4, we append (11 – 4) = 7 to the code and get the new code 487937. The remainder 7 appended to the code is called a *check digit*. If the remainder after division is 1, then (11-remainder) = (11 – 1) = 10. In such a case the character X is appended to the code. If the remainder after division is 0, then 0 is appended to the code.

In Table 14.2 we show some examples of codes and their equivalent modulus-11 codes.

Table 14.2 Illustration of Construction of Modulus-11 Code

Code	*Modulus-11 Code*
45687	4 * 6 + 5 * 5 + 6 * 4 + 8 * 3 + 7 * 2 = 111 111/11 = quotient = 10, remainder = 1. Digit to be appended (11 – 1) = 10. Use X to represent 10 45687X
68748	687480
65432	654329
69752	697524

An algorithm for generating the modulus-11 code is given below. Let $d_n d_{n-1} \ldots d_2$ be the given code ($n \leq 10$).

Algorithm for finding d_1 to be appended to the code as the least significant digit

weighted sum ← 0
for $i = 2$ to n do {$n \le 10$}
  weighted sum ← weighted sum + $i * d_i$
endfor
  r ← weighted sum *mod* 11
  (r is the remainder obtained by dividing weighted sum by 11)
if $r = 0$
  then $d_1 = 0$
  else if $r = 1$
    then $d_1 = X$
    else $d_1 = (11 - r)$
    endif
endif
Append d_1 to code.

14.3.2 Error Detection

While a code is being entered on a keyboard, a program multiplies the digits in the code by weights 1, 2, 3, ..., starting from the least significant digit. The sum of these products is divided by 11. If the remainder is not equal to 0, then an error in data entry is indicated. The algorithm for error detection is as given below:

Let the given code be

$$d_n d_{n-1} \ldots d_1$$

Algorithm for error detection

If $d_1 = X$
  then weighted sum ← 10
  else weighted sum ← d_1
endif
for $i = 2$ to n do {$n \le 10$}
  weighted sum ← weighted sum + $i * d_i$
endfor
r = weighted sum *mod* 11
if $r = 0$
  then "No error"
  else "Error in code"
endif

14.3.3 Theory of Modulus-*N* Coding

In this section we will examine modulus-*N* coding scheme and show how to pick *N* and the weights so that we can guarantee that the code will detect single transcription and transposition errors.

Let the code be

$$d_n d_{n-1} \ldots d_1 \qquad (n \le 10)$$

Let the weights used to multiply the digits be

$$w_n\, w_{n-1} \ldots w_1 > 0 \quad (n \le 10)$$

The weighted sum $= \sum_{i=1}^{n} w_i\, d_i$

(i) Detecting single transcription error. If a digit d_k is incorrectly entered as $t(t \ne d_k)$, we have a single transcription error.

$$\text{The weighted sum} = \sum_{i=1}^{n} w_i d_i - w_k d_k + w_k t$$

$$= \sum_{i=1}^{n} w_i d_i + (t - d_k) w_k$$

(ii) Let the modulus-*N* system be chosen for error detection. If there is a single transcription error, then the following condition should be satisfied:

$$\left[\sum_{i=1}^{n} w_i d_i + (t - d_k) w_k\right] \bmod N \text{ „ } 0 \tag{14.1}$$

For this to be satisfied,

$$(t - d_k)\, w_k \bmod N \ne p \,.\, N \text{ as } \sum_{i=1}^{n} w_i \,.\, d_i \bmod N = 0 \tag{14.2}$$

$$(t - d_k) \ne p \,.\, N, \text{ where } p \text{ is any positive integer} \ne 0 \tag{14.3}$$

$$w_k \ne p \,.\, N \tag{14.4}$$

$0 < |t - d_k| \le 10$ as $t, d_k \le 10$ (maximum value $t = X = 10$ and minimum value of $d_k = 0$). Also $w_k > 0$.

If *N* is a prime then Eq. (14.2) is satisfied as $(t - d_k)$ and w_k are integers. (Product of two integers cannot be a prime number).

If $N > 10$ then Eq. (14.3) is satisfied

If $0 < w_k < N$ then Eq. (14.4) is satisfied.

The smallest prime > 10 is 11. Thus, if $N = 11$ and all weights are positive and less than 11, a single transcription error will be always detected.

(iii) Transposition errors. If there is transposition error, then the code

$$d_n d_{n-1} \ldots d_k \ldots d_j \ldots d_1$$

will be incorrectly entered as

$$d_n d_{n-1} \ldots d_j \ldots d_k \ldots d_1$$

For this error to be detected, we should have

$$\left[\sum_{i=1}^{n} w_i d_i + (d_j w_k - d_k w_k + d_k w_j - d_j w_j)\right] \bmod N \text{ „ } 0 \tag{14.5}$$

For this expression to be satisfied, it is necessary that

$$[(d_j - d_k)\,(w_k - w_j)] \bmod N \ne 0 \tag{14.6}$$

Thus

$$|d_j - d_k| \neq p.\, N \tag{14.7}$$

where p is any positive integer $\neq 0$

$$(w_k - w_j) \neq 0 \tag{14.8}$$

$$(w_k - w_j) \neq p.\, N \tag{14.9}$$

Equation (14.6) will be satisfied if N is prime.
Equation (14.7) will be satisfied if $N > 10$.
Equation (14.8) will be satisfied if $w_k \neq w_j$ for all j and k.
Equation (14.9) will be satisfied if $0 < w_j < N$ for all j.

Thus any transposition error will be detected if (i) all weights are distinct, positive and less than or equal to N and (ii) N is a prime number greater than 10. $N = 11$ will satisfy this requirement.

It is thus seen that a modulus-11 check digit system will detect all transcription and transposition errors provided no two weights are equal and all weights are < 11. As all weights should be distinct, the maximum length of the code should be less than or equal to 10 digits, including the check digit.

As single transcription and transposition errors constitute 95% of all errors and the modulus-11 check digit guarantees their detection, it is a good system. Even for other errors whose detection is not guaranteed, the remainder after division by 11 will be non-zero with probability (10/11). In other words, 10 out of 11 "other errors" will also be detected. Thus, if during data entry of 100,000 codes, 100 codes are incorrectly entered, 95 are guaranteed to be detected and 50/11 = 4 will also be detected. Thus only 1 wrong code entry will be undetected.

The modulus N check can be generalized for non-numeric codes. If a code is constructed with 10 decimal digits and 26 letters of English, then the value of N should be a prime greater than 36. The number 37 is a prime. The weights should be distinct. All weights should be less than or equal to 36 in modulus-37 check system. The length of the code ≤ 36.

14.3.4 Modulus 10 Check Digit Codes

A major problem with modulus 11 check digit system is that when the remainder after division is 10 we are forced to use X as the check character. In many situations characters are considered illegal in codes and only digits are allowed. Another disadvantage of modulus 11 check digit system is that the number of digits in the code should be ≤ 10. If we use modulus 10 instead of modulus 11, the remainder after division will be at most 9 (a digit). Thus several coding schemes have been proposed using modulus 10. The simplest one is to use all weights = 1 and use mod 10 check digit. For example, if the code is: 4367865 then the check digit is

$$10 - (4 + 3 + 6 + 7 + 8 + 6 + 5) \bmod 10 = 1$$

and the coded number is 43678751.

In this code all single transcription errors will be detected. Transposition errors will not be detected. (This can be shown using the theory given in Section 14.3.3). In order to detect some adjacent digit transposition error distinct weights for adjacent digits may be used. One such method is used in European Article Numbering code. In this code 7 digits are used to code an article and the eighth digit is used as check digit. If the code is:

$$\underbrace{d_8d_7d_6d_5d_4d_3d_2}_{\text{Article number}} \qquad \underset{\text{Check digit}}{\overset{d_1}{\uparrow}}$$

The weights used are 3131313 for the article number.

If, for example, an article number is: 7548963 the check digit x is calculated as follows:

$$x = 10 - (3 \times 7 + 5 + 3 \times 4 + 8 + 3 \times 9 + 6 + 3 \times 3) \bmod 10$$

giving $x = 2$. The article code with the check digit then becomes 75489632.

This method will detect all single transcription errors. This can be easily proved using Eq. 14.1. It cannot however detect all adjacent digit transposition errors. If adjacent digits are x and y and are transposed. If the weight of x is 3 then that of y is 1. The weighted sum will be changed by

$$(3x + y) - (x + 3y) = 2(x - y)$$

If $2(x - y) \bmod 10 = 0$ then this error will be undetected. This will happen only if $(x - y) = 5$. Thus transpositions $0 \Leftrightarrow 5$, $1 \Leftrightarrow 6$, $2 \Leftrightarrow 7$, $3 \Leftrightarrow 8$, $4 \Leftrightarrow 9$, namely, 10 transpositions will be undetected. There are 100 possible transpositions of adjacent digits (as, $0 \le x, y \le 9$) and 90 of these will be detected. Thus 100% transcription errors and around 90% transposition errors will be detected.

There are other mod 10 codes also in use. One of them uses permutations of digits and is used for error detection in credit card numbers. We will not discuss it in this book.

14.4 VALIDATING INPUT DATA

Good form design and careful entry of data will reduce incorrect data from entering a file. Special measures should, however, be taken during input preparation and later special data validation programs should be written to make sure that data with errors are not stored in files.

14.4.1 Controls during Input Preparation

(i) **Sequence numbering.** Each data record is given a sequence number. This provides the information to pick the incorrect record when an error is detected in it. A missing record or an out of sequence record can also be tracked using the sequence number.

(ii) **Batch control.** Input data records consist of batches of about 50 records and each batch is numbered. One or more important fields in each record is selected and the values in the fields in all the records are manually added to form a batch control total. A batch control record is designed, which includes information such as number of records in the batch and the batch control total (also known as batch totals). This batch control record is entered at the end of each batch of records into the disk file. It is used by the data validation program to check that no records in the batch are missed and there is no data entry error in the selected field(s).

In Table 14.3 we show a batch of records and the corresponding batch control record.

Table 14.3 Illustration of Batch Control

Serial No.	*Roll No.*	*Name*	*Grade*
1	5432	K.S. Singh	A
2	5654	A.R. Kumar	A
3	5752	P. Usha	B
4	5894	K. Deo	C
5	5941	P. Jain	F
•	•	•	•
•	•	•	•
•	•	•	•
50	6452	K. Raman	B

Batch Control Record

No. of records	*'A's*	*'B's*	*'C's*	*'D's*	*'F's*
50	6	10	20	10	4

Records are read one by one by the validation program and counted. The program also counts the number of *'A'*s, *'B'*s, *'C'*s, *'D'*s and *'F'*s in the batch. The count of records and the count of *'A'*s, *'B'*s, etc., are compared with those in the batch control record. If they match there is no error. Otherwise, a data entry error is detected. The error could be in any of the 50 records. Thus a manual comparison of each record in the file with that in the form is now needed. This is feasible only if the number of records in a batch is not very large.

(iii) Data entry and verification. The same set of data records are entered by two different operators and two files are created. Corresponding records in these two files are compared by a program. Any difference in two corresponding records indicates an error. The records are then retrieved and compared with those in the data entry forms.

(iv) Record totals. Some selected fields in a record are summed and the sum is entered as a separate field in each record. The data validation program sums the fields as entered by the data entry operator and compares this sum with the manually computed hash total field. Any difference is signalled and the record is manually checked.

(v) Check digit. For important fields, modulus-*N* check digits are used to detect errors.

Other checks carried out by a data validation program are:

(a) *Character check.* Data fields containing illegal characters (e.g., a numeric field containing a character S instead of a number 5) due to data entry error are checked by the input validation program.

(b) *Radix error.* Data fields with known limits on values are checked. For example, if a field for minutes has a number greater than 60 it is a mistake. If a field represents a month it should not be greater than 12.

(c) *Range check.* Check if a field lies within allowable range. For example, an entry of 90 for number of hours worked per week by a factory worker is wrong as it is larger than the maximum allowed number of working hours per week.

(d) *Reasonableness check.* Values of fields are checked to see whether they are reasonable. For example, if the telephone charge of a customer is entered as ₹ 5000 when the customer's average charge is around ₹ 500, then it is probably an error. Such errors should be indicated by a data validation program for closer check.

(e) *Inconsistent data.* For example, an entry 31-04-2010 as a date is a mistake and should be detected. A PIN code should have 6 digits. If it has 5 or 7 digits it is an error.

(f) *Incorrect value entered.* These values are checked by using batch control totals. If it is an important data, an arror detecting code such as modulus 11 check digit should be used.

(g) *Missing data.* If a full record is missing the batch control total will indicate it. If no value is entered for a field it should be checked and indicated by the data validation program.

(h) *Data records in wrong order.* Sequence numbers in records are used to detect this error.

(i) *Inter-field relationship check.* If different fields are related in a known way this relationship can be used to check the data entered. For example, if the entry for year of study of a student in a school is 3 and the age entry is 70, then there is an error.

(j) *File header.* All input data files should have a record at the beginning, giving it an identification. Every program using a file should check the identification to ensure that it is processing the right file.

We again reiterate that careful design of controls for data entry and input validation program is essential in data processing. As volumes of data processed are large and consequences of data errors in files can be very expensive, "cleaning" input data before they get into a database is very important.

14.5 INTERACTIVE DATA INPUT

With the advent of Personal Computers, networked computers, and terminals connected to computers, considerable volume of data is input interactively not only by professionals but also by lay users such as those ordering terms from an e-commerce site. In this mode a computer is an active intelligent agent which can display on the screen a form to be filled in, or ask questions to be answered or present a pre-designed menu, giving fixed alternatives based on the application. Any of these three modes or a combination may be used in an application. The primary advantage of interactive input is the immediate response if errors are detected in input. Also, the design of the forms displayed on screens (called screen design) has more flexibility by using features of the terminals such as colour and reverse video. In interactive input design it is important to reduce key strokes, use mouse and pointer mechanisms wherever feasible, use graphical icons as appropriate, allow variations in spellings of common words, and mix upper and lower case letters in any manner. For example, the alternatives: Y, Yes, y, YES, yes must all be allowed for an Yes answer. We now explain the three main models of interactive data input.

14.5.1 Menus

In a menu a user is presented with a number of alternatives and asked to pick one. The next menu depends on the selected menu. A typical menu is shown in Table 14.4. As may be seen, a menu is selected by typing a single digit.

Table 14.4 A Typical Menu

	Select alternative and type 1, 2, or 3
1	New student
	Delete student
	Change student details

This saves key strokes and permits easy data entry. With the advent of a mouse, a cursor can be moved to one of the displayed alternatives and clicked, further reducing user's effort.

14.5.2 Template

In a menu if a user selects the alternative "New student" then a template may be displayed on the screen. A template is analogous to a form. A set of labelled fields are displayed and space for entry of data is highlighted. A typical template is shown in Table 14.5. An advantage of the template is that one can see the entire data input on the display. Further, the system is programmed to check if inputs have any errors. For example, if Roll no. is assumed to be numeric and a non-numeric character is input, the system will give an audio signal `beep' and display an error message. It may carry out a modulus check if needed. The templates are designed by using characteristics such as: colour, reverse video, highlighting field length. Brief instructions are given in the message area at the bottom of the screen.

Table 14.5 A Typical Template

STUDENT DETAILS—ENTER AS INSTRUCTED

Roll No.	[ ]	
Name	[ ]	[ ]
	LAST	FIRST or INITIALS
Dept.	[ ]	Pull down menu gives choices
Year	[ ]	
Address	[ ]	
Lines	[ ]	
	[ ]	
	PIN Code [ ]	

14.5.3 Commands

In this case data inputs are asked one by one through a series of questions which appear on the screen of the video terminal. Table 14.6 gives a typical command screen format.

Table 14.6 A Typical Command

Did you request deletion of student record?

Y

Give student's Roll no.

56789

Is the name of the student [Prafulla Sen] ?

Y

Is [Prafulla Sen] a [third year] student?

Y

Shall I delete [Prafulla Sen's] record?

Y

Observe that, based on the given Roll no., the data entry program retrieves the user's name and later the year of his study. These are shown within rectangles in Table 14.6. The user's responses are underlined.

We have seen that each method of interactive input data entry has a role to play in designing the system. In a typical problem these methods are normally combined to provide a 'human engineered system.' In Table 14.7 we give a chart showing a typical dialogue for data entry hierarchy.

Table 14.7 Hierarchy of Dialogue for Interactive Input of Data

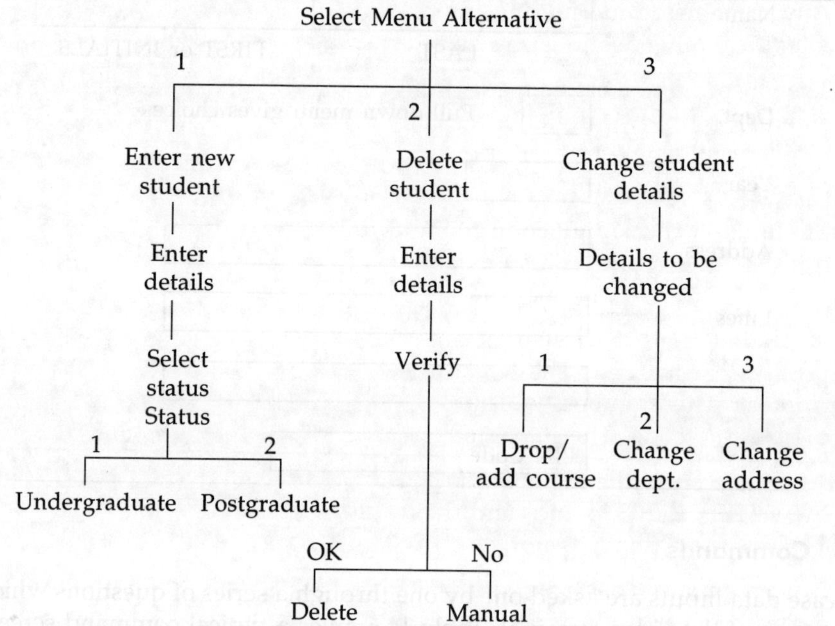

SUMMARY

1. Data may be input to a computer off-line or on-line. In on-line entry a user enters data interactively via a video terminal connected to the computer. In off-line data entry data filled in forms are entered by operators in a separate PC or a data entry machine.
2. Off-line data entry is suitable if the number of records is very large.
3. In off-line data entry, batches of data are formed and entered. They are checked by a validation program and the corrected records are stored in a file.
4. To reduce errors in input, it is essential to carefully design the forms used for entering data.
5. Important data elements are coded. Codes are necessary for unique identification, easy cross-referencing and efficient storage and retrieval.
6. There are many methods for coding. An ideal code must be concise, expandable, meaningful, comprehensive and precise. It is not possible to incorporate all these ideal features in a code.

7. Codes are classified as: (i) Serial number codes, (ii) Block codes, (iii) Group classification codes, and (iv) Significant codes. Group classification codes and Significant codes are meaningful, expandable, precise and comprehensive. They are, however, not concise. Serial and Block number codes are more concise. They are also precise and expandable but are not meaningful and comprehensive.
8. Any error made in entering important data fields such as account codes and identification codes must be detected during data entry.
9. The most common errors made during data entry are: a single digit is incorrectly entered or any two digits in the code are interchanged. These errors are called respectively single transcription and transposition errors and account for 95% of all data entry errors.
10. Modulus-11 check digit code guarantees detection of all single transcription and transposition errors. It also detects 95% of all other errors.
11. Modulus-11 check digit code is designed as follows. Given a code, the digits in it starting from the last digit are multiplied by weights 2, 3, 4, etc., and the products are added. The sum is divided by 11. The remainder is subtracted from 11. This number (which is called a check digit) is appended as the last digit of the code. The maximum number of digits in the code must be ≤ 10.
12. After data entry the digits in the code starting from the last digit are multiplied by weights 1, 2, 3, 4, etc., and the products are added. The sum is divided by 11. If the remainder is not zero then there is an error in the code.
13. It is essential to design good data validation programs to prevent data entry errors from corrupting files of input data. Validation programs need information for detecting errors. This information is provided by controls exercised during data preparation.
14. Important control mechanisms are: giving unique sequence numbers to each data record, providing a batch control record containing a count of number of records and a total of one of the fields.
15. The same data is entered by two different persons and compared to reduce transcription errors.
16. Besides this, individual data fields are checked using information on their range of allowed values, range of reasonable values, and relationships between different fields. Batch control provides information to detect incorrect values entered, missing records, and data in the wrong sequence.
17. With the advent of Personal Computers, remote terminals connected to a computer and local computer networks, considerable amount of data is entered in files interactively.
18. For interactive data input, special screens are designed on video terminals for easy data entry. Errors in data entry are instantly detected by a validation program during data entry and can be immediately corrected.

19. Common methods of interactive data input is by use of menus, templates and interactive commands for data entry.
20. A menu method is used to pick one out of many alternatives, a template method to enter new data, and a command method to add and delete data.
21. These methods may be combined to provide a user the most appropriate technique for a particular type of interactive data entry.

EXERCISES

14.1 What is the difference between on-line and off-line data entry?

14.2 Why are input data records divided into batches for off-line data entry?

14.3 What is the purpose of a data validation program?

14.4 What are the main principles used in designing forms for data entry?

14.5 A good and a bad design for entering date in a form is given in Section 14.1. What are the reasons for saying that one of them is good and the other bad?

14.6 Design a form to be used by a salesman to report to the office about the sales executed by him at different customer locations.

14.7 Why are data fields coded in an information system?

14.8 Can the name of a person be used as a code, for say, his bank account? If not, why?

14.9 What are the requirements of a good coding scheme?

14.10 Is a concise code comprehensive? If not, why?

14.11 Is a meaningful code necessarily comprehensive?

14.12 Is a comprehensive code necessarily meaningful?

14.13 Is a precise code necessarily concise?

14.14 What is the advantage of a serial number code? Why is it not normally used?

14.15 What is the main advantage of block codes?

14.16 Design a group classification code to code (i) motor vehicles, (ii) music cassettes, and (iii) books.

14.17 Is a group classification code meaningful?

14.18 Give an example of a significant code. Are significant codes expandable?

14.19 Add a Modulus-11 check digit to the codes (i) 48467, (ii) 96432, and (iii) 87646257.

14.20 Modulus-37 check is suitable for alphanumeric codes. Add a modulus-37 character to the codes (i) 4AB9W, (ii) XBY483, and (iii) CAZ4642.

14.21 The following code was entered by an operator: 449632. The last digit is a modulus-11 check digit. Is this code correct?

14.22 If a code uses hexadecimal digits, what should be N if the modulus-N check digit system is to be used with such codes? What are the allowable weights if single transcription and transposition errors are to be detected?

14.23 If modulus-11 check digit system is to generate detection of multiple identical digit transcription error (i.e., a code such as 45565 is wrongly entered as 48868), what should be the constraints on the weights?

14.24 A see-saw error is one in which one digit of the code is increased by x and another decreased by x. For example, 486732 becoming 456762. When can modulus N check detect such errors?

14.25 Credit cards have normally 16 digits. Using weights 4 and 1 alternatively starting from the most significant digit and modulus 10 error detection system find the check digit for the string 421646985432543. Will this coding scheme detect single transcription and single adjacent digit transposition errors?

14.26 Find out from the world wide web the check digit method proposed for the UID to be introduced in India.

14.27 Instead of using weights 3, 1 for alternate digits in the European article numbering system is there any advantage in using 4, 1 as weights for alternate digits. Will this code detect all single transcriptions and adjacent digit transposition errors?

14.28 Why is it useful to assign sequence numbers for data records? What are the types of errors detected by sequence numbering?

14.29 What is the purpose of a batch control record? What is the type of information contained in a batch control record?

14.30 A set of data records for student examination results has the following format:

Roll No.	*Name*	*Marks* (out of 100)			
		Paper 1	*Paper* 2	*Paper* 3	*Paper* 4

Design for these records a batch control record and a record control field and any other appropriate checks for the fields.

14.31 Give some example of fields for which a radix error check is appropriate.

14.32 What is the difference between range check and a radix check?

14.33 What are the appropriate range checks for the age of individuals in an employee file, a high school student file, and height of students in a student file.

14.34 Give some examples of fields where reasonableness check would be applicable.

14.35 Give some examples of inter-field relationship checks.

14.36 What is the main difference between menus, templates and command modes of interactive data entry? When is each of these modes appropriate?

14.37 Design a dialogue hierarchy for entering data on customers (of a manufacturer).

14.38 Design a dialogue hierarchy and the screens for a system used to reserve seats in long distance buses.

15

Designing Outputs

LEARNING GOALS

After reading this chapter a student should be able to:

1. Select appropriate devices to output information from a computer.
2. Design reports.
3. Design screens for video display units.
4. Select graphical output formats appropriate for applications.

Presenting the data processed by a computer-based information system in an attractive and usable form is very important. Very often the success and acceptance of a system depends on good output presentation. It is thus essential for a systems analyst to understand how to design output reports. With advances in computer technology, many new output devices are being introduced in the market. An analyst must keep informed about these and use new devices if they are appropriate. Currently excellent graphic displays are widely available. Speech output systems are also emerging. In this chapter we will describe the basic principles of output design.

15.1 OUTPUT DEVICES

The most important output devices are printers and video display units. Printers are used when a large volume of output is required. They are also appropriate when the output is to be distributed to many persons. When batch processing systems are used, the output is normally printed. Printers are of two broad types: impact printers and non-impact printers. In impact printers a print head strikes a printer ribbon which prints a character on paper. In non-impact printers there is no mechanical print head; characters are printed using a jet of ink or a laser beam. Laser printers produce excellent quality prints and are often used in Desk Top Publishing (DTP) system. Impact printers are of two types: line printers in which a whole line is printed at a time and character printers which print one character at a time.

Line printers are very fast (they print around 1000 lines per minute), but expensive. They are used for large volume of printing. Character printers are slower (around 80 characters per second). Currently the cost of printing a page with a character impact printer is cheaper compared to an inkjet or laser printer. They are used for low volume printing when multiple carbon copies are required. Inkjet printers are used to print letters and single copy documents. Laser printers are becoming cheaper and thus very popular.

Video display units use flat panels for display. They have a keyboard which allows entering data. Twenty four lines, each 80 characters long, can be displayed on the display unit. Video display units are used in on-line systems to display results and answers to queries. Colour display units can display a large number of colours. The information displayed on a video display unit may be printed using an inkjet or laser printer or plotted using a plotter. For more detailed description of these output devices the reader is referred to the book by Rajaraman [21].

A new type of output unit entering the market is an audio output unit. Such an output device announces a result using a loudspeaker. Currently the most common audio output gives digits or numbers. Short words are also synthesized by the computer. The pattern of bits corresponding to a word is stored in an audio output buffer storage. A speech synthesizer converts this to spoken words. Currently the speech output of synthesizers are good. The most common audio output unit answers enquiries such as a request for telephone number, balance in one's bank account, and instructions to plant operators, arrival and departure times of trains etc. Pre-recorded dialog systems which use telephones are also being widely employed.

15.2 OBJECTIVES OF OUTPUT DESIGN

In order to select an appropriate output device and to design a proper output report format, it is necessary to understand the objectives the output report is expected to serve. This can be found by asking the following questions:

1. Who will use the report?
2. What is the proposed use of the report?
3. What is the volume of the output?
4. How often is the report needed?
5. Where and when are the reports required?

If the report is for top management, it must be summarized, highlighting important results. Graphical outputs such as bar charts and pie charts convey information in a form useful for decision making.

If the report is for middle management, it should highlight exceptions, that is, anything not normal. For example, in a stores management system, information on whether items are rapidly consumed, or not consumed for a long time should be highlighted. Exceptions convey information for tactical management.

At the operational level all relevant information should be printed. For example, in a payroll system, pay slips of all employees have to be printed each month. In a mailing system all address labels have to be printed periodically.

An important rule to be followed is to provide the volume of information appropriate for use by a person when he needs it. Further, the total volume of printed output should be minimal. Irrelevant reports are worse than no reports. Short reports must be sent to the intended recipient for display on his or her desktop/laptop computer. Before finalising report format it is essential to create several of them, show them to the intended users and obtain their feedback.

15.3 DESIGN OF OUTPUT REPORTS

A report normally has the following structure:

1. A *report heading* which appears only on the first page of the report. It would be desirable to include date and version number.
2. A *page heading* and subheadings which appear at the top of each page of the report.
3. A set of records with some common characteristics may be grouped together. Such a group is called *control group*. A heading, called the *control heading* may be given for this group.
4. Abstract of information contained in each such group of records is printed at the bottom of the group with a description label called *control footing*.

Using Table 15.1 we will explain the terminology.

1. The report heading appearing once at the start of the report is:

   INDIAN INSTITUTE OF SCIENCE—ACADEMIC ROLL LIST
   5-9-2001 VERSION 1
2. The page heading which will appear on top of each page is:

   ROLL LIST OF STUDENTS SEM 1/2000–2001
3. The control heading and subheadings are:

   LIST OF STUDENTS IN AEROSPACE ENGG. DEPT.

   ROLL NO. NAME

   Another control heading and subheadings in Table 15.1 are:

   LIST OF STUDENTS IN CHEMICAL ENGG. DEPT.

   ROLL NO. NAME
4. Lines containing detailed information given in the report appear below the appropriate control headings. In Table 15.1 the line

   97101115 A B BHATTACHARYA

   appearing below the heading:

   ROLL NO. NAME

   is called a *detail line*.

Table 15.1 Illustration of Terminology Used in Describing Reports

Report heading →	INDIAN INSTITUTE OF SCIENCE—ACADEMIC ROLL LIST	
	5-9-2001	VERSION 1
Page heading →	ROLL LIST OF STUDENTS SEM 1/2000–2001	
Control heading →	LIST OF STUDENTS IN AEROSPACE ENGG. DEPT.	
Detail lines →	ROLL NO.	NAME
	97101115	A.B. BHATTACHARYA
	97101125	A.K. CHANDRA
	97101135	R.K. GUPTA
	⋮	⋮
	97201198	P.R. RAO
Control footing →	TOTAL NO. OF STUDENTS IN AEROSPACE ENGG. DEPT. = 68	
	LIST OF STUDENTS IN CHEMICAL ENGG. DEPT.	
	ROLL NO.	NAME
	97102105	A.C. ARVIND
	97102121	L.S. BHATIA
	⋮	⋮
	97102131	P.R. ZAVERI
Control Footing →	TOTAL NO. OF STUDENTS IN CHEMICAL ENGG. DEPT – 63	
Final control Footing →	TOTAL NO. OF STUDENTS IN SEM 1/2000-2001 = 852	
Report Footing →	END OF IISc BANGALORE ROLL LIST FOR SEM 1/2000–2001	

5. Abstract of the information contained in a control group appears at the end of the control group and is called the *control footing*. In Table 15.1 the lines

   TOTAL NO. OF STUDENTS IN AEROSPACE ENGG. DEPT = 68

   TOTAL NO. OF STUDENTS IN CHEMICAL ENGG. DEPT = 63

   are control footings.

   There is another type of control footing called *final control footing*. This abstracts a desired information for the entire report. In Table 15.1 the final control footing is

   TOTAL NO. OF STUDENTS IN SEM 1/2000–2001 = 852

6. Information written at the end of each page is known as *page footing*.
7. Information printed at the end of a report is called *report footing*. In Table 15.1 the report footing is

   END OF IISc BANGALORE ROLL LIST FOR SEM 1/2000–2001.

To summarize, we see that a printed report is divided into the following group of lines, called *report groups:*

First group	:	Report heading
Second group	:	Page heading
Third group	:	Control heading
Fourth group	:	Detail heading and lines
Fifth group	:	Control footing
Sixth group	:	Page footing
Seventh group	:	Final control footing
Eight group	:	Report footing

Having decided what report groups should appear in a report, an analyst decides the layout of the report in a print chart which depicts the number of columns and lines in a report. A sample is shown in Figure 15.1. The chart has a number of columns normally equal to that in a printer. The report headings, etc., are entered on the chart by the analyst to examine the form in which the printed output will appear. The most important consideration in designing the print chart is proper form layout for easy readability. The print chart helps in selecting appropriate headings, entering headings, picking control headings, etc. Once a form layout is determined using a print chart, it may be handed over to a programmer who may use an appropriate *report generator* program. For example, COBOL has a report generation section with its own rules of syntax and semantics. In the print chart the number of columns reserved for each detail line to be printed by the program, including control footing and report footing, is indicated. One may use a convention used in a language such as COBOL to describe the format of the individual fields. In Figure 15.1, for example, the convention used is 9 for digits, X for characters, B for blank, Z for digit or leading zeros and $ for currency symbol. Several report generating systems (e.g., Crystal Reports) are now available to prepare and print reports.

The general principles to be used in designing output reports are:

1. The design must be such that it can be read from left to right and from top to bottom.
2. The most important item, such as the key field, should be the easiest to find.
3. All pages must have a heading and a page number. The date on which the report was prepared should also be printed. Version number should be indicated.
4. All columns must be labelled with meaningful labels.
5. Too many details should not be given.
6. Control footing abstracts information about groups of detail lines and must be effectively used.
7. Similarly, page and report footings give summaries and are useful.
8. If a signature is required at the end of a report there should be space and printed line provided for it.

Reports are often sent to government agencies such as the tax department periodically. The formats for such reports are specified by government agencies and must be strictly followed. Some common forms needed particularly by Government agencies are now available in the internet (See http://www.FormsIndia.com)

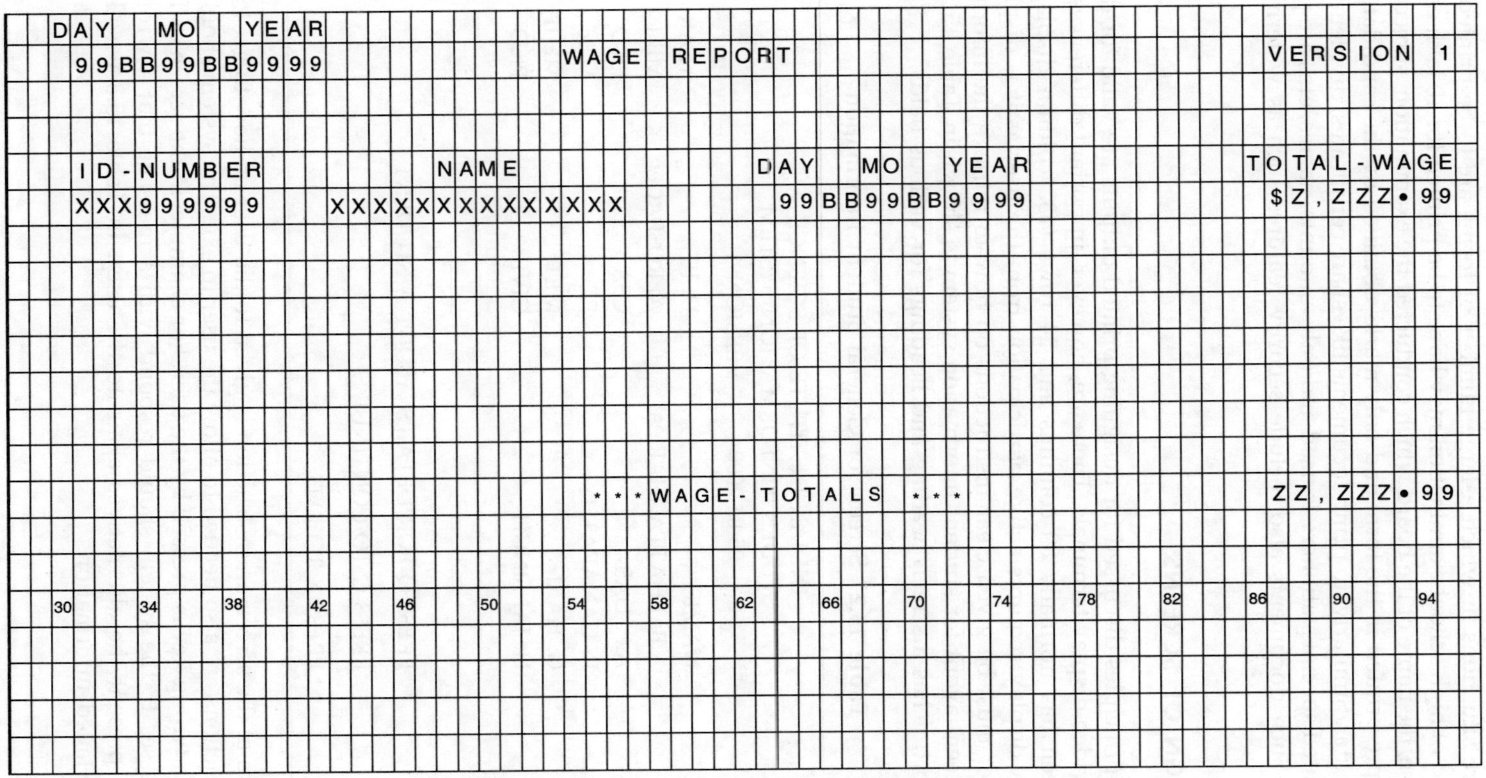

Figure 15.1 Print chart for the wage report example.

There are situations in which a large number of forms are to be printed, e.g., the marks record of high school examination may have to be printed for 10 lakh candidates. In such a case, the name of the board which conducted the examination, subjects offered, etc., can be pre-printed in the form and only marks obtained by the candidates printed by the computer. Similarly, to print documents like share certificates, dividend warrant, etc., pre-printed special stationery is used and only amounts, names, etc., are printed by a program. Such documents also include security features such as watermarks and holograms.

15.4 DESIGN OF SCREENS

Most of the principles discussed for designing printed outputs are valid for displaying outputs on video display terminals. There are, however, important differences. A video display screen has normally 80 columns and 24 rows (132 column display is now available). A display requires active participation from the user as it is displayed on-line. The user must be given clear instructions on how to retrieve the information he needs. A screen layout for student information system is shown in Table 15.2. As in a printed report, it has a screen heading and headings for various fields.

Table 15.2 Screen for General Student Information

INDIAN INSTITUTE OF SCIENCE
STUDENT INFORMATION SYSTEM
05-05-2000 VERSION 1

ROLL NO.	NAME	DEPT.	YEAR
9501325	A.B. BHATTACHARYA	AEROSPACE	ME I
9602415	A.P. DAS	CSA	Ph.D
9602325	P. GANAPATHY	EE	M.Sc
9701425	G. HARI	MET	ME II
9702112	H. JAI SINGH	CIVIL	Ph.D
______	____________	______	______
______	____________	______	______

PRESS D FOR DETAILS ABOUT A STUDENT

PRESS C TO CONTINUE

PRESS X TO EXIT

At the bottom of the screen an instruction is given to the user on how to get to the next screen. Instructions are also needed to allow the user to exit. The system should be able to recover if a wrong key is pressed by the user. For example, in the screen of Table 15.2 if a user presses E, the system should respond with a message that a wrong key is pressed and present the alternatives again to the user. In Table 15.3 we illustrate a screen for detailed information.

Table 15.3 Screen for Detailed Student Information

INDIAN INSTITUTE OF SCIENCE

STUDENT INFORMATION SYSTEM

05-05-2000
VERSION 1

ROLL NO.	9701425	
NAME	G. HARI	
YEAR	ME II	
GUARDIAN	P. GANESHAN	
ADDRESS	41 OLIVER STREET	
	MYLAPORE	
	MADRAS 600 004	
	PRESS D FOR MORE DETAILS	
	PRESS ANY KEY TO RETURN	

Instead of pressing keys it is more convenient to provide a Graphical User Interface (GUI) with "buttons" which when pointed to by mouse movement and clicked will take the specified actions. Thus in the screen of Table 15.2 the following buttons may be provided at the bottom of the screen.

Details	Continue	Exit

Besides tabular reports, text reports are also used by businesses. Such reports must have a heading and date. They must be left justified and use upper and lower case letters (e.g., Do not use all upper case letters). It is not essential to right justify. It is better to avoid hyphenation. Also abbreviations shoud be avoided unless it is universally used. It is better to give full expansion of an abbreviation when it is used for the first time, for example, Local Area Network (LAN).

15.5 USE OF BUSINESS GRAPHICS

All video terminals nowadays allow display of graphics and different types of charts. These are called *business graphics*. Business graphics presentation normally consists of facilities to draw graphs showing trends, bar charts, pie charts and maps. Each of these has a role to play. Graphs normally show the relationships between two variables, typical time along the X-axis, and variables such as sales, student enrollment, and power consumption along the Y-axis. This type of presentation is used when growth or decline of some variable with respect to time has to be summarized (see Figure 15.2). Graphs have the advantages of condensing and presenting the information in a form where its significance is easily comprehended.

Bar charts are normally used to show the distribution of the number of items having specified characteristics within a specified range. For example, a bar chart showing the number of employees in different age groups in a company is shown in Figure 15.3.

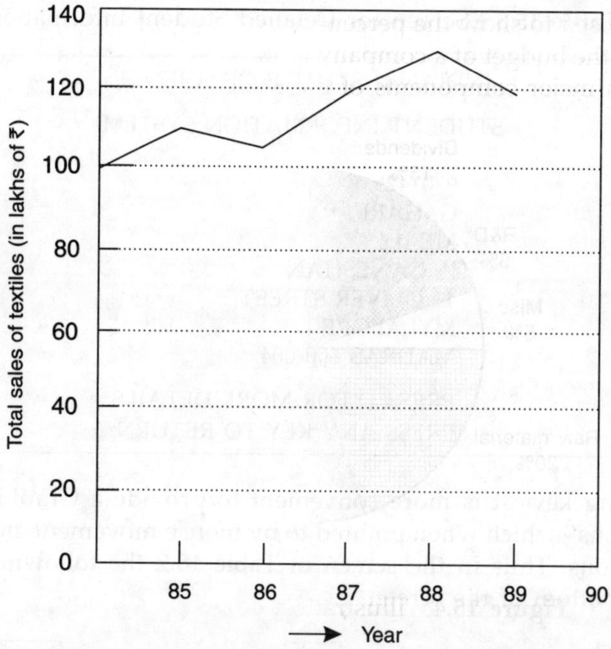

Figure 15.2 Illustration of use of a graph.

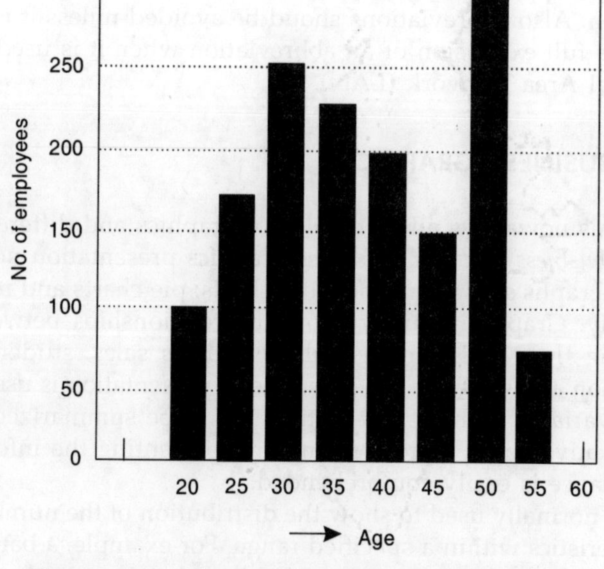

Figure 15.3 Illustration of use of bar chart.

Pie charts are used to show the percent or part of a total used by different activities, for example, how the budget of a company is allocated. A pie chart immediately focuses one's attention on major components of the budget (see Figure 15.4).

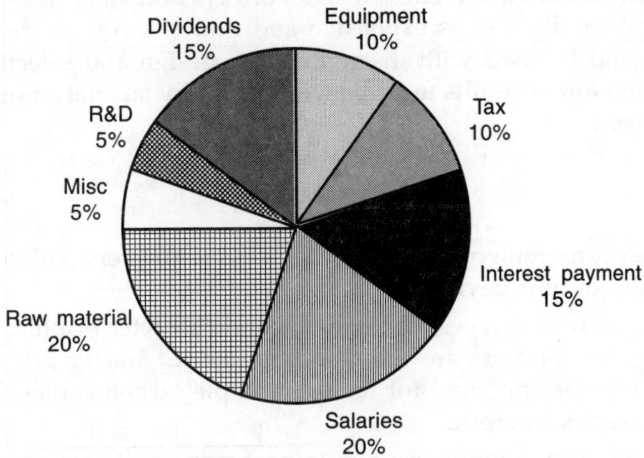

Figure 15.4 Illustration of use of pie chart.

Maps are useful to show the spread of sales offices or markets or maintenance shops in a geographic region. For example, Figure 15.5 shows a map of major spare parts dealers of a company. From the map it is immediately seen which geographical areas are well covered and which are not.

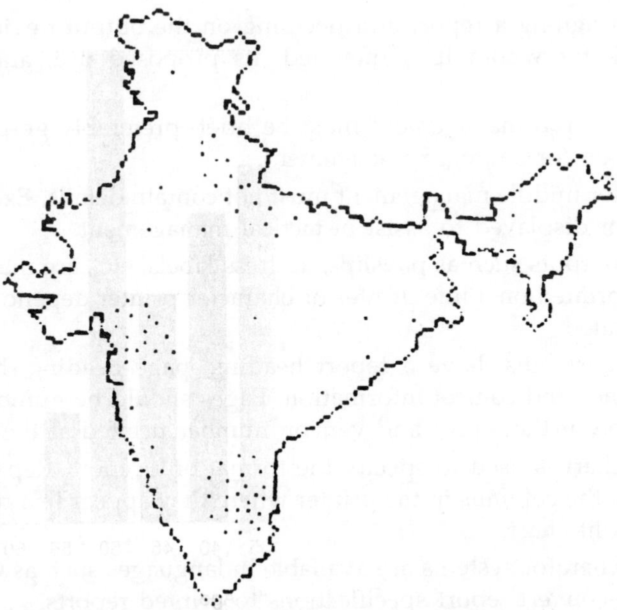

Figure 15.5 Illustration of use of a map in depicting geographical distribution of spares outlets of a company.

We can, of course, combine these forms. For example, in a map, a bar chart can be shown for each region of a country. One more development now is the availability of colour and animation on video display terminals. This further enhances the variety of output presentation. For effectiveness it is not always necessary to have colour. Further, hard copy of colour displays is expensive and multiple copies are difficult to make. Thus colour should be used with discretion. The design and selection of appropriate forms for presentation of results must be worked out by an analyst in consultation with user's management.

SUMMARY

1. The most commonly used output devices are printers, video display terminals and audio output devices.
2. Printers are of two types, line printers and character printers. For large volume printing, line printers are most appropriate. Character printers are useful for low volume printing and for taking multiple carbon copies. They also provide simple graphics output.
3. Inkjet and laser printers are non-impact printers. Laser printers provide high quality printing. Both inkjet and laser printers can display graphics output.
4. Video displays are normally used for on-line display of information. They are interactive and a user has to be instructed how to get appropriate information.
5. Audio outputs are becoming popular with the advent of inexpensive speech synthesizers. Currently audio output is used in enquiry systems to answer simple queries.
6. Before designing a report and deciding on the output device to be used, one must ask for whom it is intended, its proposed use, and its volume and frequency.
7. Reports for top management must be brief, preferably graphical, and convey information for strategic management.
8. Reports for middle management must not contain details. Exceptions should be printed or displayed to assist in tactical management.
9. Routine outputs such as payslips, address labels, etc., are massive and periodic and are printed on a line printer or character printer depending on the volume to be printed.
10. Every report must have a report heading, page heading, heading for fields, details lines and control information. Pages should be numbered and the date of creation of the report and version number printed at the top of the report.
11. A print chart is used to specify the format of a report. Report headings, page headings, the columns in the printer where they appear in a report are specified by the print chart.
12. Report generator systems are available in languages such as COBOL. These are useful to convert report specifications to printed reports.
13. The layout of information on video display screens is also specified using similar principles. The main difference is the need to instruct a user to select appropriate displays.

14. With current technology Graphical User Interfaces are easy to design. They should be used wherever they are appropriate.

15. Graphical outputs are easy to obtain on video terminals, inkjet and laser printers. Graphs, bar charts, pie charts and maps are easy to draw and should be used whenever appropriate.

EXERCISES

15.1 What is the difference between impact and non-impact printers? Give an example of each type of printer.

15.2 What is the speed range of a line printer? What is the speed range of a character printer?

15.3 For what types of outputs are video display units appropriate? What is the main difference in characteristics between video display units and printers?

15.4 What type of output will you use for the following:

(i) Strategic management
(ii) Tactical management
(iii) Operational management

15.5 When is a detailed printed output appropriate?

15.6 What is the difference between off-line and on-line output presentation?

15.7 What are the various headings used in a report? Why are they required?

15.8 What is a print chart? How is it used? Prepare a print chart for stores data processing system.

15.9 Why are control footings used in printed reports?

15.10 What is a report generator?

15.11 List the important points to keep in mind while designing the format of a report.

15.12 What is the main difference between the design of an output screen for a video terminal and the page layout for a printer?

15.13 What do you understand by business graphics?

15.14 Pick the appropriate graphics presentation for the following applications:

(i) Relative enrollment of students in various departments in a University.
(ii) Growth of student strength in a department over a period of 10 years.
(iii) Sales percent of a product in 6 regions of a country.
(iv) Proportion of total revenue of a state from direct taxes, indirect taxes, public loan, land revenue, income from public sector companies, and miscellaneous receipts.
(v) Engineering colleges established in different parts of a country.

15.15 When is colour presentation of graphics useful? Give examples.

16

Control, Audit and Security of Information Systems

LEARNING GOALS

After reading this chapter a student should be able to:

1. Explain the need for controls in an information system.
2. Design controls for an information system.
3. Plan auditing methods for an information system.
4. Devise testing methods for an information system.
5. Explain methods used to ensure security of information systems.

In this chapter we will discuss three related aspects which are important in the design of information systems: (a) control, (b) audit, and (c) security. Control is a method to ensure that a system processes data as it was designed to. Audit is a check of the system by an auditor who wants to make sure that the system is built as per specifications and the processed results are correct. Audit also has to ensure that there are no errors in both entering and processing data. Before a system is released for general use, it should be tested. Testing is essential to make sure that a system performs as per specifications. Security is concerned with the protection of the data resource and the programs against accidental or intentional destruction or modification or disclosure to unauthorized persons.

16.1 CONTROL IN INFORMATION SYSTEMS

Any data processing system should not only process data as per specifications, but must also control that processing. Controls are essential to ensure the reliability of the reports produced by the system. They must be built into the design of the system. Some simple spot checks of outputs for their correctness are not sufficient. Information systems should be designed to handle massive volumes of data. For example, the processing of high school examination results in a State such as Uttar Pradesh will have to handle over a million records. To reliably process such large volumes of data, formal controls are essential. Methods should be devised to control the flow of data in and out of the data processing system, and ensure that all data entering the system are correct and included, and no duplicate data enter the system. If some record is omitted by mistake, a method should be devised to find the location of such an error without having to search the whole system. It should be remembered that business systems, particularly those which deal with money, are subject to audit. Records dealing with individuals, customers, etc., should be accessible and any enquiries should be answered.

Controls do cost more programming time, processing time, and memory space. These considerations should not limit the extent of control. Necessary controls should be planned and established during the initial stage of the design. The potential damage to the reputation and finances of an organization is too great compared to the cost involved.

16.1.1 The Objectives of Control

The primary objectives of controls are:

1. To make sure that the data entering the computer are correct.
2. To check the clerical handling of data before these reach the computer.
3. To provide a method of auditing the steps in a procedure to detect quickly where an error has occurred in the procedure.
4. To ensure that accounting for tax purpose or other legal requirements is carried out according to the law.
5. To guard against fraud that may affect the financial standing or reputation of the business.

16.1.2 Control Techniques

(i) Organization measures. It is necessary to assign well-defined responsibilities for input preparation, delivery, output use, and operation and maintenance of systems. Specific responsibilities for reporting any changes in a system must be assigned. All changes must be monitored. Performance and recording of operations should be in different hands. For example, payment to a vendor should be balanced with an amount payable entry. This entry should not be made by the person who will perform the balancing. In a warehousing system, for example, data on quantity of various items received by the warehouse should independently match with the quantity for which payment is made, based on suppliers' invoices.

(ii) Input preparation controls. This was discussed in Section 14.4. We had described the following controls:

(a) Sequence numbering
(b) Batch controls
(c) Data entry and verification
(d) Record totals
(e) Self-checking digit.

(iii) Processing controls

(a) *Proof figures.* These are used to check an important arithmetic operation in a procedure. This check satisfies the dual purpose of detecting any data entry/process error or a hardware error. A proof figure is an additional data element entered in each record. For example, assume the following record structure:

Item no., qty. supplied, cost per unit

Another field known as a proof cost is appended to the record. The modified record is:

Item no., qty. supplied, cost per unit, proof cost.

Proof cost is manually calculated for each record in the following way: Assume that the highest cost which is ever expected is H. Then

Proof cost = H – cost per unit

As part of the procedure, the following additional calculations are performed:

```
total cost ← 0; total qty. ← 0; check total ← 0;
for all records do
   total qty. ← total qty. + qty. supplied
   total cost ← total cost + qty. supplied * cost per unit
   check total ← check total + qty. supplied * proof cost
endfor
   if total qty. * H = total cost + check total
      then print "No error"
      else print "Error"
   endif
```

If an error message results then it could be due to a data entry error or an arithmetic error. The location of the error may be found by breaking the **for** loop into a number of sets and calculating the proof totals for each **for** loop.

(b) *Two-way check.* The same quantity can be calculated in two different ways and compared. For example, in a payroll system for each record the system calculates:

Gross pay, deductions, net pay

The gross pay, deductions and net pay can be accumulated separately and one may check if (sum gross pay – sum deduction) = sum net pay.

(c) *Relationship check.* When a known relationship between two data elements exists and one of the elements is correct, it is possible to check the other element by using a relationship. For example, if a discount is known for a set of sales then the procedure can sum all sales and sum all rebates and check if rebates total equals (discount * sum of all sales).

If sum of sales = 150,000 and discount is 5% then the rebate total should equal 7500.

(d) *Check point and restart.* A check point procedure is a check routine which is executed at specific intervals during the execution of a large data processing program. The point where the check routine is run is called a *check point*. The purpose of a check point is to find out if processing has been carried out correctly upto the check point. At check points, quantities such as control totals and proof figures can be checked. If the processing is correct the status of the machine is recorded by writing this on a disk. The normal procedure is then resumed.

Check point procedure divides a long job into a series of short ones. Each short one is independent and is checked independently. If the check is correct then enough information is written on the disk to enable automatic restarting of processing from this check point. If the check is incorrect the last part of the process is discarded; errors, if any, are corrected and the system is restarted from the previous check point.

A restart procedure restores the status of the computer as it existed at the specified check point. Restoring status implies restoring the values of all CPU registers, restoration of accumulated totals, re-establishing switches and counters, restoring the file status of secondary files, and so on.

The proper use of check point and restart procedure in a program improves the operational efficiency of a data processing system. If power failure, or hardware or software failure occurs, then the system can be restored to the last check point. The processing done between the last check point and the system failure is the only work that is wasted.

Restart procedure also allows intentional interruption of a running program. Programs may be intentionally interrupted to accommodate priority jobs, emergency maintenance, change of shift, etc.

16.2 AUDIT OF INFORMATION SYSTEMS

Auditors (both internal and external) of organizations have the responsibility of ensuring that computer-based financial and related information systems are reliable, include all records, and are protected against frauds. The main difference between manual systems and computer-based systems is the non-availability of intermediate records in computer-based systems for inspection. The problem is complicated by the processing rules being incorporated in a stored program whose functioning and reliability are not immediately obvious to an auditor who is not a computer professional.

The auditor follows one of the following methods: (a) auditing around the computer, (b) auditing through the computer, and (c) auditing with the computer.

Auditing around the computer means checking only the inputs and outputs of a computerized system. Auditing through the computer implies that the computer program itself is checked. Auditing with the computer involves writing special programs to check the system. The existing system is used and sometimes a synthetic data input is used to check the correctness of the system.

16.2.1 Auditing around the Computer

In auditing around the computer, the computer and the programs running in it are bypassed. The input fed to the system is reconciled with the output. For example, accounts receivable is reconciled with the opening balance, receipts and payments, using only the input and output data records.

16.2.2 Auditing through the Computer

An auditor has to establish an *audit trail* which allows examining selected intermediate results. If an error is found in the output, the audit trail is expected to lead to the source of the error. Batch totals and check point procedures localize the source of an error and provide an audit trail.

If an auditor is consulted at the time of system design he will provide his requirements which can assist in incorporating proper checks and audit trails. Specific audit and control procedures comprise the following steps:

- Provide facility to trace a variable value through processing steps and print intermediate values when requested.
- Selectively print records and transactions in the system which meet certain criteria assigned by the auditor. The criteria may be to print inactive accounts with large balances, highly active accounts, etc.
- Produce a detailed journal of all transactions meeting certain criteria and their effect on account balances in the master file.
- Examine credit and debit balances in an accounting system and report if the system does not balance both.
- Examine the adequacy of input and batch controls.

Besides this, an auditor should ensure that (a) logs are kept of all runs of systems to fix responsibilities when a fraud is suspected, and (b) systems are adequately documented to check the logic of procedures and data flows.

An auditor may also keep a set of input data for testing a system. Test results are compared with expected results. This is a useful method to check if an unauthorized amendment to a program has been made.

16.2.3 Auditing with the Computer

Special audit package is used to check a system. The main features of an audit package are:

- Extracting data based on specified criteria for inspection.
- Totalling specified data items from records based on some criteria, e.g., total debit of all customers from Maharashtra.
- Introducing a procedure to check selected discounts, interest charges and payments.
- Examining totals when key values change from one group to another.
- Matching with an independent file some master file data to detect out of sequence or missing records.

16.3 TESTING OF INFORMATION SYSTEMS

When a system is developed, it is hoped that it performs properly. In practice, however, some errors always occur. The main purpose of testing an information system is to find the errors and correct them. A successful test is one which finds an error. The main objectives of information system testing are:

- To ensure that during operation the system will perform as per specifications.
- To make sure that the system meets users' requirements during operation.
- To verify that the controls incorporated in the system function as intended.
- To see that when correct inputs are fed to the system the outputs are correct.
- To make sure that during operation, incorrect input, processing and outputs will be detected.

The scope of an information system test should include both manual operations and computerized operations. Information system testing is a comprehensive evaluation of the programs, manual procedures, computer operations and controls. Information system tests may be classified as program tests, string tests, system tests, pilot tests and parallel tests. These are now discussed.

(i) **Program tests.** These are designed to test the logic of programs. Program logic is normally very complicated and it is practically impossible to test all the paths taken by a program. Normally individual modules are tested. Test data are generated to test all logical paths in the module. A very good tool to determine the paths to be taken is to make a decision table listing of all condition tests. All logically possible rules in the decision table can then be determined and used to generate a complete set of test data. Use cases also provide good test data.

Most common errors in programs occur at boundary points in tests. Test data should test all boundary points where there is a transition from true to false or vice versa in the conditions being tested in the program. Another common error is made in counting. Thus the program must be tested with counts which are one more or one less than what is specified. Some tools are available which analyze a program and create a flow chart equivalent to it. Such a tool is useful in generating test data.

(ii) **String tests.** These are carried out on a set of related program modules. The purpose of a string test is to ensure that data are correctly transferred from one program in the string to the next. For example, in a student result processing system, the grade processing module for post-graduate students must use valid output from the post-graduate student registration module.

(iii) **System tests.** These are used to test all programs which together constitute the system. In a payroll system, for instance, all programs in the system such as calculation of bonus, overtime, and costing program (which uses the output of the payroll system) are to be tested together. Testing is conducted using synthetic data. Both valid and invalid transactions are used in this test. For example, a non existent employee code may be used in a transaction to see if it is rejected. Similarly, some unreasonable data may be used to check whether the input controls function as expected.

(iv) **Pilot tests.** A set of transactions which have been run on the present system are collected. Results of their processing on the existing system are also kept. This set is used as test data when a new computer-based system is initially developed. The two results are matched. The reason for any discrepancy is investigated to modify the new system.

(v) Parallel runs. The aim here is similar to pilot tests. In this, however, both existing and new computer based systems are run simultaneously for a period of time and the results from the two systems are compared. This is a good method for complex systems but is expensive.

Preliminary tests on a system before it is released for general use are known as alpha tests. This is followed by a beta testing period when the system is released for general use and carefully observed for malfunctions. After this phase it is put to general unrestricted use.

16.4 SECURITY OF INFORMATION SYSTEMS

Data entering a data processing system and the program processing the data must be kept secure. By security we mean protecting the data and programs against accidental or intentional destruction or disclosure to unauthorized persons. It is the responsibility of a systems analyst to ensure the security of both data and programs. The following requirements should be met to ensure security:

1. The data and programs must be protected from theft, fire, disk corruption and other types of physical destruction. Duplicate copies are kept in a fire-proof vault in a place away from the data processing centre. This is particularly important for financial data.
2. Data should be reconstructable in case of loss despite precautions. Backup copies of master files and transaction files are kept in a fire proof almirah.
3. The system should be tamper proof. Password or biometric (e.g., thumb print) system and file security keys are used to bar unauthorized access. If password system is broken by a clever programmer a secrecy transformation (called encryption) can be used to transform the stored data. Even if the data is accessed, it will not be meaningful as it has been encrypted. One simple data transformation would be to map every character in a file to another character by using a secret mapping table. For example, a table may transform 1 to 8, 2 to 7, 3 to 5, ……., A to W, B to U, C to X, … . This is called transcription. It may be combined with transposition in which characters are shifted by a specified number of characters. The transformation method used is kept secret. To obtain the original data the inverse transformation is applied. A person who gains unauthorized access cannot understand the data. An authorized person will get back the original data by applying the inverse transformation.
4. Any person gaining access to a file should be identified. Thus attempts to access data is logged and identity is also recorded. This will inhibit potential data lock breakers.
5. Only authorized persons should be allowed to change data. Password or biometric system is used to prevent unauthorized access. Every access, authorized or unauthorized, should be logged by the systems. Any change should be monitored by the system.

Information systems have now become vital in day-to-day operations of organizations. For example, a failure of servers or network in a banking or railway reservation system will bring the whole organization to a halt. It is imperative thus to design systems which are *resistant to failure*. One method is to duplicate servers and communication systems. They should be in two different geographic locations. All transactions are duplicated on both the systems so that if one of them fails, the other system takes over. This is a very expensive solution and should be used only if a business cannot tolerate a failure of even a fraction of a second. Not many systems are of this type. If an organization can tolerate a few minutes failure periodic check points every few minutes would be sufficient. Apart from failures, a more serious problem is a disaster. *Disasters* make it impossible to restore a system immediately, for example, fire, floods, terrorist attacks or earthquake may make entire systems non-operational. Disasters are rare but all organizations must have a plan to recover from them. Backup copies of all files should be taken at the end of each day or every few hours, depending on how critical the data is, and kept at a different site. Transferring files using communication systems has become feasible and thus a disaster recovery site can be in a different geographical location. For example, the main servers of a major Indian bank are located in Mumbai and the disaster recovery site is in Chennai.

Besides security, another concern of a system designer is to ensure the *privacy of data*, particularly individual's data in a database. Privacy refers to the rights of individuals and organizations to ensure that data about them is not revealed to others without their consent. This issue of privacy goes beyond the merely technical issue of data security. It is a moral, ethical and legal issue to be tackled by society through appropriate legislation. With massive databases on individuals, which store tax returns, medical data, personal data, etc., an unscrupulous person or even a government may use it to harass an individual unless data privacy is ensured.

Another recent problem which affects file security is the prevalence of *computer virus*. A virus is a small program written by a mischief maker and spreads by copying itself from one computer's secondary memory to another's secondary memory. For example, a common *virus* copies itself from a pen drive in which it resides to a PC's disk thus spreading the virus. Some viruses store a counter and every time an infected file is copied the count decreases by 1. When the count reaches zero, a program in it destroys files stored in the hard disk.

Virus also spreads through computer networks and replicates itself on many computers connected to the network. It is thus essential for a security system to protect files from viruses. One physical control is not to allow free copying of files from unknown sources and not allowing indiscriminate use of pen drives. Anti virus programs are available in the market which examine files and remove viruses if any from them. A warning is issued to a user when this is found and saves the user's disk from getting infected.

There is no fool-proof technique of protection against viruses in networks. Strict access controls, logging of all accesses, and examining suspicious programs in detail are the only means. The threat of viruses may decrease if virus propagators are convicted and jailed when caught. This is already happening in the West. When computers in an organization are connected to the Internet one has to be concerned with attempts to access the system by hackers (i.e., programmers who are irresponsible vandals). A computer with a special program called a firewall is installed at the gateway of the company's network to the Internet to inspect and deny access to hackers.

SUMMARY

1. Controls in an information system are essential to ensure the reliability of its reports.
2. The primary objectives of controls are to make sure that all data to be processed are included and they are correct. Controls are essential to check clerical handling of data and to trace the source of the error when an error is detected.
3. Controls are also essential to guard against intentional frauds.
4. Control methodology includes organizational measures, input preparation controls, and extensive editing of input data.
5. Besides control on input data, controls on processing are also needed. Processing controls use proof figures, two-way checks and relationship checks.
6. A check point procedure splits a large processing job into smaller runs. At each break the major results and computer status are stored in a secondary storage device. Restoring from a check point is thus easy.
7. Check point—restart facility provides an audit trail and improves the operational efficiency by providing convenient restart points from where a program interrupted by power failure, hardware failures, etc. can be resumed.
8. Auditing of computer-based systems is performed by auditors to ensure that computer-based financial and related information systems are reliable, include all records, and are protected against fraud.
9. Auditing can be done through the computer by running specially developed test programs to extract data records for spot checks and perform cross-checks on input and processed output files.
10. Auditing of the program may also be done to ensure that proper control schemes have been implemented in the program. The data processing system is run with specially designed data sets to check whether the system conforms to specifications.
11. Data and programs of an information system must be protected from theft, fire, disk corruption, and loss due to other reasons. If data or program is lost in spite of precautions, it must be reconstructed.
12. A system is tested before release to ensure that it conforms to specifications during operation, meets user requirements, controls function effectively, and its outputs are correct.
13. The scope of a system test includes manual procedures, computer programs, computer operations and controls.
14. Systems must be protected from unauthorized changes by controls on access. Critical data may be transformed by using a secrecy transformation (called encrption) so that data, even if accessed, are meaningless.
15. Continuous functioning of systems in spite of faults or disasters should be ensured by implementing appropriate control measures.

16. Privacy of individual's data stored in a database must be ensured. This is a moral, ethical and legal issue.
17. A recent problem in data security is the advent of computer virus. Virus is a mischief-maker's program which reproduces itself and corrupts data stored on disks.
18. Precautionary measures against virus infection are: not allowing copying of data/programs of unknown origin and using anti-virus programs resident in the computer to check all incoming files.

EXERCISES

16.1 Why are controls necessary in an information system?

16.2 What are the objectives of control?

16.3 A college has 1500 students whose final examination results are declared using computer processing. There are 5 subjects, each carrying 100 marks. Classes are awarded as follows: Marks 60 or above I Class, 50 or above II Class, below 50 Fail. Devise an appropriate control scheme for processing results.

16.4 What organizational measures are necessary to ensure proper control of an information system?

16.5 What do you understand by a proof figure? Design a proof figure for the example given in Exercise 16.3.

16.6 Enumerate the controls used for ensuring correct processing.

16.7 What is check point restart procedure? What is its purpose?

16.8 Why should information systems be audited?

16.9 What is an audit trail?

16.10 What is the difference between control and audit?

16.11 Enumerate useful audit procedures.

16.12 What is an audit package? Enumerate some of the important features of an audit package.

16.13 What is the purpose of security measures in an information system?

16.14 What are the security measures taken to protect important data files?

16.15 What is the difference between security and privacy? Do secure systems ensure privacy?

16.16 What is the difference between security measures and control measures?

16.17 Is a password system sufficient to ensure security of access to files?

16.18 How can privacy be ensured in an information system?

16.19 What do you understand by a computer virus?

16.20 How can you protect a computer installation from computer viruses?

16.21 Why are system tests necessary?

16.22 What are the objectives of system tests?

16.23 What is the difference between a pilot test and a parallel test?

16.24 How are program test data generated?

16.25 Explain how information systems are kept operational if there is a disaster.

16.26 Explain how the continuity of operation of an information system can be ensured if a system failure such as disk crash occurs.

17

Electronic Commerce

LEARNING GOALS

After reading this chapter a student should be able to:

1. Define e-commerce and evaluate its advantages and disadvantages.
2. Classify e-commerce systems as B2C, B2B or C2C.
3. Explain the need for a layered architecture of e-commerce system and describe the functions of each layer.
4. Define Electronic Data Interchange and explain its role in B2B e-commerce.
5. Describe how security is ensured in e-commerce.
6. Compare various methods of encrypting messages and explain the need for digital signature.
7. Explain various electronic payment systems used in e-commerce.

During recent years a revolution has taken place in the way business is done. This revolution is primarily due to the convergence of computers and telecommunication technologies and the emergence of a number of Internet Service Providers called ISPs. These service providers facilitate connection of computers to the Internet—the world wide network of computers. The Internet connects the computers of many companies and also connects individuals. It is very much like the world wide telephone network but is much cheaper and allows exchange of not only documents but also multimedia information—namely audio, graphics and video information. The Internet has spawned a number of innovations in doing business between organizations, between individuals and organizations and between individuals and individuals. These are commonly known as Business to Business (B2B), Business to Customer (B2C), and Customer to Customer (C2C) transactions. These transactions constitute Electronic Commerce and is abbreviated as e-Commerce. The methods of analyzing and designing information systems we have discussed so far in this book are applicable to e-Commerce but have to be enhanced to account for the fact that data flow into and out of an organization is getting further automated. In other words, human intermediaries in data handling are being dispensed with and consequently the flow of physical documents is reduced. This reduces the overall time and cost of doing business. From the systems design angle this places more responsibility on system designers to detect and correct errors in data flow at the source of the data. Special actions are needed to ensure security of transactions. In this chapter we will first explain what is e-Commerce, we will then describe why e-Commerce has become a fast growing activity and finally describe the various components of an e-Commerce system and how they interact to enable businesses to provide innovative services.

17.1 WHAT IS e-COMMERCE?

e-Commerce is concerned with transactions carried out between organizations, businesses or individuals on-line using their respective computers which are connected by a telecommunication system. The transactions include orders sent to vendors to supply items, invoices sent by vendors, payment usually made by debiting an organization's account with a bank and crediting to the vendor's account, cash payments made using what is known as electronic cash (e-cash). The important point is that all transactions are carried out electronically using the network connection. Transactions may be between two commercial organizations, between government department and a commercial organization, between a customer and a commercial organization or between two individuals. There are a variety of e-Commerce applications. Some of these are:

- Retail stores such as book stores, music stores, toy stores, etc., selling goods.
- Auction sites using which an individual buyer and seller can buy and sell goods.
- Co-operating businesses connected using their own private telecommunication network carrying out transactions in a semi-automated way.
- Banks connected to their customers providing services such as deposits, payments, and giving information on status of an account.
- Railways/airlines/cinema theatres permitting booking tickets on-line using the Internet and paying for them using credit cards.
- Filing tax returns with government agencies on-line using the Internet and obtaining an immediate acknowledgement.
- Use of the Internet to promote marketing, advertising, sales and customer support.

EXAMPLE 17.1 Business to Business (B2B) e-Commerce

We will now take the example of ordering goods from a vendor for which a document flow diagram was given in Chapter 5 (Figure 5.6) which we reproduce with slight modifications as Figure 17.1. In this block diagram each solid line represents a paper document and each dotted line physical item flow. If this system is to be changed into one which fully integrates computation and communication which is the basis of e-Commerce, the following should be done:

1. Each of the blocks at the customer's site should be replaced by a Personal Computer. In other words, the receiving office clerk, the purchase office clerk, inspection office clerk, stores office clerk and accounts office clerk, each will work with a PC using which they can view documents, enter data or carry out specified transactions.

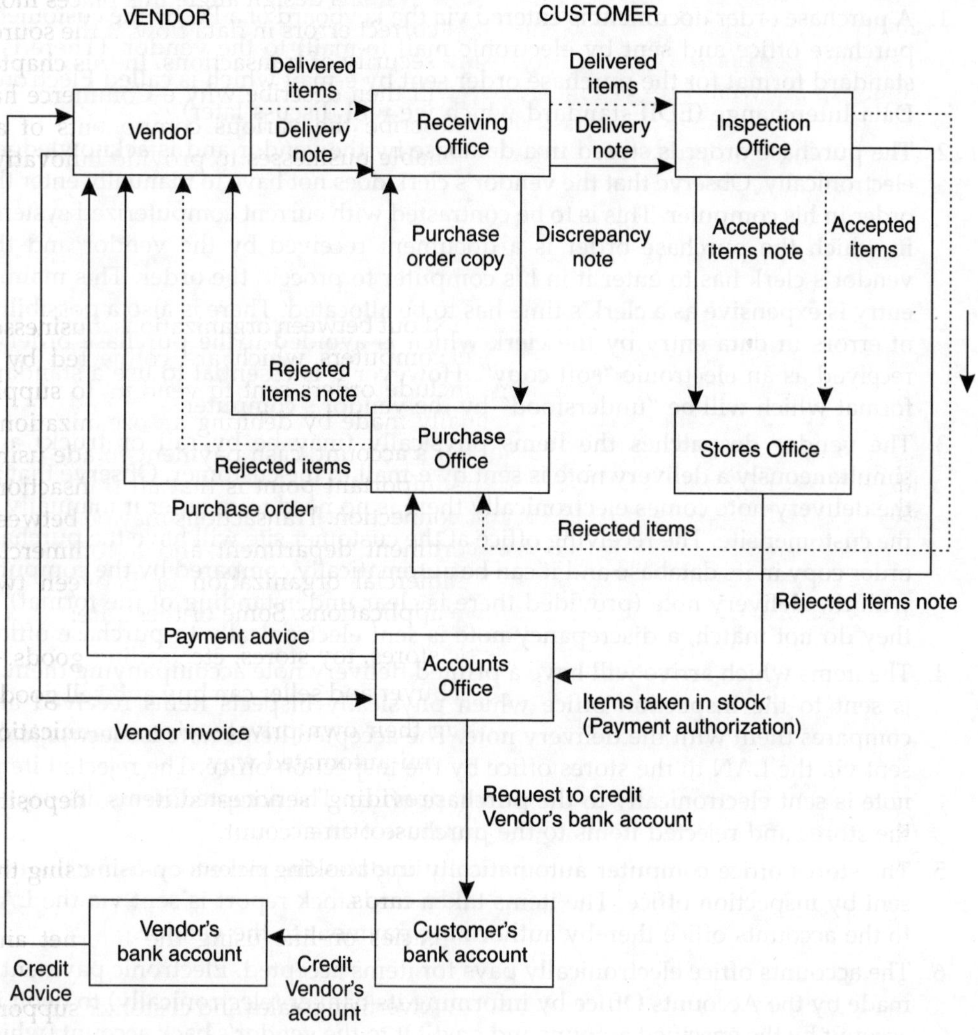

Figure 17.1 Block diagram of document, item flow and funds transfer in a vendor supply system.

2. All these computers will be interconnected by a network. Such a network is called a Local Area Network (LAN).
3. Information may be sent/received between offices of the customer using the LAN.
4. The vendor will have his own LAN to which a computer is connected for carrying out transactions with customers.
5. The vendor's LAN is connected to the customer's LAN either through a Public Switched Telephone Network (PSTN) or by a private communication line.

The above steps will enable the vendor and the customer to transact business electronically. The business transaction will proceed as follows:

1. A purchase order document is entered via the keyboard of a PC by the customer's purchase office and sent by electronic mail (e-mail) to the vendor. (There is a standard format for the purchase order sent by e-mail which is called Electronic Data Interchange (EDI) standard which we will discuss later)
2. The purchase order is stored in a database by the vendor and is acknowledged electronically. Observe that the vendor's clerk does not have to manually enter the order in his computer. This is to be contrasted with current computerized systems in which the purchase order is a document received by the vendor and the vendor's clerk has to enter it in his computer to process the order. This manual entry is expensive as a clerk's time has to be allocated. There is also a possibility of errors in data entry by the clerk which is avoided if the purchase order is received as an electronic "soft copy". However it is essential to use a standard format which will be "understood" by the vendor's computer.
3. The vendor despatches the items physically (may be by rail or truck) and simultaneously a delivery note is sent by e-mail to the customer. Observe that as the delivery note comes electronically there is no need to re-enter it manually at the customer site. The receiving office at the customer site will have the purchase order copy in its database and it can be automatically compared by the computer with the delivery note (provided there is clear understanding of the format). If they do not match, a discrepancy note is sent electronically to purchase office.
4. The items which arrive will have a printed delivery note accompanying them. It is sent to the inspection office which physically inspects items received and compares them with the delivery note. The accepted items note is electronically sent via the LAN to the stores office by the inspection office. The rejected items note is sent electronically to the purchase office. The accepted items are sent to the stores and rejected items to the purchase office.
5. The stores office computer automatically updates the inventory using the note sent by inspection office. The items taken into stock report is sent via the LAN to the accounts office thereby authorizing payment to the vendor.
6. The accounts office electronically pays for items accepted. Electronic payment is made by the Accounts Office by informing its banker (electronically) to debit its account by the specified amount and credit it to the vendor's back account (which maybe with a different bank). This is known as Electronic Funds Transfer (EFT) which is also an important aspect of e-Commerce. Observe that it is not necessary to print a cheque and mail it.

From the above example we see that the following hardware and software would be needed for establishing e-Commerce between businesses:

1. Each of the businesses must have a LAN interconnecting its offices. The offices themselves should have computers for data entry/receipt, comparison, etc. The system may be a distributed client/server type system with each office being a client and the databases being stored in appropriate servers. The internal system architecture of the business is not a major issue. However, the protocol used by the LAN is normally the same as that used by the Internet, namely, TCP/IP. The organizational computer network using this protocol is called an *Intranet*. Besides using TCP/IP protocol intranets also have one or more World Wide Web servers.

2. The two intranets must be interconnected. There are two alternatives. One is to connect each of the intranets to the Internet. This is an inexpensive solution but may be insecure unless special precautions are taken. The other solution is to connect the intranets to constitute what is known as an *Extranet*. Extranet is a private interconnection of the intranets of business associates. Each intranet supports web pages which can be accessed by the members of the extranet.
3. A method of paying for goods or services received electronically. This implies that the business partners must know one another's bank account details. Further, funds transfer must be secure. No one should be able to illegally access a business partner's account and transfer money.
4. We have assumed that documents are interchanged by e-mail. This is acceptable between close business associates if security and authentication of documents sent and received is ensured.

EXAMPLE 17.2 Business to Customer (B2C) e-Commerce

We will now take another example of e-Commerce involving an individual and a shop selling goods. This is called Business to Customer (B2C) e-Commerce. One of the earliest such shops was a bookstore known as amazon.com set up in USA which sells primarily books. There are many such companies in India now. One of them is called Fabmart.com which first sold music cassettes/CDs and now books and several other items. We will now explain how these Internet based shops popularly known by several names such as Virtual-shops, Cyber-shops, Dot-com shops, E-Stores, function. All these have the following common characteristics:

1. Customers have access to the Internet from their homes or work places and wish to purchase items sold by these shops. Primary reason is convenience as one can shop at any time from home and items will be delivered. Customer normally knows the world wide web address of the shop or uses a browser to locate it. These shops have a world wide web address ending with .com (hence called Dot Com Shops) and are connected to the Internet. For example the address of a music and books shop in India is www.fabmart.com.
2. The customer logs on to the web address of the shop. If he wants a specific item he states that by keying it in a specified place in a form displayed on his terminal. If the item is a book and he wants to look at books on the same subject he gives the subject and the shop searches its database and gives a list of available books on that subject. The shop may also give the contents page of each of the books, review of the books and cost of the books.
3. If the customer wants to buy one or more books he points to the book details shown on his VDU using his mouse and clicks. The vendor's computer enters the prices, adds them, provides discount if any and shows the net amount payable. The customer is then asked to enter his credit card number and address to which the items are to be shipped. Option is also given to the customer to pay cash on delivery.
4. If the payment is by credit card, the vendor electronically (i.e., via a communication line) sends credit card details of the customer and the amount of bill to the credit card company's computer.

5. If the customer's credit is OK the credit card company authorises the sale. The e-shop can now proceed to fulfill the order.
6. The e-shop's computer sends an order acknowledgement accepting the order and asks the customer for shipping instructions. Based on the shipping instructions the e-shop specifies the mode of transport and delivery period.
7. The e-shop normally does not stock the items. It orders them from a distributor (normally electronically). After they are received at vendor's warehouse, the items are packed and despatched to the customer's address.
8. The credit card company sends a bill to the customer. It also credits the vendor's bank account by the amount of the bill. The various steps involved in this transaction are shown in Figure 17.2.

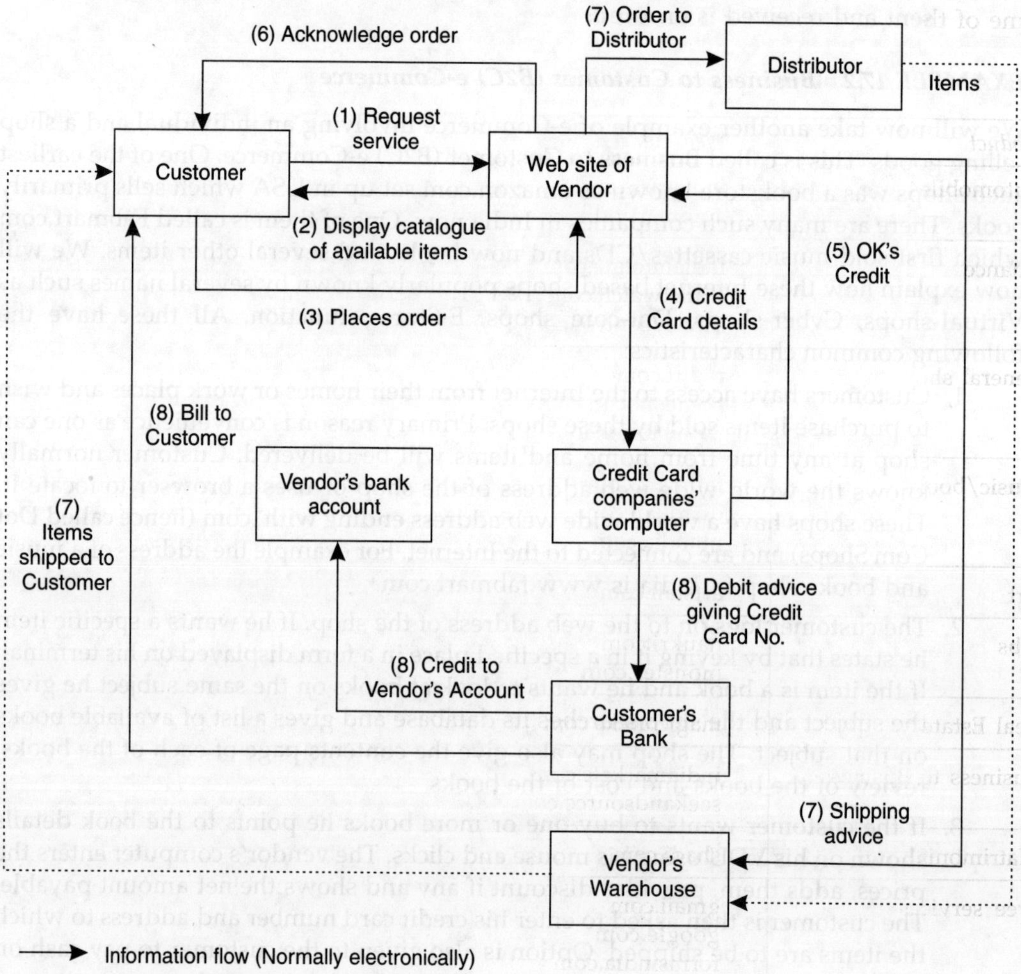

Figure 17.2 Business to customer e-Commerce.

EXAMPLE 17.3 Customer to Customer (C2C) e-Commerce

The last example we will take is one in which two individuals want to sell/buy items. The items are usually used items, collector's items such as stamps/coins or antiques. The seller posts the description of the item and the expected price of the item on a website maintained by a company which acts as a middle man or broker. An individual who logs on to this site looking for items may be interested in the item for sale. He then offers to buy the item and he may quote a price. The price is mutually settled by exchanging messages by e-mail. The broker then arranges to collect the items from the seller and despatch it to the buyer. He collects a fee from the buyer and the seller for his services. The primary advantage of this transaction is that the Internet enables two individuals in distant places to come together to buy and sell using an intermediary's web address.

There are a large number of web sites in India for various goods and services. We list some of them in Table 17.1. This list is representative and subject to change.

Table 17.1 Some Web Sites which are Popular in India

Product	*Web Address*	*Service Provided*
Automobiles	indiacar.com automartindia.com	New and used car sales Used cars
Finance	indiainfoline.com walletwatch.com moneycontrol.com	Investment advise Personal finance Personal finance
General shopping	rediff.com homeshop18.com indiabazaar.com	Portal with several services General goods General goods portal
Music/books	fabmart.com dhadkan.com phindia.com	Music, books and other goods Music Books by PHI Learning
Art	artstall.com	Virtual arts gallery for paintings
Jobs	naukri.com monster.com	Find a job
Real Estate	magicbricks.com	Flats, houses etc.
Business to Business	indiamarkets.com seekandsource.com	Listing of businesses, suppliers, business directory
Matrimonial	shadi.com	Find a partner
Free services	gmail.com google.com formsindia.com	Free e-mail site. Free Internet search All types of forms for tax returns, etc.

17.2 ADVANTAGES AND DISADVANTAGES OF e-COMMERCE

We saw in the last section that a large number of goods and services are now being sold/ bought using the worldwide web. The major advantage to a customer using this mode of shopping are:

1. One can buy/sell items from anywhere sitting at home provided an Internet connection is available.
2. The shopping can be done 24 hours a day, 365 days in a year—An Internet based shop never closes!
3. One can avail of services such as financial services, legal services, medical advise etc., from appropriate portals.
4. Wide variety of goods (particularly items like books and music) are accessible easily without spending time and money in physically visiting and searching in many shops.
5. Anonymous friendly advice may be available on items one may like to buy/rent.

The advantage which accrue to a business are:

1. With a website a business can reach out to a world-wide customer base at a very low cost.
2. Order processing cost is reduced as manual data entry is reduced. Business is also carried out faster as all documents are exchanged electronically.
3. Inventory size is reduced as transaction time is reduced.
4. Funds transfer is faster.
5. A large number of potential business partners can be quickly explored with appropriate search engines and e-mail correspondence.
6. There is no need to maintain a business premises with showroom in a locality where rent is high.

The major disadvantages of e-Commerce are:

1. Currently Internet access is not widely available in India.
2. Communication infrastructure is expensive and not very reliable particularly in small towns in India.
3. Payment by credit card requires faith in the system security. As of now there is fear of entering credit card number on the Internet for paying for goods and services in India.
4. Electronic Data Interchange standards have to be in place before Business to Business e-Commerce can increase. Small businesses may find it difficult to conform. This is, however, being alleviated by using XML for data interchange.
5. Many persons go shopping for social contacts, touch and feel before buying items. e-Commerce will de-personalize transactions.
6. A major concern is security of transactions on the Internet. Spies or hackers can steal and misuse credit card numbers if appropriate care is not taken.

7. Shopping portals will be vulnerable to attacks by hackers unless special precautions are taken. One type of nuisance is called denial of service in which a large number of frivolous enquiries are posted to a portal making it inaccessible to legitimate customers.
8. Portals have to be protected from virus attacks and other electronic vandalism and espionage by erecting fire walls and special security systems.
9. Customer's privacy may be lost if regular log is kept of his/her buying habits.

In spite of these disadvantages e-Commerce is bound to rapidly increase due to its convenience. Some of the security and privacy issues are not severe in Business to Business e-Commerce using private networks connecting them.

In order to set up e-Commerce sites a number of technologies are required which we will discuss in the next section.

17.3 e-COMMERCE SYSTEM ARCHITECTURE

In this section we will discuss the technology, namely the hardware and software, required for e-Commerce. The system architecture for e-Commerce consists of several layers, each layer providing some services. At the bottom is the physical network which connects the computers participating in e-Commerce and at the top level is the e-Commerce applications. We will now examine each layer and see what services they provide. The layered architecture is shown in Table 17.2.

17.3.1 Physical Network

If computers are to communicate with one another they should be physically connected. Most businesses have a Local Area Network (LAN) connecting all their computers. The LAN usually connects machines with unshielded twisted pair (UTP) of wires. LANs using UTP can communicate at the rate of 1 Gigabits/sec even though 100 Mega bits/sec is more common. The number of computers that can be connected to a segment of a LAN is limited to around 16. If a large LAN with more than 16 computers is required, it is created by connecting LAN segments using an electronic circuit known as *bridge*. Besides UTP other interconnections use fibre optic cables (for higher speed LANs) and wireless connection to connect mobile computers to LANs. Those interested in detailed information on physical networks may refer to the book by Stallings [29].

When two businesses want to communicate with one another their LANs are connected using what is known as a *router* and a *modem* to the telephone network provided by Bharat Sanchar Nigam Ltd. (BSNL). This network is known as Public Switched Telephone Network (PSTN) or Plain Old Telephone System (POTS). PSTNs are not very secure. Thus if businesses want to closely collaborate they have their own private leased communication lines.

17.3.2 Logical Network

The single most important technology which has enabled the growth of e-Commerce is the *Internet*. The Internet connects tens of millions of computers spread all over the world

Table 17.2 Layered Architecture for e-Commerce

Logical Layers	*Services in each Layer*	
Application Layer	Customer to Business e-commerce Business to Business e-commerce Customer to Customer e-commerce	
Middleman Services	Mail order houses, hosting services Payment schemes, e-cash Mailing services Value Added Networks Digital Signature Certification Authority	
Secure Messaging	Encryption EDI Firewalls	
World Wide Web Services	HTTP Hypertext transfer protocol	
	structured documents	HTML Hypertext Markup Language XML Extensible Markup Language
	compound documents	OLE, opendoc
	Search Engines Software agents	
Logical Network	Internet Intranet Extranet	TCP/IP Protocol
Physical Network	PSTN (Public Switched Telephone Network) Wireless Optical fibre LAN Bridges Routers	

enabling them to exchange information and share resources. The major applications of Internet are exchanging electronic mail, exchanging files (text as well as multimedia), storing information in a form which allows others connected to the Internet to access it and remote logging into a computer and using it to run programs. For two computers in different locations to communicate it is necessary that

(i) each one must have a unique address.

(ii) messages originating from one computer must be routed to the destination via a Public Switched Telephone Network. It may be local, national or international PSTNs.

(iii) There must be software in each computer to format the messages using commonly agreed rules so that the messages are properly re-routed and interpreted. This commonly agreed set of rules is called the *Internet Protocol* (IP).

The unique address required by a computer in order to access the Internet is called its IP address. The IP address is a 4 byte address and is expressed in what is known as the dotted decimal format, e.g., 202.42.128.3. The IP address for a business or an individual is provided by the Internet Service Provider (ISP). IP addresses are an important and scarce resource particularly because the number of computers connected to the Internet is rapidly growing. The IP address is converted into a string of characters for ease of remembering and grouped into *domains*. For example, in the address:iisc.ernet.in, the top domain is the country name abbreviated *in*, the ISP is *ernet* which is the host of *iisc*. IP addresses are controlled by an international authority and a hierarchical organization of addresses allows this authority to decentralize assigning of addresses. For example, *ernet* is given a range of IP addresses and it allocates a set of addresses to *iisc* which in turn allocates addresses to various departmental servers. The clients connected to the departmental servers are then given their unique address by the department.

The Internet breaks up a message sent from a source to a destination into a number of *packets*. A packet consists of two parts, the part containing the information which is called the *payload* and a part called the *header* (see Figure 17.3) which is used to route the packet. The header will consist of the source and destination addresses, the serial number of the packet, error detection bits and other control bits. Messages are broken into

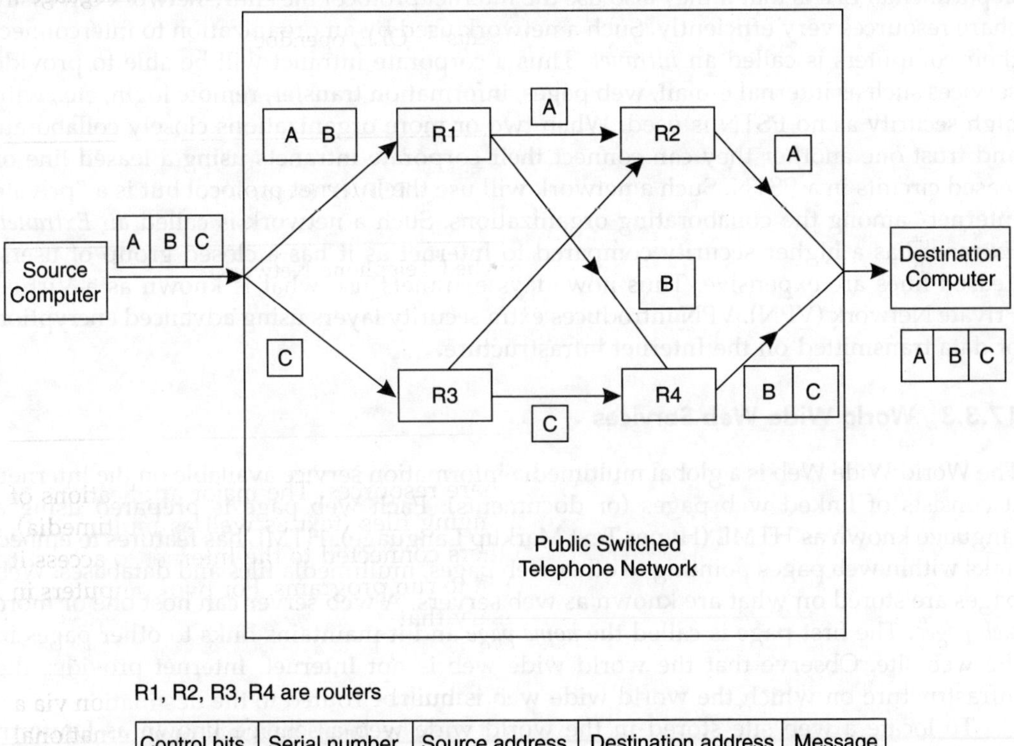

Figure 17.3 Packet communication used in Internet.

packets as it reduces the cost of transmission and improves the reliability of transmission. The cost is reduced as a number of packets (may be belonging to different messages) can be assembled and sent along any free communication channel. The packets are stored in routers along the path and forwarded to another router when the communication link is free. This is called *packet switching*. It is also more fault-tolerant because if a line is not working the stored packet can be sent along another line which may be working. Observe that different packets belonging to a message may travel along different paths. They are finally assembled at the destination using the serial number of each packet. The major disadvantage of packet switching is that the time taken for a message to reach a destination cannot be predicted. This is not a disadvantage for applications such as e-mail, file transfer, etc., but is a disadvantage for real-time messages such as voice conversations and video transmissions. The internet protocol has now been improved to allow real-time data transmission also. The routing of data as packets in the Internet is illustrated in Figure 17.3. One of the major advantages of Internet is its ability to connect computers from any manufacturer and LANs using different technologies together into a uniform access system by enforcing that the computers use a software layer conforming to the Internet protocol known as TCP/IP (Transmission Control Protocol/Internet Protocol). It was realized by many organizations with a large number of computers and departmental LANs that if they also use the Internet protocol the entire network can easily share resources very efficiently. Such a network used by an organization to interconnect their computers is called an *intranet*. Thus a corporate intranet will be able to provide services such as internal e-mail, web pages, information transfer, remote login, etc., with high security as no PSTN is used. When two or more organizations closely collaborate and trust one another they can connect their corporate intranets using a leased line or leased circuits in a PSTN. Such a network will use the Internet protocol but is a "private internet" among the collaborating organizations. Such a network is called an *Extranet*. Extranet has a higher security compared to Internet as it has a closed group of users. Leased lines are expensive. Thus nowadays extranets use what is known as a Virtual Private Network (VPN). VPN introduces extra security layers using advanced encryption of data transmitted on the Internet infrastructure.

17.3.3 World Wide Web Services

The World Wide Web is a global multimedia information service available on the Internet. It consists of linked web pages (or documents). Each web page is prepared using a language known as HTML (Hyper Text Markup Language). HTML has features to embed links within web pages pointing to other web pages, multimedia files and databases. Web pages are stored on what are known as web servers. A web server can host one or more *web pages*. The first page is called the *home page* and it maintains links to other pages in the web site. Observe that the world wide web is not Internet. Internet provides the infrastructure on which the world wide web is built.

To locate a web site stored in the world wide web a scheme known as Uniform Resource Locator (URL) is used. An example of a URL is given below:

http://www.freesoft.com

In this example http specifies the protocol to be used. In this case it is hypertext transfer protocol. This is the protocol used for web search and leads to the home page stored in the world wide web server of freesoft.com. There are other protocols used in the Internet for other services. For example ftp:// is used for transferring files from one computer to another connected to the Internet. ftp stands for *file transfer protocol*. For example

ftp://freesoft.com/<filename>

will transfer the contents of the specified file to your computer if you have access permission from the server. www.freesoft.com preceded by :// is the address (called domain name) of a computer (called a server) which is permanently connected to the Internet. The computer may be located anywhere in the world.

The information on a web page can be retrieved by a customer (or user) using a *browser* program which runs on his/her desktop computer connected to the Internet. There are many browsers, the most popular of which are Netscape, chrome and Internet Explorer. The URL is entered in the location field of the browser screen. The browser program connects to the specified web server, interprets the html markups and displays the document on the browser screen. Browsers have excellent Graphical User Interface (GUI) which simplifies access to web pages to anyone. Most organizations maintain a web site on a server in their organization or on a server which is rented by a service agency. Hosting of organization's web pages has now become an important business. These businesses keep a large number of powerful servers on their network with reliable connection to the Internet. They create the web pages of the organization based on specifications given by them and continuously update them on request from the contracting organizations. Web presence is now essential for any business as it publicises its activity. Besides organizations, individuals also create and maintain web page to "sell themselves." In Figure 17.4, we give the structure of a typical web site which has a home page and links to other local or remote web pages.

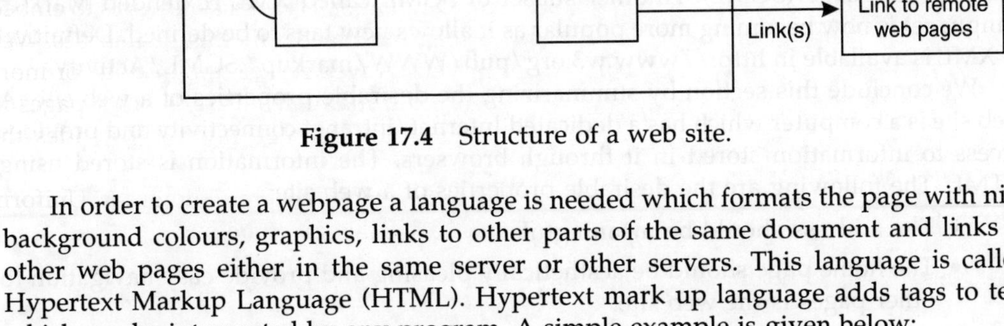

Figure 17.4 Structure of a web site.

In order to create a webpage a language is needed which formats the page with nice background colours, graphics, links to other parts of the same document and links to other web pages either in the same server or other servers. This language is called Hypertext Markup Language (HTML). Hypertext mark up language adds tags to text which can be interpreted by any program. A simple example is given below:

```
<HTML>
<HEAD>
<TITLE>          </TITLE>
</HEAD>
<BODY>
<H1> Analysis and Design of Information Systems </H1>
<P> This is <B> third edition </B> of <I> Rajaraman's </I> book. </P>
</BODY>
</HTML>
```

will display the following:

> **Analysis and Design of Information Systems**
> This is **third edition** of *Rajaraman's* book.

Observe the various commands introduced with the text. Some of them are:

- Document delimiters such as <HTML>, <HEAD>, <TITLE> and <BODY>
- Section heading <H1>. More levels are available
- Paragraph and other spacing commands such as <P> above
- Character attributes such as <B> for bold face, <I> for italics
- Graphic images to be displayed with the document
- Listing using bullets or sequence numbers
- "Anchor" commands which specify text or images that can be clicked on to another HTML document either in the same server or another server.

Observe that when you design a web page you can pick words and tag them with anchor commands. When these words are clicked the tag activates a link to the specified page, graphics file, audio or video file. As the use of inter and intranets is increasing most documents are now created using HTML format. Standard word processor outputs can be converted to HTML format using tools. There are also specialized tools available to create web pages.

HTML is a subset of a much larger standard language known as Standard Generalized Markup Language (SGML). Another subset of SGML called XML (Extended Markup Language) is now becoming more popular as it allows new tags to be defined. Definition of XML is available in http://www.w3.org/pub/WWW/markup/.SGML/Activity.

We conclude this section by summarizing the desirable properties of a web site. A web site is a computer which has a dedicated Internet/intranet connectivity and provides access to information stored in it through browsers. The information is stored using HTML. The following are the desirable properties of a web site:

- The address should be meaningful
- The home page should be aesthetically pleasing and provide easy navigation to other pages is the web site.
- It should contain meaningful and current information

- The scope and coverage of information in the site should be clearly specified
- It must have good help and search features
- It must have links to related sites
- It must be easy to navigate using the links
- It should provide facilities for users to give their comments.

The following web site give good guidelines for authoring web sites.

- www.w3.org/provider/style

The number of web sites in the world wide web runs into tens of millions and is continuously growing. Information in the web is poorly structured and contains very useful information along with poor quality unauthenticated information. Finding relevant documents is not easy. There are many tools known as *search engines* which aid users in their search. These engines (which are actually search programs) receive a user query, systematically explore the web to locate documents, evaluate their relevance and return a rank ordered list of documents to the user. Among well known search engines are Yahoo (www.yahoo.com), Bing (www.bing.com) by Microsoft and Google (www.google.com). There are so many search engines being developed and used that there is a website which gives the latest news on search engines (www.searchenginewatch.com). One method of finding an e-store is to use a search engine to locate it. For example, if you want to find an e-store selling electronic goods you may use a search engine such as Google and type in: e-stores India electronic goods. This will retrieve a large number of sites.

17.4 ELECTRONIC DATA INTERCHANGE

As was pointed out earlier in Business to Business e-Commerce all documents such as purchase orders, invoices, shipping notes are exchanged electronically. There are no paper forms. To enable electronic exchange of documents a standard has to be agreed upon. Electronic Data Interchange (EDI) translates information in printed or hand written forms to standard electronic formats and transmits them. When such "electronic forms" are received they are to be interpreted correctly by the recipient's computer program and used. Observe that there is no manual intervention. The major advantage is elimination of paper forms and re-entry of data at multiple points. This reduces cost and increases reliability.

Let us now look at the steps a company X should follow to establish an EDI partnership with company Y (see Figure 17.5).

1. The first step is to agree on a standard format for commonly used documents such as request for quotation, response, purchase order, invoice, payment advice, etc. The formatting information should include headings of various fields, their length, character types to be used, their ordering, units of quantity fields, etc. Remember that company X may have to deal with many other companies and it is desirable to have a universal standard for forms. Attempts have been made to evolve such a standard. Currently two standards are available. They are ANSI X.12 standard proposed by the American National Standards Institute and EDIFACT (Electronic Data Interchange For Administration, Commerce and Transport) standard proposed

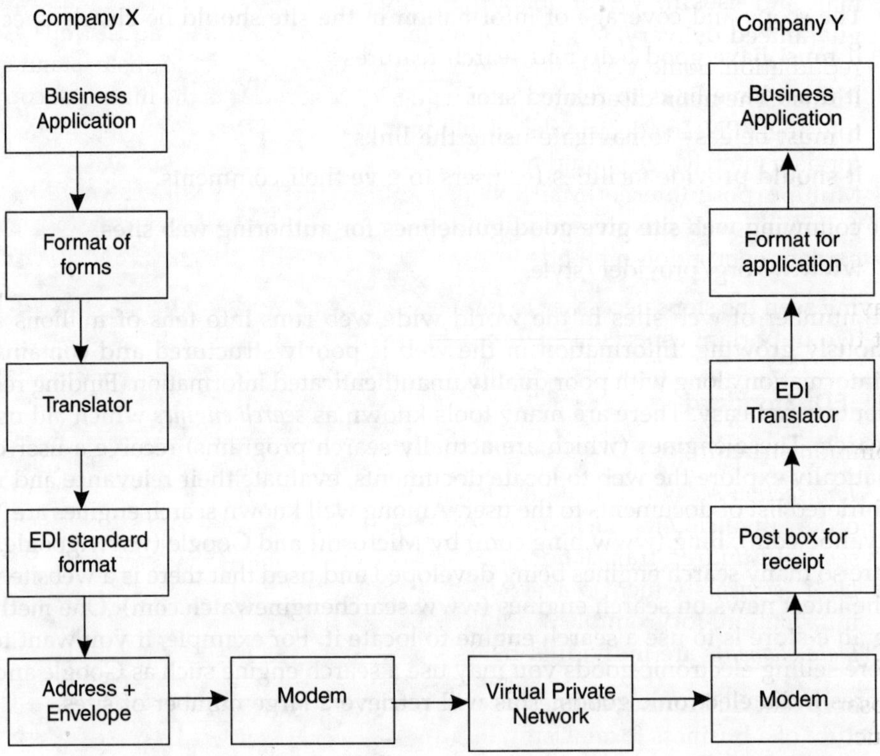

Figure 17.5 Steps to be followed for Electronic Date Interchange.

by the United Nations Economic Commission for Europe. The EDIFACT standard is an international standard and may ultimately become the commonly accepted standard for international trade. It is the standard accepted by the Govt. of India.

2. Having agreed on the EDI format standard, programs should be written to translate data received in EDI format to a form acceptable by the application program using this input. Similarly the sender should translate documents created by application program to EDI format before transmitting it to the intended receiver. Many organizations use XML to format forms.
3. The next decision is how the data is to be sent/received between companies. A virtual private network (VPN) using the Internet is the preferred method nowadays. The important points are to ensure reliable, guaranteed and secure receipt of electronic documents by the intended receiver, acknowledgement of receipt, non repudiation (i.e., a sender after sending a document cannot later maintain that he never sent the document), and tracing transactions (later when required). Private networks called Value Added Network services (VAN) offer their networks to provide EDI services at a cost. VANs provide post boxes for each of its subscribers who want to use their services. A sender addresses a form to be sent to a receiver and deposits it in an out box with VAN. The VAN service software takes this and delivers it to the addressee. The addressee picks up the

mail and uses it. VANs operate 24 hours a day, 7 days a week and provide guaranteed delivery, security, provides acknowledgement, audit trails and non repudiation. Some VANs may even provide conversion of forms to standard EDI format. The main disadvantage of using VAN services is the high cost compared to using VPN. If a decision is taken to use the VPN then an appropriate standard for data interchange should be used. The recommended standard is called Secure/ Multipurpose Internet Mail Extensions abbreviated S/MIME. S/MIME uses a combination of private and public key encryption and digital signature. We will explain encryption methods in the next section.

Having seen the steps necessary to implement EDI between companies we will briefly look at two important aspects of EDI implementation.

17.4.1 EDI Standard

The common structure of EDI messages contain:

1. *Transaction set* which is equivalent to a business document such as a purchase order, invoice, etc. Each transaction set has several data segments. There are several hundred transaction sets defined in the standard.
2. *Data segments* which is a group of data elements that together convey some information, for example a purchase order line, company name and address.
3. *Data elements* are individual fields such as price of an item, quantity, etc.

Observe that EDI format defines a serial ordering of data elements corresponding to 2D structure of a business form. Using EDI standard is expensive for Small and Medium Businesses (SMBs). Thus the current trend is to use XML based business forms.

17.4.2 e-Mail based Exchange of EDI

This is simple and widely supported. One would usually use Security enhanced Multipurpose Internet Mail Extension (S/MIME) to enclose EDI data within an e-mail messages. The cooperating companies (or organizations) which agree to trade using EDI should establish the following:

- The e-mail addresses for sending/receiving EDI messages. Addresses for other communications related to EDI transfer.
- Method of encrypting messages, digital signatures and acknowledgement of received mail (we will discuss encryption and digital signatures in the next section).
- Which EDI standard format will be used (ANSI or EDIFACT) or XML is to be used.

Instead of e-mail two organizations may decide to use ftp (file transfer protocol) to exchange EDI messages in which case there must be an agreement on

- Directory and file naming conventions.
- File encryption protocols and keys.
- ftp log-in names, machine names and passwords.
- EDI message format.

For large data sets ftp would be preferable.

17.5 SECURITY IN e-COMMERCE

As e-Commerce normally uses PSTN and is often between entities who are not known to one another, a number of security issues become important. These are:

1. Transactions take place from a computer (connected to a company's LAN) to another company or customer using the Internet. As the Internet connects several networks one has to make sure that an unauthorized person does not gain access to the company's confidential information. Both hardware and software solutions are needed to ensure this.
2. The communication between companies should be protected from snoopers.
3. When a company receives a message it should be able to make sure from whom it has come. In a written communication there is a signature and a company seal which authenticates the letter. An electronic mail communication should have a Digital Signature so that the receiver knows that it is from an authorized business. It should also ensure that the authentication of a digital signature must be maintainable in a court of law in case of disputes.

In this section we will discuss these three issues.

17.5.1 Network Security using a Firewall

A common method of protecting an organization's LAN from unauthorized access is to place what is known as a *firewall* between the LAN and the external network (see Figure 17.6). A firewall is a computer or a router which controls and monitors all traffic between the LAN and the Internet. A good firewall should allow employees of an organization to have full access to approved Internet services, while granting access to the organization's LAN only to specified outsiders.

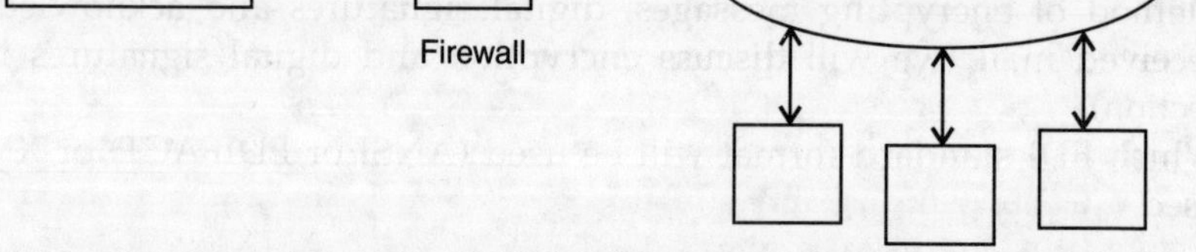

Figure 17.6 Firewall to protect organization's computers.

The simplest firewall is what is known as a *packet screening* and filtering router. The router is a hardware device which can be programmed to allow access to the network only from a set of computers connected to the Internet. These computers are identified by their IP address (i.e., Internet address). Other screening rules may be to allow access only to one IP address in the LAN which may host a specified web page.

Another type of protection is called a *proxy application gateway* which is a special server program which runs on the firewall machine or other internal computers which have full Internet access. It is primarily used for web browsing and obtaining other Internet services to members of an organization. The proxy gateway would allow employees of an organization to access only designated addresses outside, limit use of certain browsers and disallow use of some protocols with known security holes.

At a higher security level a company may use what is known as a *hardened firewall host* which is a stripped down computer in which proxy application gateway programs are installed. It will require both inside or outside users to connect to the trusted applications in it before connecting to any machine. This type of firewall is configured to prevent unauthenticated interactive log-ins from the *external world.* The major advantages of a hardened firewall host are

- Security processes are concentrated in one machine
- Names of systems on LAN, e-mail addresses, etc. are hidden from outsiders
- Network service management is simplified by locating services such as ftp, e-mail, search engines, etc., in the firewall machine.

Depending on the level of security required one may opt for one of the above types of firewall.

17.5.2 Data Encryption with Secret Keys

When data travels over public networks (PSTN) there is a danger that an unauthorized person may capture it and use it causing harm to the sender/receiver. Thus one should encode (or scramble) the data before sending it on a public network so that even if an outsider is able to read it he will not be able to understand it and use it. One should also take precautions to prevent unauthorized persons from accessing a database. If, by some means, he is able to access it, the data stored should be in an encoded, i.e., scrambled form so that he cannot read and use it to harm the organization. For example, sensitive databases are those containing credit card numbers, passwords, financial data, etc. Encoding or scrambling data to make it difficult to decode is called *encryption.* Encryption is a transformation of a data in any form (text, audio, video, graphics) into another form (called a cryptogram or ciphertext) which cannot be understood. In order to understand the data one needs a *key* which is used to decrypt the message. Messages to be encrypted are also known as *plain text* and encrypted messages are known as *cryptograms* or *ciphertext.*

All ciphertexts are obtained by applying two types of transformations on the plain text. These are transpositions and substitutions. Transposition and substitutions are perfomed using keys. Let us take an example of how encryption is carried out. Consider a plain text:

THIS IS A MESSAGE

which is to be encrypted. It is first transformed using transposition. The transposition rule is applied to blocks of characters. For example, a transposition key (41253) can be interpreted as:

1. Write the message in blocks of 5 characters. For the above plain text after blocking we get:

   THISI SAMES SAGE

2. Replace the first character in the block by the 4th character, second character by first character, third character by the second character, fourth by the fifth character and the fifth by the third character. After applying this transposition we get:

   STHII ESASM SAGE

   (Observe that the block smaller than 5 characters is left as it is). The next transformation is substitution. The substitution rule with key = 4 may replace each character by a character which is 4 letters away from it in the alphabet. For example A will be replaced by E, B by F and finally Z by D (by assuming that alphabets are wrapped around). If the substitution transformation is applied, the message will become

   WXLMM IWEWQ WEKI

The *algorithm* on how the encryption is done can be publicized without compromising the encrypted message as a person not knowing the transposition and substitution keys will not be able to easily decode the message. This general method of encryption is illustrated in Figure 17.7.

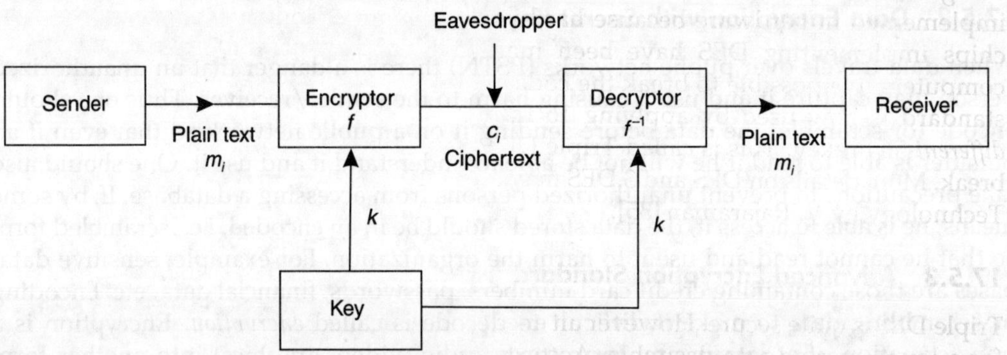

Figure 17.7 Use of secret key for encryption.

We can formalize the system as follows:

Let

$$\text{plain text} = (m_1, m_2, m_3, \ldots, m_n)$$

$$\text{cipher text} = (c_1, c_2, c_3, \ldots, c_n)$$

where $c_i = K(T_i\,(m_I))$

T_i is transposition of the ith block and K is substitution.

This general idea is used in a very popular encryption method called the Data Encryption Standard (DES) introduced by IBM in 1975 and standardized by U.S. Government in 1977. It was reasonably secure, i.e., trying out all possible keys exhaustively to break the code will take too long even using the most powerful computer of 1977. DES has been modified to what is known as Triple DES to ensure security with today's (2011) faster computers. DES applies transformations on blocks of 64 bits corresponding to binary encoding (may be ASCII) of message text. The plain text is exclusive ORed with the key to obtain the ciphertext. If the key is exclusive ORed with the ciphertext we get back the original plain text as shown below:

M = Plain text	01101100	11011000	11011010
K = Key	10101111	00101100	01011011
$E = M \oplus K =$	11000011	11110100	10000001 (Encryption)
$E \oplus K =$	01101100	11011000	11011010 (Decryption)

This general idea is used in DES.

DES encrypts 64 bit blocks. This is encrypted with a 56 bit key. First the 64 bits are permuted with a secret key. The resulting block is divided into two 32 bit blocks (L_i, R_i) which are the left and right half of each block. The following complex procedure is applied 16 times

$$L_{i+1} = R_i;$$

$$R_{i+1} = L_i \oplus f(R_i, K_i)$$

where K_i is the secret key used in the ith round. The resulting block is again permuted using the secret key to obtain the final encrypted block. DES was designed to be implemented in hardware because hardware implementation is fast. Integrated circuit chips implementing DES have been marketed. However with increasing speed of computers it is possible to break the code as it uses only a 56 bit key to encrypt. A new standard is now used by applying 16 rounds of DES three times, each time using a *different* 56 bit key. This is called Triple DES (abbreviated 3 DES) and is very difficult to break. More details on DES and 3 DES may be found in the book Essentials of e-commerce Technology by V. Rajaraman [35].

17.5.3 Advanced Encryption Standard (AES)

Triple DES is quite secure. However an encryption method which is expected to be secure for a longer period was desirable. Accordingly the National Institute of Standards and Technology (NIST) of U.S.A. decided to introduce a new standard by inviting cryptographers world wide to propse encryption algorithms. The conditions to be satisfied by the algorithm were:

- It must be an open algorithm disclosed to all with no payment. Security must be ensured only by the key and not the algorithm.
- It should be a symmetric key algorithm.
- It should be a block cipher encrypting blocks of at least 128 bits.
- It should be simple enough to be implemented in hardware including embedded chips in credit cards.
- It should use keys of length 128, 192 or 256 bits depending upon the extent of security required.

Many proposals were received. The one designed by two Belgian cryptographers V. Rijmen and J. Daemon was accepted and is currently known as Advanced Encryption Standard (AES). The algorithm is applied on blocks of 128, 192 or 256 bits with 128 or 256 bit encryption keys depending on the level of security required.

The encryption method we have discussed so far are called *symmetric key* or *private key* encryption as the encryption and decryption use the same key known to the two parties exchanging messages. If a business conducts transactions with 1000 other business, 1000 different keys should be distributed and the keys should be kept in a secret directory. If the directory is compromised a new set of keys should be distributed. The keys should be distributed using a different secure channel of communication, e.g., by registered post. The key distribution problem is a challenge in a symmetric key encryption method.

17.5.4 Data Encryption with Public Key

The public key cryptography allocates two keys to each organization wanting to communicate with one another. One of the keys is called a *public key* of the organization as it is available to any one wanting to send a ciphertext to that organization. The organization has another key which it uses to decrypt the ciphertext it receives to recover the plain text, this is shown in Figure 17.8. A popular public key system is known as RSA system as it was invented by three Professors Rivest, Shamir and Adleman. The procedure works as follows:

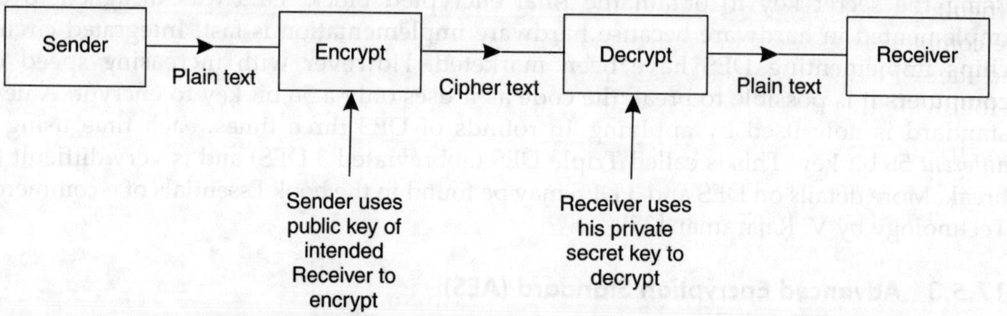

Figure 17.8 Public Key Encryption System.

1. If an individual or organization R wants to use RSA it picks two sufficiently large prime numbers p and q and finds $n = p \times q$.
2. Next $\phi = (p - 1) \times (q - 1)$ is computed.
3. A number e is picked which is relatively prime to ϕ and is declared as R's *public key*.
4. A number d is then found which satisfies the relation

$$d\ e\ mod\ (\phi) = 1.$$

   The number d is R's *private key*.
5. An organization S wishing to send a message to R encrypts a text t using R's public key, using the formula shown below

$$c = t^e\ (mod\ n)$$

   Observe that S must also know the value of n.

6. When R receives the ciphertext c it decrypts it by using the equation

$$c^d \pmod{n} = t$$

Let us take an example

1. R picks two primes $p = 41$ and $q = 43$

$$n = p \times q = 41 \times 43 = 1763$$

2. $\phi = (p - 1) \times (q - 1) = 40 \times 42 = 1680$
3. Picking a number which is relatively prime to ϕ we get

$$\text{R's public key} = e = 19$$

4. $d\ e \bmod (\phi) = 1$ gives $d = 619$

$$\text{R's private key} = 619$$

5. Let message to be sent be 1511

$$\text{ciphertext } c = (1511)^{19} \bmod 1763 = 909$$

6. Message recovery: $(909)^{619} \bmod 1763 = 1511$

The reason why this works is outside the scope of this book. One may refer to the book of Denning [30] cited in the references. The strength of this method lies in the fact that given a number n which is a product of two large primes it is difficult to factor n and get the two primes. Compared to DES the RSA encryption technique is computationally difficult. Thus for large plain texts RSA is not applied in practice. Instead RSA and 3 DES are combined as shown in Figure 17.9. The plain text is encrypted using 3 DES or AES and the secret key necessary to decrypt it is sent using RSA. There are two advantages in following this procedure. They are:

1. Encrypting plain text using 3 DES is faster
2. The secret key used in 3 DES can be unique for each plain text as it is encrypted using public key encryption and sent along with each plain text.

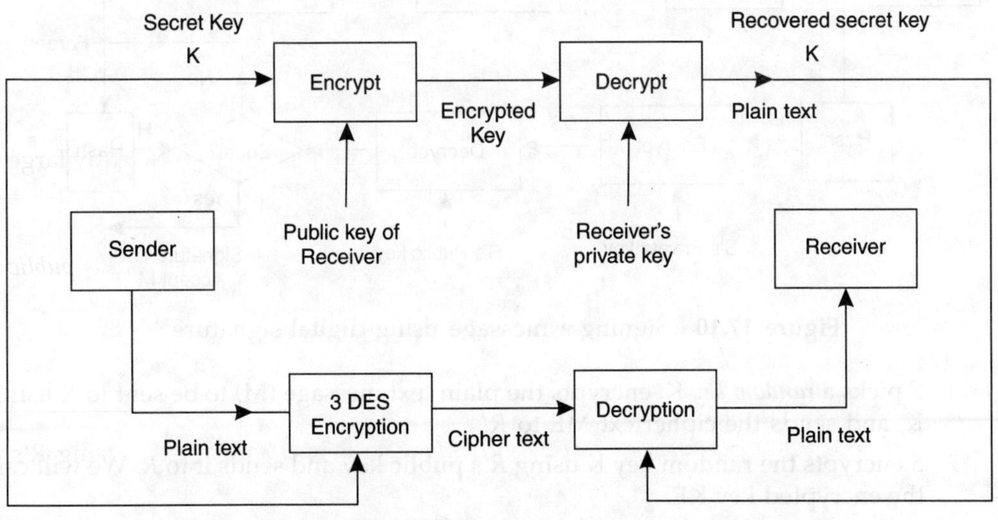

Figure 17.9 Combining public key and secret key systems.

17.5.5 Digital Signature

The RSA system is used primarily to protect from unauthorized listeners messages sent on a public network. There are two problems which may occur. They are:

1. If *S* sends a message to *R*, there is no surety that *S* indeed sent the message. If *R* has to be sure that the message was sent by *S* then *S* must sign it. Physical signatures are unique and can be verified. We need similar method of signing an e-mail message so that *S* cannot later on say that he never sent the message. In other words, *S* should not repudiate after sending, say, a purchase order to *R*.
2. Further, no one should be able to act as though he is *R* and receive messages intended for *R*. The public keys of all potential participants in e-Commerce are known. If *W* somehow is able to convince *S* that *R*'s public key is his, *S* will be sending messages intended for *R* to *W* and *W* can read it using his private key. There is thus a need for a third party to authenticate public keys of all the participants.

We will first explain how a digital signature system works (see Figure 17.10). Assume that a sender *S* wants to send a message to a receiver *R* and sign it. The following steps are carried out by S:

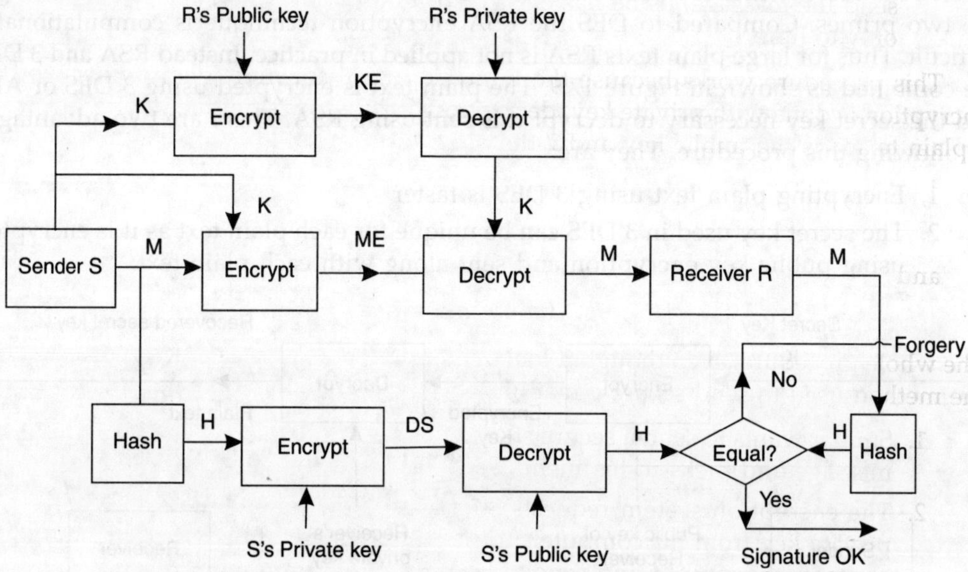

Figure 17.10 Signing a message using digital signature.

1. *S* picks a *random key* K, encrypts the plain text message (M) to be sent to *R* using K, and sends the ciphertext ME to *R*.
2. *S* encrypts the random key K using *R*'s public key and sends it to *R*. We will call the encrypted key KE.
3. *R* will be able to decrypt KE using his private key and get K.

4. Having obtained K, *R* can decrypt the ciphertext ME sent by *S* and gets M.
5. Now *R* has to be convinced that *S* sent the plain text. This can be done only if *S* signs the message. Signing of the message is done as follows:
6. S creates a unique abstract of M, namely H = *f*(M) by using a hash function *f*. The function *f* must be such that two different messages do not lead to the same hash. In other words if M1 and M2 are distict $f(M1) \neq f(M2)$. The number of bytes in H must be much smaller than the number of bytes in M. Hashing is done primarily to reduce the size of the signature. (A hashing method called MD5 (Message Digest 5) is popular). Also hashing the message M ties H to M. In other words, the signature will use H which is tied to the document being sent.
7. H is encrypted by S using his *private key* and transmitted to R. This is his digital signature DS.
8. As R already has M he can hash it using the known hash function *f* to obtain H.
9. When DS is received by R, he decrypts it using the *public key* of the sender S.
10. The decrypted value must be H. If it is not then it is a fake message. If it is H then R is convinced that it is signed by S. S cannot repudiate (i.e., say that he did not send the message) as he has encrypted H using his *private key* which is known only to him.

This procedure works because the RSA algorithm is symmetric. In other words if encryption is done with private key, decryption can be done with the public key. If *m* is a plain text, *e* is the public key and *d* the private key.

$$(m)^e \bmod n = m'$$

$$(m')^d \bmod n = m$$

and

$$(m)^d \bmod n = m''$$

$$(m'')^e \bmod n = m$$

The whole procedure, even though it looks complicated, is very easily implemented. It is the method used to send secure e-mail. It would look like the following:

1. Sundar wants to send a secure signed message to Ramu. He composes the e-mail message and picks in his menu; send mail "signed and encrypted".
2. The encryption system requests Sundar's password and uses it to retrieve his private key from his computer. It also retrieves the public key of Ramu. It now encrypts the mail M and sends ME, KE and DS to Ramu.
3. When Ramu receives the mail and tries to read it, his terminal prompts him saying it is encrypted and signed. He uses his private key and requests the system to decrypt the message. To authenticate the message he looks up Sundar's public key and requests the system for authentication which it does using the algorithm stored in it.

The second question we raised at the beginning of this section was about the authenticity of public keys. This is done by some organizations which issue public key certificates after verifying the credentials of an organization or individual. Thus if a customer A wants to do business with a merchant B electronically, B can send an e-mail to the certification authority requesting certification of A's public key, e-mail identity, etc. Once the certifying authority certifies the public key, transaction can proceed. The certification authority takes on legal responsibility in case of disputes on identity. Governments approve a number of organizations in the country as certification authority. For details a student may refer to [35].

17.6 ELECTRONIC PAYMENT SYSTEMS

Payment is an important component of e-Commerce. In normal commerce there are many modes of payment each with its own advantages and disadvantages. The most common payment especially for low value purchases is cash. Credit cards are the next preferred mode of payment by customers to merchants. If the customer is a trusted party, merchants normally accept cheques. Settlements between businesses or organizations are normally by cheque. Thus in e-Commerce also we have to be concerned with cash payments, credit card payments and cheque payments. We will explain only credit card and electronic clearance of pay orders. e-cash is not very mature and we will not discuss it.

17.6.1 Credit Card Payment

There are many competing systems for payment by credit card. We will describe one protocol called Secure Electronic Transaction (SET) protocol which consists of the following steps (see Figure 17.11),

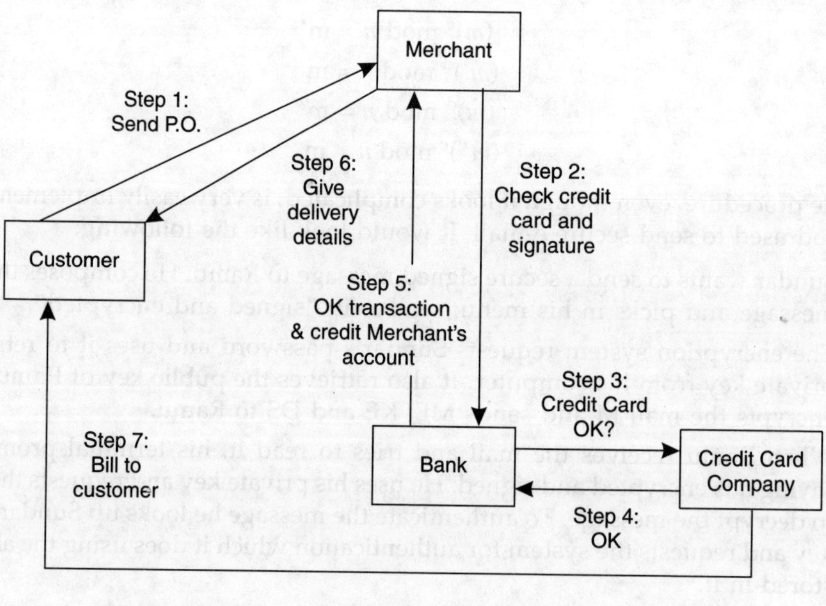

Figure 17.11 Secure Electronic Transaction Protocol.

Step 1 Customer sends a purchase order with an encrypted credit card number to the merchant. The credit card number is encrypted using the Internet's Secure Socket Layer (SSL) using https protocol.

Step 2 The merchant sends the amount of purchase with the customer's credit card number (encrypted) to the credit card company's bank. The bank's computer decrypts the credit card number, authenticates it and checks credit balance on the credit card.

Step 3 If it is OK it checks with the credit card company to see if card is OK.

Step 4 Credit card company OKs the transaction and signs it. It shows customer's transaction details in customer account.

Step 5 The bank when it receives the signed authorization from the credit card company sends a message to the merchant to honour the purchase order. The bank credits the merchants account by appropriate amount.

Step 6 The merchant accepts customer's purchase order and informs him about delivery details.

Step 7 At the end of the month the credit card company sends a consolidated bill to the customer.

17.6.2 Electronic Payment in B2B e-Commerce

This system uses the Electronic Clearing System for cheques which has been standardized by the Reserve Bank of India which acts as an Automated Clearing House (ACH). In order to use this a vendor, purchaser, their respective banks and ACH must all have certified public keys.

The steps followed in e-clearance of payment are as follows (Figure 17.12):

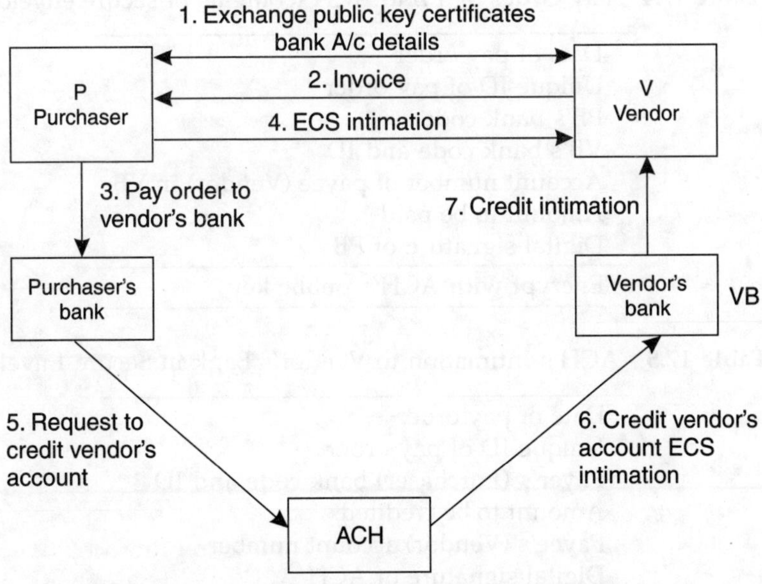

Figure 17.12 Electronic clearance of payment.

1. The purchaser and vendor exchange their public key certificates.
2. The vendor sends invoice with amount payable to the purchaser.
3. The purchaser checks invoice and authorizes payment. A pay order with details given in Table 17.3 will be sent to his or her bank PB encrypted with PB's public key.
4. The purchaser intimates the vendor that ECS payment has been initiated.
5. PB checks accounts balance of the purchaser and if OK debit his or her account and send the information given in Table 17.4 to ACH encrypted with ACH's public key. It is assumed that ACH has certified public keys of all its member banks.
6. ACH debits PB's account specified and credit VB's account. It sends information given in Table 17.5 to VB to credit the vendor's bank account.
7. VB credits the vendor's account with payment received. It intimates the vendor that ECS payment has been received and give payer's (purchaser's) ID.

Table 17.3 Pay Order to Purchaser's Bank (contents of secure envelope)

Date on which to be paid Unique ID of pay order Purchaser's account number Vendor's account number (Payee) Amount to be paid Digital signature of purchaser (Payer)
Encrypt with PB's public key

Table 17.4 Pay Order by PB to ACH (contents of secure envelope)

Date of pay order Unique ID of pay order PB's bank code VB's bank code and ID Account number of payee (Vendor) in VB Amount to be paid Digital signature of PB
Encrypt with ACH's public key

Table 17.5 ACH's Intimation to Vendor's Bank in Secure Envelope

Date of pay order Unique ID of pay order Payer's (Purchaser) bank code and ID Amount to be credited Payee's (Vendor) account number Digital signature of ACH
Encrypt with VB's public key

The encryption used is normally RSA public key encryption of a unique symmetric session key for each transaction between the two parties. The symmetric key is encrypted using Triple DES encryption or AES. For digital signature MD5 hashing is commonly used. Normally banks have their own intranet (using either VPN technology or leased lines) connecting all their branches. Bank intranets are connected to one another using VPN which is a cheaper alternative to leased lines. ACH is usually connected to bank intranets using leased lines.

We do not discuss electronic cash payment in this book. Those interested may refer to the book by V. Rajaraman on e-Commerce [35].

17.7 CONCLUSIONS

We have seen in this chapter that with the emergence of world wide telecommunication networks and the Internet, a big change is taking place in the way organizations do business. Methods of sending purchase orders, payment for goods and services, security and confidentially are all being changed due to the emergence of e-Commerce. A system analyst designing an information system for an organization should now be aware of the procedure adopted in e-Commerce. We have thus given a broad overview of this important topic. Many of the procedures and standards are still emerging and not yet standardized. It is however clear that e-Commerce will have a profound impact on information systems design and implementation. Another important trend is e-commerce using mobile devices such as mobile phones, laptops, tablets etc. There are over 500 million mobile subscribers in India (2010) and it is growing at approximately 8 million per month. Mobile e-commerce (abbreviated m-commerce) is growing rapidly. The student should thus keep track of developments. The World Wide Web is a good source of getting the latest information and trends.

SUMMARY

1. The sharing of business information, maintaining business relationships and conducting business transactions by using telecommunication networks is usually defined as Electronic Commerce.
2. e-Commerce is normally categorized as Business to Business (B2B), Business to Customer (B2C) and Customer to Customer (C2C).
3. The major advantages of e-Commerce are anytime, anywhere transaction, reduction in cost of transactions, reduction in time to market products, faster inter-business transactions and faster transfer of funds.
4. The major disadvantages of e-Commerce are poor security of transactions unless special precautions are taken, loss of privacy, lack of legislation to settle international disputes and menace of hackers.
5. e-Commerce architecture consists of the following layers:

   The lowest layer in the physical network which may be a LAN connected by unshielded twisted pair wires, Public Switched Telephone Networks, WAN using optical fibre, etc. The next layer is the logical network such as Internet, intranet and extranet (all of which use TCP/IP protocol). Over this is the world wide web and services such as web pages, browsers, and search engines. Above this is the security layer which deals with encryption, digital signatures, etc. On top of this are Electronic Data Interchange and Electronic payment services. All these layers are necessary to write application systems.

6. At the physical level most organizations use local area networks using unshielded twisted pair cables and Ethernet protocol. The cheapest method of interconnecting organizations is to use Public Switched Telephone Network.
7. Internet is the most important logical network which enables e-Commerce. The Internet protocol called TCP/IP is universally used. Organizational private networks which use TCP/IP protocol are called intranets. Organizational intranets are often connected by a (virtual) private secure communication link. Such a network is called an Extranet.
8. World wide web is a global multimedia information service available on the Internet. It supports web pages .
9. Web pages are located using a scheme known as Universal Resource Locator (URL) which is its address. Browsers are used to locate web pages.
10. Web pages are created using a language known as hypertext markup language (HTML). Words can be picked and tagged to connect the page with other related pages.
11. It is imperative for every organization to have a web site consisting of a home page and related web pages in today's Internet world to publicize their functions. The pages must be attractively designed and updated regularly.
12. Electronic Data Interchange (EDI) is essential in e-Commerce. EDI replaces printed forms by a standard electronic format which can be interpreted correctly by computer programs of cooperating businesses.
13. EDI standards have been published by American National Standards Institute and by United Nations Economic Commission for Europe. The United Nations standard is called EDIFACT and may become the standard for international commerce. Small and medium businesses use XML based EDI.
14. Value Added Network Services provide electronic post boxes to their clients to exchange EDI documents. They guarantee security and delivery of documents. They also provide services to convert an organization's forms to standard format such as EDIFACT.
15. If Internet is used for exchanging business documents Secure Multipurpose Internet Mail Extension (S/MIME) standard is recommended.
16. An organization's intranet is connected to the Internet via a proxy server or a hardware unit. This is called a firewall. Firewall protects an organization's computers from unauthorized intruders.
17. Messages exchanged between organizations using the Internet can be easily tapped by eavesdroppers. It is thus necessary to scramble them to prevent eavesdroppers from understanding the messages. It is done by encrypting messages.

18. Message (plain text) is encrypted by transposing characters of the plain text by a specified permutation and substituting characters by other characters. The encrypted text is called ciphertext.
19. This general idea is used in a standard encryption method called Digital Encryption Standard(DES). DES encrypts 64 bit blocks with a 56 bit key. Currently a system called Triple DES (3 DES) which applies DES with three different 56 bit keys is used as it is more secure.
20. A system in which the encrypting and decrypting keys are same is called a symmetric key system.
21. A new encryption standard called Advanced Encryption Standard (AES) has now been standardized for symmetric key encryption. It uses 128 or 256 bit encryption keys on blocks of 128, 192 or 256 bits.
22. The main problem with a symmetric key system is the need to distribute the key securely to all participating businesses. Symmetric key encryption/decryption is fast.
23. Two-key based system called RSA system does not require distributing secret keys. It has two keys for each participant in the communication, a private key and a public key. If A wants to send a message to B, A encrypts the message using B's public key. B decrypts it using his private key. Thus there is no key distribution problem. It is, however, slower than the symmetric key system.
24. RSA system is based on the fact that it is difficult to factor the two prime components from their product, particularly, when the prime numbers are large.
25. In RSA system, a message encrypted with a private key, can be decrypted with the corresponding public key. This is used in digital signature.
26. In order to sign a message the sender hashes the message with a known algorithm to get a message digest MD. MD is encrypted with the sender's private key and sent to the intended receiver. Let us call it MD_e. The message itself is encrypted with a symmetric key and sent. The recipient decrypts the message and computes the message digest MD using the known hashing algorithm. He then decrypts the encrypted message digest MD_e using the sender's public key. If (MD_e) decrypted = MD then the message is not a forgery as only the sender knows his private key. This signature ties the signature to the message and cannot be repudiated by the sender.
27. To ensure that public keys of organizations do belong to them there are certification authorities which check the legitimacy of organizations and issue public key certificates.
28. In e-Commerce payments are primarily made as credit card payments or cheque payments.
29. Credit card payments are made using a protocol called Secure Electronic Transaction (SET protocol). It uses RSA system and digital signatures.
30. Cheque payments are made between organizations using electronic clearing service and public key certificates issued by a certifying authority.

EXERCISES

17.1 Define e-Commerce. What are the different types of e-Commerce?

17.2 Explain B2B e-Commerce using an example of a book distributor who stocks a large number of books, which he distributes via a large network of booksellers. Assume that the distributor has stocks of books of a large number of publishers and booksellers order books as and when their stock is low. Distributors give 1 month's time to booksellers for payment.

17.3 Explain B2C e-Commerce of a customer reserving airline tickets from his home or place of work.

17.4 Explain C2C e-Commerce with an appropriate example.

17.5 List the advantages and disadvantages of e-Commerce.

17.6 Explain the system architecture of e-Commerce by looking at it as a set of layers with the physical network at the bottom layer and applications at the top layer.

17.7 Define Internet. Why is Internet important in e-Commerce?

17.8 What do you understand by an IP address? How is an IP address represented numerically?

17.9 What is a domain name? How are e-mail addresses assigned?

17.10 What is packet switching? What are the advantages of packet switching? Why is packet switching used in Internet communication?

17.11 What is an intranet? Give an example of an intranet.

17.12 What is an extranet? Give an example of an extranet.

17.13 What is the world wide web? In what way is it different from the Internet?

17.14 What is the role of HTML in web page design? Give a HTML code to display

Introduction to e-Commerce
This is a **new book**

17.15 What is a URL? Explain the various parts of a URL?

17.16 What do you understand by EDI? Is EDI used in B2C or B2B e-Commerce? Why is EDI important in e-Commerce?

17.17 What are two major EDI standards used in e-Commerce? Which is the standard accepted for Government transactions in India?

17.18 What is VAN? What services do VANs provide? What are the advantages and disadvantages of VAN?

17.19 If Internet is to be used for EDI which mail standard is used?

17.20 If e-mail is to be used to exchange EDI between two businesses what are the points on which they should agree?

17.21 Why is security important in e-Commerce? What are the security issues to be taken into account while designing a security system for e-Commerce?

17.22 What is a firewall? What are the functions of a firewall?

17.23 What is packet screening? Which hardware device performs packet screening?

17.24 What is a proxy application gateway? What are the functions of this gateway?

17.25 What is a hardened firewall host? What are its functions? In what way is it different from proxy application gateway?

17.26 Given a plain text:

THIS IS A SAMPLE SENTENCE FOR ENCRYPTION.

Apply the permutation (231564) and the substitution: (letter → letter + 6) and obtain the cipher text.

17.27 What is DES? Explain what DES does when the following hexadecimal plain text is input to a DES hardware.

A1907FBCD986543201FED14E890ABCA5

17.28 What do you understand by symmetric key cryptography? What are the main advantages and disadvantages of symmetric key cryptography?

17.29 What is AES? Compare it with DES.

17.30 What is public key encryption? In what way is it different from private key encryption? Why is it important in e-Commerce?

17.31 What are the main differences between DES based encryption and RSA based encryption? Is it possible to combine these two systems? If so explain how?

17.32 Given two prime numbers 23 and 41 design a RSA system. Explain with an example how it works.

17.33 What is a digital signature? Why is it necessary in e-Commerce? What are the necessary conditions a hash function used in digital signature should satisfy?

17.34 Give a block diagram of a system for transmitting a signed purchase order from business 1 to business 2.

17.35 What is a public key certifying authority? Why is a certifying authority required in e-Commerce? How does a certifying authority performs its tasks?

17.36 What types of electronic payment systems are required in e-Commerce? Why are there different types of payment systems? Explain the necessary characteristics of each type of payment system and give an example each of where it is used.

17.37 Explain SET protocol used in credit card transactions. What is the main interesting aspect of SET protocol which gives confidence to customers transacting business using the Internet?

17.38 In using SET protocol who has to keep a database of public keys of all customers? How is the customer assured that he will not be double charged for the same item purchased?

17.39 What are the main differences between electronic cheque payment and credit card payment in e-Commerce? Explain cheque transaction protocol used in e-Commerce.

18

System Design Example

LEARNING GOALS

After reading this chapter a student should be able to:

1. Analyse a fuzzy description of a problem and obtain a clear specification of requirements to be met by a computer-based system to solve the problem.
2. Integrate various tools learnt in earlier chapters to develop an information system.
3. Prepare documentation of a computer-based information system.

In this chapter we will analyze a case study of designing a system. Also, we will integrate various methods we learnt in this text so far. We will then show how to document the system.

18.1 A SYSTEM FOR JOURNAL ACQUISITION

National Centre for Science Information (NCSI) subscribes to 300 journals to supplement the journals bought by the main library of the state. It is desired to automate the journal acquisition and subscription system.

We will follow the steps given in Table 18.1 and document the system as suggested in this table. The first step in design is fact gathering. The organizational chart of NCSI is simple and is shown in Figure 18.1. The Chairman of NCSI determines the operational policies in consultation with an advisory committee. The day-to-day operations of the NCSI are supervised by a Senior Scientific Officer. The person in charge of journal acquisition and control is a scientific assistant who knows computing. A programmer is assigned to work with the scientific assistant to computerize the system.

The persons interviewed are the Chairman, the Senior Scientific Officer and the scientific assistant. The main questions asked were:

- Why is a computer-based system being considered?
- What are the basic objectives of the proposed system?
- What is the volume of data to be processed?

Table 18.1 Steps to Follow in the Design and Documentation of an Information System.

1. **Fact Gathering and Requirements Specification**
   - Hierarchy chart of organization
   - List of persons to be interviewed
   - Questions to be asked
   - Records of discussions
   - Feedback documents
   - Outputs of interview
     - Document flow diagram
     - Outputs which will be produced by the system
     - Rough data flow diagram
     - Requirements to be met
     - Prioritized refined objectives
2. **Feasibility Analysis**
   - Refined quantified goals
   - Resource needs—People; machines
   - Cost—Rough estimate of hardware and software costs
   - Benefits—Quantified and qualitative
   - Time estimate for system implementation
3. **System Specifications**
   - Data flow diagram
   - Structured specifications of procedures
   - Broad specification of data needs
   - Hardware specification
4. **System Design**
   - Database design
   - Data encoding for input
   - Input menu (if any)
   - Editing procedures
   - Control and audit procedures
   - Test plan
   - Program structure chart
   - Output design
   - Screen design
   - Procedure parameterization
   - Procedure description (refined)
5. **Implementation Plan**
   - Users' manual
   - System manual
   - Parallel runs
   - Users' feedback
   - Review

- What is the frequency of processing?
- What are the benefits expected from the system?

Interview with the Chairman and the Senior Scientific Officer led to the following observations:

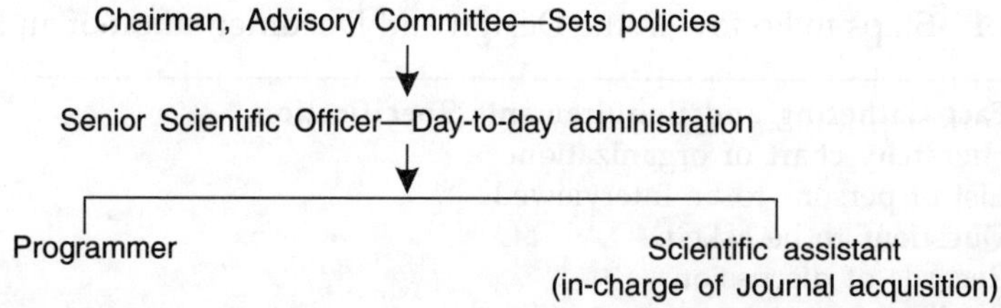

Figure 18.1 Organizational chart of National Centre for Science Information.

About 300 journals are subscribed by the Centre, with periodicity varying from half yearly to monthly. Most journals are quarterly. The journals mostly come from abroad by surface mail and take approximately two months to reach NCSI from the date of despatch. Not all journals are published on time and delay in publication is noticed. Often journals are lost in the mail. If a journal is lost and a request for replacement is sent promptly, the publishers normally send a duplicate copy free of charge. If a request for replacement is not sent promptly, then it would be difficult to get a replacement copy free or even by paying its cost. Most journals are renewed in October as a volume begins with its first number in January. Subscription renewal has to be done promptly to avoid missing issues due to delayed payment. As foreign exchange payment is involved, renewal and payment take time and advance action is required. Currently two scientific assistants are responsible for journal procurement and follow-up and they handle approximately 50 issues per week. It is found from experience that some mistakes are made due to a large number of journals coming in a week. It is necessary to have systematic processing which has now become feasible with the advent of inexpensive Personal Computers (PCs) and thus it is felt that a PC-based solution would be cost effective and beneficial.

The basic objectives of the system are:

- To follow up regularly receipt of journals, promptly detect delay in receipt of an issue and send a reminder to the journal publisher.
- To request the publisher to send invoice for journal subscription well ahead of time.
- To renew subscriptions to journals promptly.
- To print out each week a list of journals received and display it so that users know of the arrival of journals.
- To answer queries about the status of subscription, receipts, missing issues, etc., of any journal in the collection.

The volume of data to be processed is about 50 journals received per week. Thus the number of transactions to be entered are 50 per week. It is a very low volume application. Besides regular entry of 50 transactions per week, there may be a dozen queries per week. Per year 300 journal renewal subscriptions are paid. This, however, gets bunched in a month or two, typically in October and November of each year. The main benefits are:

- Prompt follow-up with publisher to get replacement of missing issues.
- Prompt subscription renewal, thereby avoiding subscription cancellation and delay in receipt of journals.

- Ability to answer queries regarding subscriptions quickly.
- Ability to easily analyze budget for journals and find yearly cost, allocation of subscription in different fields of science and trends in expenses.

18.2 DOCUMENT AND DATA FLOW DIAGRAMS

The document flow diagram is shown in Figure 18.2. The outputs which will be produced are:

- List of journals received each week.
- Reminder to publishers for journals not received and for invoices not received.
- Journal renewal with subscription to publishers.
- Indents to purchase section for journal procurement.
- List of completed volumes of journals to library for binding.

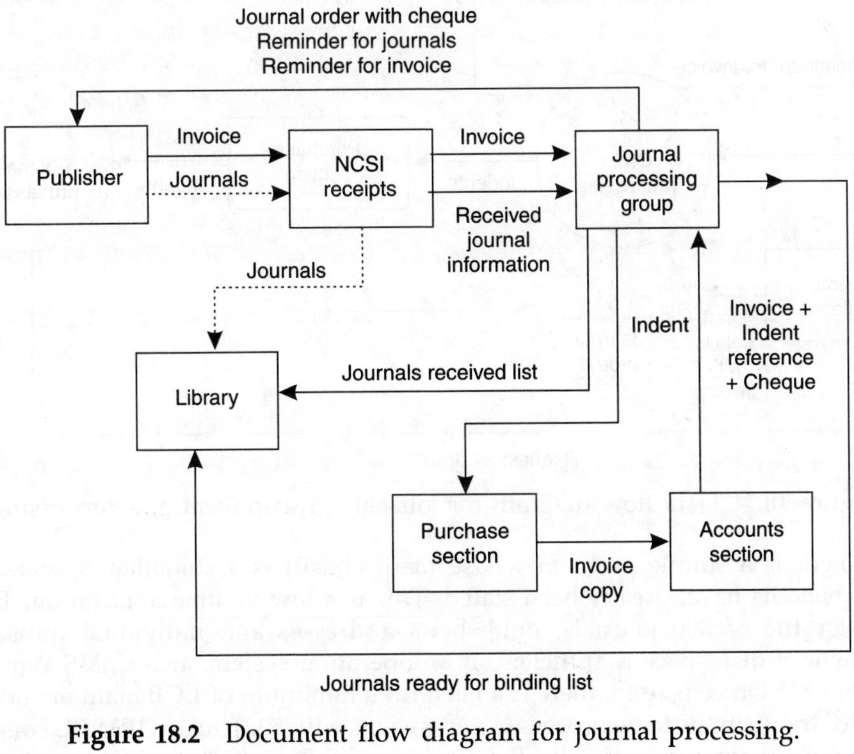

Figure 18.2 Document flow diagram for journal processing.

The rough data flow diagram is shown in Figure 18.3. The requirements to be met and objectives of the system have been stated earlier.

18.3 FEASIBILITY OF THE SYSTEM

The next step in system development is feasibility study. Computerizing journal

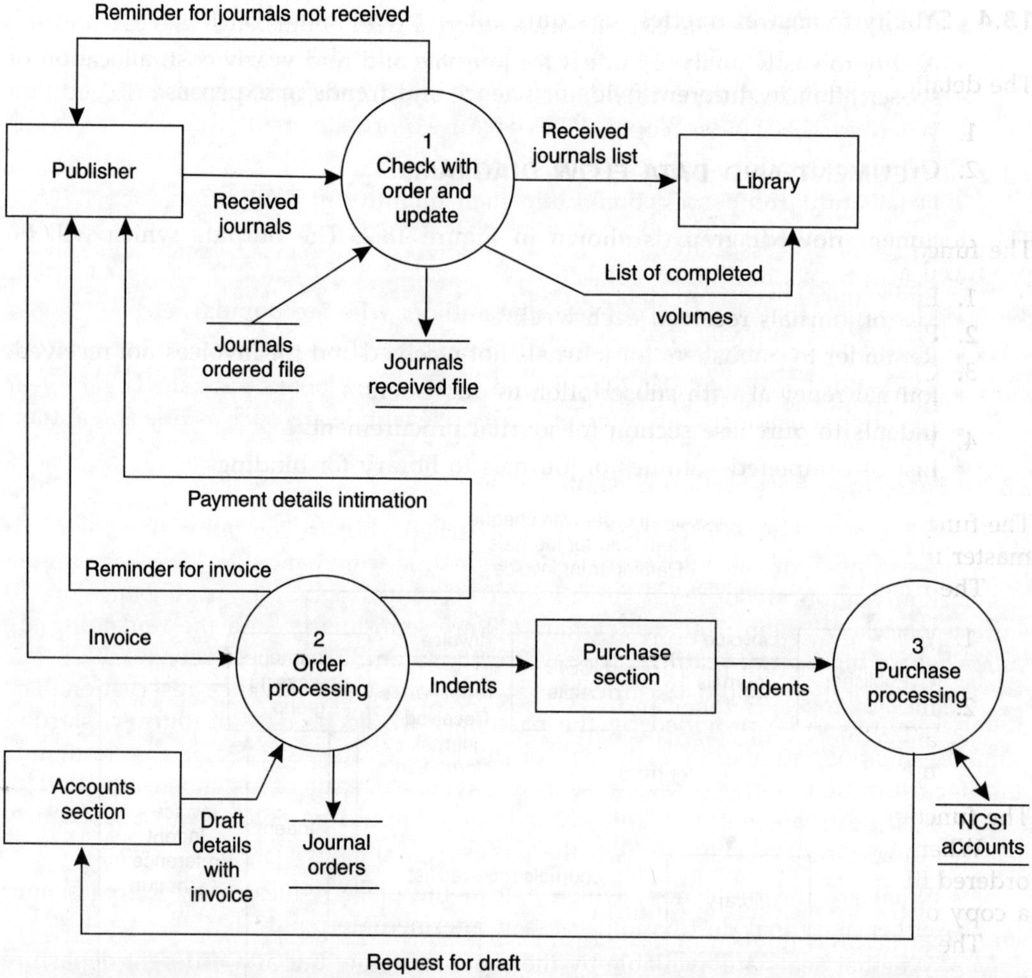

Figure 18.3 Data flow diagram for journals procurement and processing.

procurement is a simple problem whose main benefit is a smoother operation. The detailed benefits have already been stated. This is a low volume application. To store the master file of 300 journals, publisher's addresses and individual transactions, 1 Megabyte of disk space is sufficient. If an operating system such as MS Windows 7 along with MS Office is used, there is a need for a minimum of 1 GB main memory. The number of transactions to be processed per week is only 50. Thus an IBM PC compatible with 1 GB main memory and 320 GB disk space and 80 cps printer is adequate for this application. The cost of such a system is around ₹ 30,000. Such a system can also be used for other applications such as word processing and keeping track of the books in the NCSI library.

The system can be implemented by a programmer within six months. This time includes codification of journals, system design, implementation, testing, and trial run for a month.

18.4 SYSTEM SPECIFICATION

The detailed data flow diagram is given in Figure 18.4. The functions of Process 1 are:

1. When a journal is received, check with journal ordered master.
2. Generate details of journal received (journal identity, name, volume and issue no., month) and send details to Process 2.

The functions of Process 2 are:

1. Enter journal received detail in journals received file.
2. Print list of journals received during a week and send to the main library.
3. Check list of journals expected from journal master with the journals received file. Send the list of journals expected and not received to Process 3.
4. When all issues of a journal expected during an year have been received, send the information to journal binding section of the library.

The function of Process 3 is: using the journals not received list, use journal publisher master file to generate reminders to the publishers and request replacement.

The functions of Process 4 are:

1. When an invoice is received from a publisher, generate an indent. Send it to the purchase section for further action.
2. In September, October and November check using journal master file whether an invoice is received from the publisher for journal due for renewal in that month. If not, send request for invoice to the publisher.

The function of Process 5 is as follows:

When a bank draft with invoice copy is received from accounts, update journal ordered file, enter payment information in payment detail file, and mail bank draft with a copy of the invoice to the publisher.

The structured English description of each process is now given:

Procedure for Process 1

for each journal received **do**
  Input journal identification code, volume and issue no.
if journal id code is in journal database **then**
  Retrieve journal details from database and send details to Process 2;
else
  Give message "Journal not in journal database. Check identification code entered";
endfor

Procedure for Process 2

for each journal details sent by Process 1 **do**
  Enter details in journal received file;
  Print details of journals received in weekly list;
endfor
  Send list of journals received during week to the library.

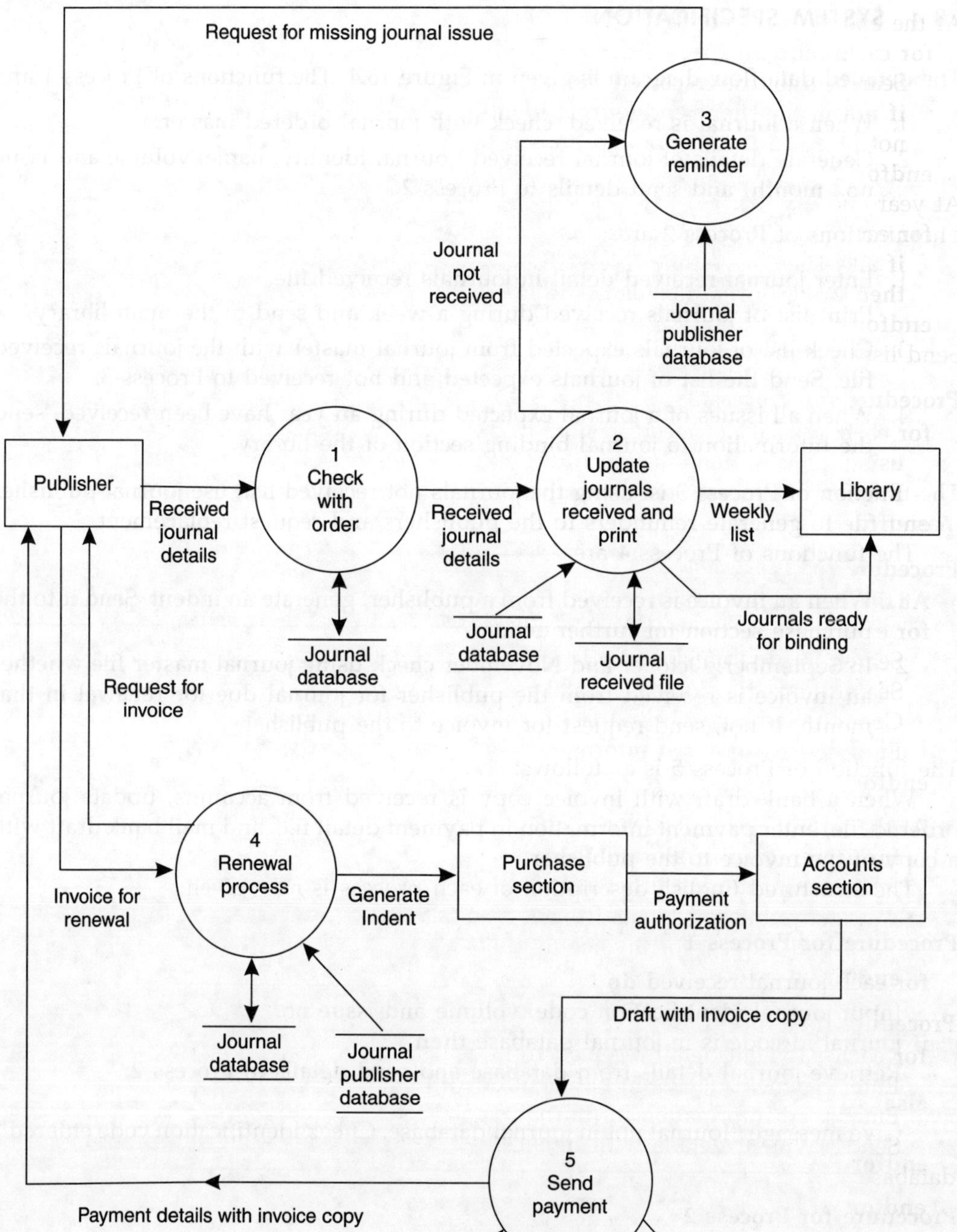

Figure 18.4 Detailed data flow diagram.

```
At the end of each month
  for each journal expected during month as found in journal database do
    Search in journal received file;
    if not in file then send journal identification, volume no., and
    not received transactions to Process 3;
  endfor
At year end
  for each journal in journal received file do
    if all issues of volume received
    then print journal details in binding list
  endfor
Send list to library.
```

Procedure for Process 3

```
  for each journal not received note from Process 2 do
    using journal identification find publisher details from Journal-
    Publisher database; print letter requesting replacement to publisher;
  endfor
```

Procedure for Process 4

```
  At the end of the month
  for each invoice received from publisher do
    Search Journal database and find journal details;
    Search Journal—Publisher database to find publisher details;
    Generate indent and send to purchase section;
    Enter journal ordered information in Journal database;
  endfor

  At the end of September and October
    for each record in Journal database do
      if journal not yet ordered then
        get publisher details from Journal—Publisher database;
        send request for invoice to publisher;
    endfor
```

Procedure for Process 5

```
  for each payment detail received from accounts section with invoice copy do
    Enter payment details in payment detail file;
    Enter paid information in Journal database;
    Send payment details with invoice to publisher using data in Journal-Publisher
database;
  endfor
```

18.5 DATABASE DESIGN

(i) **Data about journals.** Journals identification (id.), Journal code (ISSN), journal name, vol. no., periodicity, month of first number, parts per number, subject code, language, publisher identification, time allowed for delivery, year of entry.

(ii) Data about subscription details. Journal identification, date of invoice, yearly subscription, currency of subscription, date when subscription sent, subscriber code, amount in rupees, subscription amount in foreign currency, invoice no., order no., draft no., draft date, bank name.

(iii) Data about publisher/agent supplying journal. Publisher identification, publisher's name, address, agent or publisher, id of journals published.

The entity-relationship diagram for the application is shown in Figure 18.5

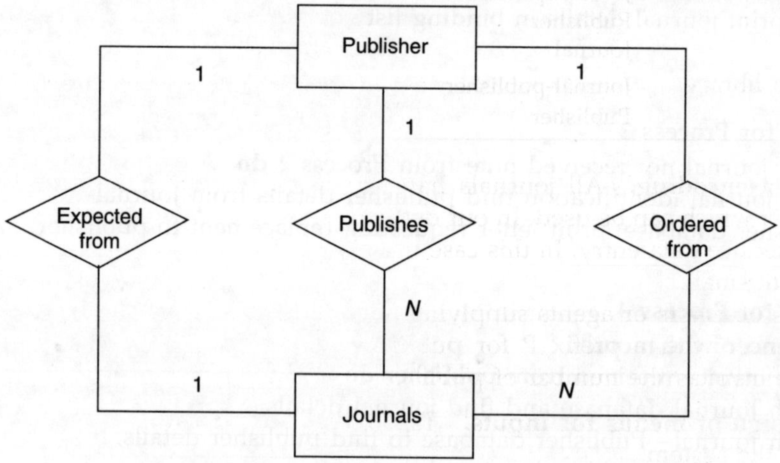

Figure 18.5 E-R diagram for database.

The publisher-journal relationship is one-to-many. One publisher may publish many journals, but a journal is published by only one publisher. The normalized relations are given below. (Key fields are shown in italics).

Journal: *Journal id,* journal name, journal ISSN, subject code, language, key words, numbers per volume, parts per volume, month of first number.

Publisher/Agent: *Publisher id.,* publisher's name, address, agent or publisher, publisher type.

Journal-publisher relation: *Journal id., publisher id.*

Journal expected: *Journal id., vol. no.,* first no. month, periodicity, expected delay.

Journals received: *Journal id., vol. no.* of issues in volume, no. of issues received, last issue no. received, year.

Journal payment details: *Journal id., vol.,* order no., order date, publisher's invoice no., invoice date, foreign currency amount, rupee amount, draft no., draft date, bank name.

After designing the database, we revert back to the detailed data flow diagram of Figure 18.4 to specify the relations needed by each process in the DFD. We give this as Table 18.2.

Table 18.2 Relations Needed for Processes in the Data Flow Diagram

Process	*Relations to be Read*	*Relations to be Written*
1.	Journal expected	
2.	Journal Journal expected	Journal received
3.	Journal-publisher Publisher	
4.	Journal-publisher Publisher Journal	
5.	Journal-publisher Publisher	Journal payment details Journal expected

(iv) Data encoding. All journals have a standard international code known as ISSN which can be used. In our design a simple serial number is used as a code to reduce data entry. In this case this is acceptable as the number of journals is quite small.

Publishers or agents supplying journals are also encoded as a simple serial number with a prefix P for publisher and A for agent. The simple code is sufficient as the number of publishers is very small.

(v) Design of menus for inputs. Tables 18.3–18.7 give various menus designed for the system.

18.6 CONTROL, AUDIT AND TEST PLAN

As the volume of transaction is low in this example, no extensive control and audit procedures are needed. The program should check whether duplicate copies of an issue of a journal have been received by mail, and whether the invoice amount and order amounts match. The total of order amounts and draft paid amounts should also be compared.

Table 18.3 Menu for Picking Processing Needed

AUTOMATED SERIALS CONTROL SYSTEM
NATIONAL CENTRE FOR SCIENCE INFORMATION

MAIN MENU

A> ORDER PROCESSING
B> JOURNAL MAINTENANCE
C> PUBLISHER DATABASE MAINTENANCE
D> QUERY
E> EXIT
ENTER YOUR OPTION

Table 18.4 Menu for Order Processing

MENU A: ORDER PROCESSING

1 ADD/DELETE JOURNAL
2 REQUEST FOR INVOICE OR REMINDER FOR INVOICE
3 INPUT INVOICE
4 PRINT INDENT LIST
5 ENTER PAYMENT DETAILS
6 PRINT PAYMENT DETAILS
7 EXIT
ENTER YOUR OPTION

Table 18.5 Menu for Journal Maintenance

MENU B: JOURNAL DATABASE MAINTENANCE

1 UPDATE JOURNAL INFORMATION
2 ENTER JOURNAL RECEIVED DETAILS
3 PRINT REMINDERS
4 EXIT
ENTER OPTION AND PRESS RETURN

Table 18.6 Menu for Maintaining Publisher Details

MENU C: MAINTAINING PUBLISHER DATABASE

1 ADD PUBLISHER NAME
2 DELETE PUBLISHER NAME
3 MODIFY PUBLISHER DETAILS
4 EXIT
ENTER OPTION AND PRESS RETURN

Table 18.7 Menu for Enquiry or Printing

MENU D: ENQUIRY OR PRINTING

ENQUIRY	PRINT OPTION
DISPLAY	
A RECEIVED JOURNALS	1 LIST JOURNAL YEARWISE
B LIST JOURNALS BY KEYWORD	2 LIST JOURNAL ALPHABETICALLY
	3 LIST NEW JOURNALS
	4 LIST JOURNAL BY LANGUAGE
	5 LIST JOURNAL BY KEYWORD
	6 LIST JOURNAL BY SUBJECT
	7 PUBLISHER LIST
	8 MAILING LIST
	9 BINDING LIST

ENTER YOUR OPTION AND PRESS ESc

Test Plan. We use the process description to evolve a test plan.

1. Enter a wrong journal code and see the response.
2. Enter a correct journal code and see the response.
3. Enter wrong journal volume and see the response.
4. Enter a few journal codes expected in a month and see the response.
5. Create a journal not received transaction and see the result of processing.
6. Try giving year-end message and see the effect.
7. Create payment details transaction and enter to see the effect.

The input screens themselves must have prompts and messages for wrong data entry.

18.7 IMPLEMENTATION PLAN

The steps to be followed to implement the system are:

1. Codify journals being subscribed.
2. Create a master database of journals with details.
3. Codify publishers and agents who supply journals.
4. Create a master database of publishers with addresses.
5. Create journal id., publisher id., database
6. Based on knowledge of delays in delivery, enter in journal expected relation the expected delay.
7. Create a sample set of weekly, monthly and year-end transactions.
8. Create transactions corresponding to receiving draft from accounts for subscription payment.
9. Design screens for data input and query answering.
10. Test each process using sample data.
11. Integrate all processes and databases.
12. Create user manuals.
13. Create systems manuals.
14. Design user's feedback forms and a system review plan.

(i) User's manual. One page instructions is prepared on how to use the system. As the system is interactive with screens and prompts, no extensive user manual is needed.

(ii) Systems manual. This contains all the documentation so far stated, namely, document flow diagram, data flow diagram, E-R diagram, screens, output, etc.

The complete program listing with comments in the program is part of the system documentation. This documentation is essential for the person in-charge to maintain the system.

(iii) User's feedback form. This form should be designed to obtain user's opinion on ease of use, ease of data input, quality of screens and system response time. A user may be asked to rate each of these as excellent, good, fair or poor.

(iv) Review plan. The system is to be reviewed at the end of one year.

18.8 CONCLUSIONS

In this text we have studied various tools used to design a computer-based information system. The tools are primarily intended to assist in system design. The system design process itself is quite involved and requires, besides the knowledge of tools, sound commonsense and judgement to decide the level of detail to which the design effort should proceed. One can learn a lot through experience. The main trait of a good systems designer is the ability to design the system to accommodate change and be receptive to users' needs.

SYSTEM DESIGN EXERCISES

These are exercises for the student to design small systems following the guidelines given in this chapter. The examples are briefly stated. Make suitable realistic assumptions without oversimplifying your model. Your solution should include, as far as possible, all points specified in Table 18.1.

18.1 Book Publisher's Assistant

This system attempts to help a book publisher in maintaining details about his books, customers, receipts/payments, etc. A book may be published afresh or it could be a new edition or just a reprint. The copyright may belong to either the author(s) or the publisher. The publisher may reprint any number of copies if he holds the copyright. If the author holds the copyright, he may publish only as much as is asked for by the author. A publisher is expected to pay royalty on books sold each year.

Various book sellers, libraries, institutions, etc., order books. If the required number of copies are available the publisher sends the books and updates his stock level. If they are not available, he may send a partial consignment provided it is acceptable to the customer. Payments may be either by cash/credit, subject to suitable credit limits. The publisher may decide to reprint a popular book if he holds the copyright. Otherwise he approaches the author for consent. The author may also revise it and the publisher brings out a new edition.

Assume that the publisher has published roughly 10,000 books of various titles in various subjects and that roughly 100 new books are published every year and that 100 new editions are also brought out each year. Assume that 200 reprints of various books are also published. The minimum number of copies published is 1000. On an average about 5000 copies of each book are published.

Your system should provide the following information:

- A list of various customers, libraries, etc., placing orders has to be maintained, with relevant details.

- For each customer the list of books ordered with details such as cost, address of customer, date of order, date of delivery, qty. delivered, etc.
- For each book relevant details such as author, date of publication, price, edition no., copyright holder, copies allowed for export, etc.
- Details of reprints, new editions of books.
- Details of current stock, books and their quantity in publication, etc.
- Books published by an author.
- List of books in a given subject.
- The books that get sold out fast, the authors who are popular, etc.
- Details of pending payments by customers, pending payments to authors, etc.
- The publisher would also like to send a list of new books published every year to various customers/libraries, etc., with relevant details such as title and author.

18.2 Passenger Reservation System

Assume that ABC Bus Corporation has approached you for computerizing their bus reservation system for various destinations originating from Bangalore. The Corporation has currently about 1000 buses spread over 60 routes operating from Bangalore to various places. Of them, 700 are regular, 200 are semi-luxury, and the remaining are super deluxe buses. The seating capacity are 48, 42 and 36, respectively.

The Corporation would like to have at least one week advance reservation. The details required to be provided by the passenger would be date of journey, starting point, destination, number of seats (half or full), concession required, if any, route no., and departure time of the service. If seats are available, seats of his choice are allotted. However, if seats are not available, your system should be able to provide enquiry services in a variety of ways to help the passenger take a decision.

- What are the route nos., which halt at the place requested by the passenger either as a final destination or as an intermediate point?
- Whether seats are available by the same route no. but at a different departure time.
- Whether seats are available by a different route no. on the same day.
- The earliest date on which seats are available for a place by a given route no.
- The cost of travel by various categories of buses (namely, ordinary, deluxe, etc.) to a destination by various routes.

The above are some of the questions that could be asked by the customer.

Two seats are reserved for VIPs and/or emergency quota. However, if nobody turns up for the same, upto half an hour before the departure, the seats are allotted to those in the waiting list according to priority.

Concessions are given in respect of students during summer vacation. Thus they need to pay only 50% of the actual fare. Employees and their eligible family members may travel free once a year, by producing the appropriate identity cards. Your system must verify that the total distance travelled by such a person does not exceed 3000 km. Seats are not allotted for half-ticket but a seat may be allotted for two half-tickets issued.

Your system should provide a facility for cancellation. 10% of the fare shall be forfeited if the cancellation is done at least two days prior to the journey. 20% shall be forfeited if the cancellation is done one day prior to the journey but not more than two days, 25% if done 8 hours but not more than a day and 30% if done at least half an hour in advance but not more than 8 hours and 40% for less than half an hour and up to half an hour after departure of the bus. In all other cases the amount is forfeited. The reservation fee of ₹ 5 is always non-refundable. Journey can be postponed/advanced subject to availability of seats by the same route at most once at no extra cost.

Your system would have to maintain a database of various routes originating from Bangalore and the fare by each of them for various categories of buses to various destinations. Frequently the Corporation revises the fares and your system should accommodate this. Assume that more than one counter is simultaneously active for reservation/cancellation.

18.3 Electricity Billing

KKN Electricity Board is interested in computerizing its operations, of which customer billing is one. Assume for simplicity that the computerization is restricted to each office on a stand-alone basis and that too only for customer billing.

Customers are classified into four major groups: domestic, commercial, agricultural, and industrial. The following figures for Nelamangala divisional office may be taken as an Index:

Category	*Connections*
Domestic lighting	50000
Domestic AEH	10000
Domestic 3 phase	2000
Commercial single phase	10000
Commercial 3 phase	4000
Industrial LT	1000
Industrial HT	250
Agricultural	1000

Each connection has associated characteristics and rates. The charges are given below for the various connections for the consumption of electricity.

Category	*Rate* (₹)	*Minimum charges* (₹)
Domestic	1.65	50
Agricultural	0.50	25
Commercial	2.30	150
Industrial	2.20	1000

Power cut may be imposed and quotas fixed for users. For exceeding the quota, penalty is charged at twice the normal rates. The allowable quota is calculated based on the connected load and the percentage of power cut. For example, consider a domestic AEH connection of 2.7 kW. The normal unrestricted consumption is calculated as 2.5 times the connected load of 2.7 kW. Hence the unrestricted consumption per month is 2.5 * 2.7 * 30 ≃ 200 kWh or 200 units. For a power cut of 50%, the permitted quota is 50% of 200 = 100 units. If a user with the above characteristics exceeds this and uses 140 units (domestic AEH), his dues would be calculated as 100 * 1.65 + 40 * 1.65 * 2 = ₹ 297. Penalty for not paying on time shall be calculated as 10% interest charges on the balance overdue, subject to a minimum of ₹ 25.

(i) When meter does not record. If the meter is not recording, a flat rate for consumption based on the connected load and type of connections shall be made.

Category	*Units/Month*
Lighting	Connected load * 60
AEH	Connected load * 60
Commercial	Connected load * 120
Industrial LT	Connected load * 240
Industrial HT	Connected load * 300

For agricultural connections a flat rate of ₹ 50 shall be charged.

(ii) Demand charges. In the case of industrial consumers the charge is based not merely on the consumption but also on the power factor/demand. For example, if the power factor is consistently below 0.8, penalty charges will be levied. Find out more about this and see how you can incorporate this.

Based on the above the customer's bill is to be computed. The customer is permitted to pay without the bill also, as for example, loss of bill, bill not yet delivered to him, etc. The amount he pays is based on the previous month's consumption. The balance is to be adjusted next month, and interest also is to be paid, if he has paid less. The customer may pay either by cash or by cheque. This is to be noted down and in case of cheque, actual date of realization of the cheque should also be noted.

The customer's deposit should be at least three times the average of the last three months. If this is violated the deposit should be increased.

Your system should maintain details about customers, connections, receipts from customers, consumption, etc.

(iii) Fresh connection/cancellations. Fresh connections are granted throughout the year and the details should be noted down, including date of connection. Similarly, connections can be cancelled though this happens rarely. Every connection is assigned a unique number called the R.R. number. Your system should be able to answer the following queries:

- Which consumers violate quota and use more power?
- Which consumers pay always late?
- Who are the consumers whose deposit has to be increased?
- Which are the fresh connections given last month, with their numbers and details?
- Who are the consumers having overdue charges above the fixed amount?

- What percentage of receipts comes from Industrial HT?
- What percentage of power is consumed by Industrial HT?

18.4 Fixed Deposits, Shares, Debentures

Consider a public limited company that raises funds by issuing shares, debentures and also by accepting deposits from the general public/institutions, etc. The company wishes to computerize this operation. Fixed deposits could be for varying terms such as 1, 2 or 3 years with interest rates as shown in the table. It also has another scheme of deposits where the total amount is paid at maturity consisting of the principal plus the interest compounded quarterly. The following table illustrates this.

	Fixed deposits interest paid periodically			*Maturity payable interest + principal on maturity*		
Category	*Rate of interest (%)*			*Amount repayable on maturity (₹) for a deposit of ₹ 1000*		
	Period of deposit (years)			*Period of deposit (years)*		
	1	2	3	1	2	3
General public	10	11	12	1100	1232	1405
Employees	10.5	11.5	12.5	1105	1243	1424
Institutions	9.75	10.75	11.75	1097	1127	1396

Furthermore, the interest may be paid quarterly/half yearly/annually in the case of fixed deposits. Deposits may be renewed on maturity. Whenever the company announces acceptance of deposits, public may apply for the same. Deposits are accepted in multiples of ₹ 1000. The list of depositors and their details have to be properly stored. Your systems would have to

- Maintain the list of depositors' names, addresses, amount deposited, period, rates, type of deposits, etc.
- Send dividend warrants or interest periodically to the depositor if he has opted for the same.
- Send maturity notice to depositor on maturity and based on depositor's decision, renew/reinvest/repay the deposit.
- Answer who are the depositors with a deposit of ₹ 50,000 or more.

Similarly, the company also raises debentures periodically. They may either be convertible or non-convertible. Convertible debentures can be converted to equity shares on completion of specified period. Assume a rate of interest of 13%. Debentures are issued for periods of 1, 2 or 3 years. Your system would have to maintain all the details as in the case of deposits. In addition you would have the following additional problems:

- Convertible debentures can be converted to equity shares at maturity. This has to be ascertained from the depositor and action taken accordingly.

Similarly, the company also offers shares periodically. Both equity shares and preference shares may be issued. The dividend is declared at the end of the year. You would have to keep track of the dividend rates, dividend issued to the share holders, etc. Also, allotment of shares is done by lots. The prospective customers are grouped into classes based on the amount applied for. For example, a person who applies for more than 10,000 shares will be given firm allotment while those who apply in the range 5000–10,000, 1 in 5, etc. The following tabular data illustrates this:

Amount applied for	*No. of persons allotted*
No. of shares > 10,000	Firm
5000 < No. < 10,000	1 in 5
1000–5000	1 in 10
100–1000	1 in 100
< 100	1 in 1000

Use the above data as a guideline in determining allotments. Can your system handle the following?

- Different type of holders: varying trends
- Which type of holding is preferred: shares/debentures/deposits?
- Aggregate holding of an individual

18.5 Hostel Information System

Assume that your college hostel authorities have approached you for developing a computerized information system that can handle hostel room allotments and keep track of them.

You are quite familiar with the hostels. As a first step, catalogue the details such as various hostels, their capacities, types of rooms, i.e., single/double/spacious double/married students' apartments, etc. Assume that all men's hostels put together have about 800 rooms while there are about 1200 men students. Assume a suitable figure for ladies hostels also. Since the number of rooms are less than the number of students, some students have to share a room. Assume that research students are given priority, followed by M.E. and B.E. students. Married students are given married students' apartments, if available; else they fall into general pool.

Find out how many students join/leave every year in each of the college programmes. A student is allotted a room on joining as follows: If the student is married and married students' apartment is available and all others senior to him in his category have been allotted an apartment, then he is allotted an apartment; else he gets only single/double/spacious double room.

- If the student is unmarried, or he is married but no married students' apartment is available, he gets either a single/spacious double/double room, i.e., if a single room is available and all others senior to him in his category (viz., Research/ M.E./B.E) have been allotted the same, he also gets a single room; else he has to be given a spacious double room. If that also is not available then allot him a double room. Priorities based on course (Research, M.E., B.E.) and within that the date of joining is to be maintained so as to give a student a better room as and when it is available. Your system should maintain the list of hostels, rooms and details and should be able to,

- give the lists of available rooms, those occupied, those likely to become free, etc.,
- intimate students as and when their priorities increase about the new rooms which are available and handle room allotments,
- prepare a list of students who do not vacate their rooms even after the last date of vacating their rooms after their course is over, and
- allow students to change rooms within the same class either mutually or to any other vacant room of the same class.

18.6 Hotel Room Booking

This project addresses the information requirements of a hotel that allows its rooms to be rented by customers. The hotel has various rooms in the first, second and third floor. Rooms may be regular/deluxe/suite type and can be either single/double/triple. However, suite types does not have single rooms. Further, rooms may or may not be air conditioned. But all deluxe/suite types are always air conditioned. The number of rooms and booking rates are shown in the following table (Rates may change):

Rates per day (₹) and No. in each floor

	Regular				*Deluxe*				*Suite*			
	Rate	No. of rooms			Rate	No. of rooms			Rate	No. of rooms		
AC Single	150	50	30	20	250	30	40	40	–	–	–	–
Non-AC Single	100	75	–	–	–	–	–	–	–	–	–	–
AC Double	250	20	30	30	400	40	20	20	500	10	30	20
Non-AC Double	175	30	10	10	–	–	–	–	–	–	–	–
AC Triple	350	10	15	15	600	05	10	10	700	03	05	07
Non-AC Triple	275	10	05	05	–	–	–	–	–	–	–	–
Floor		G	F	S		G	F	S		G	F	S

A customer shall be charged on a per-day basis. Minimum of one day shall be charged to the customer. The number of days shall be computed based on the check-in time and check-out time. If the actual time is less than one day, it shall be considered as one day. For example, if a customer checks in at 9.00 hours on 21.2.1990 and checks out at 12 hours on 22.2.90, he shall be charged for two days, whereas if he had checked out before 9.00 hours on 22.2.90, he would be charged only for one day.

An advance of one day's charge shall be paid at the time of check-in for the type of room to be occupied. Your system should assist in enquiry. The following are some of the typical questions that could be asked:

- What are the various rooms and their details such as type/whether occupied, etc.?
- What are the rates applicable to each type of room?

- What is the status of a given room, and details of occupancy?
- What are the free rooms of a particular type such as regular/single deluxe/ suite on a particular floor?

The customer requests a room of a particular type. If available, it is allotted; else, enquire and find out the vacant rooms of other types/in same/different floor; if acceptable to the customer accept the advance and book the room. Enter customer details into the system.

Advance booking may be done on request. Customers shall provide details such as:

Name, address, type of room, date from, date to, payment details.

If rooms are available they may be allotted. Advance payment remains the same. 50% of the advance paid shall be forfeited in case of cancellation at least 8 hours in advance; else, the whole amount shall be forfeited.

When a customer checks out, his bills should be settled. In addition to the room charges there may be extras such as food and beverages, telephone, Internet services used by him, and extra beds used. Whenever the customer uses a service, it is noted down with details of his name, room no., service, date, service cost, etc.

Besides, there are three conference rooms, C1, C2, C3, which are charged on an hourly basis.

	C1	C2	C3
Charge/hour (₹)	300	400	500
Capacity	40	55	60

These can also be booked in advance. 50% of the estimated charges shall be paid as advance. 15% of the advance paid shall be forfeited in case of cancellation provided it is done at least 8 hours before; else 25% of the advance shall be forfeited. Accommodation can be extended beyond the duration for which it was booked provided there is no reservation for the next hour. In order to allow for smooth change of parties, a gap of one hour shall be provided between two reserved sessions. For example, if the room has been booked from 9.00 hours to 12.00 hours the next reservation shall be allowed starting only from 1.00 hours.

Extras may be charged such as refreshments provided to the participants in the conference, etc., apart from any telephone/fax services. The rates mentioned above may be revised from time to time. Your system should be capable of handling it.

18.7 Book Acquisition and Cataloguing

Assume that the central library is interested in computerizing its operations, of which book acquisition and cataloguing form a subsystem. Books are continuously being acquired and added to the library. Currently it has 2 lakh volumes. Books are either acquired directly from publishers or from booksellers. Periodically publishers and booksellers send a list of new books to the library. Departments also send in their suggestions. Based on this, the library sends a list to the publishers/booksellers. The publisher then sends a set of sample books for inspection. The library committee scrutinizes the books and draws a list of books to be bought. Books not approved for purchase are returned.

Assume that 10 bookstores and two publishers have been identified. A bookstore is generally a supplier of a particular publisher. However, he may also be a supplier of several publishers. For example,

Book store 1:	McGraw-Hill dealer
Book store 2:	Narosa, PHI Learning
Book store 3:	Wiley
Book stores 4, 5, etc.:	Other publishers.

Book acquisition section. Sample copies for inspection as well as books against firm orders come here. If the book has been ordered, then it is given a classification number. This number is sometimes marked by the publisher on the book itself. Otherwise it is determined by identifying the subject being dealt with by the book. Note that this system of classification is based on an international standard and one cannot give an arbitrary classification number. For example, in the Dewey decimal coding used commonly in libraries, 000 represents Computer Science. The subdivision 05 within this represents an area of Computer Science such as programming languages. Thus 000.05 would cover books in this area of Computer Science.

Next, the book is given an accession number. This is simply a serial number indicating the unique number of the book in the library. Thus two books cannot have the same accession number. The book is then catalogued. Both subjectwise and authorwise cataloguing has to be done. For multiple authors cataloguing has to be done for each author.

Every consignment received from a publisher/bookseller either as a sample or against a firm order has to be checked/verified to see that it matches the delivery slip, and the details of books received, date, publisher, order no., if any, are all noted. Similarly, books sent back after inspection should also be noted for details.

Books are released and go to the display section. They stay there for some time before they can be issued on loan. However, books can also be sent to departmental libraries based on departmental requests. If so the details are noted. Books are marked as reference or reference texts are not normally loaned. Else they may be loaned after display.

Can your system answer the following?

- List of bookstores/publishers and their details
- Books requested by depts., but no action taken yet
- Books requested by depts. and request sent to publishers
- Books not yet acquired, but samples received from publishers/bookstores
- Pending payments to publishers
- Given details of a book title/author, does the library have it?
- Books on a given subject of a given publisher brought out in the last two years
- List of acquired books and details in the last two years (subjectwise/authorwise/publisherwise)
- List of books sent to departmental libraries.

18.8 Library Issues and Receipts

Assume that the central library is interested in computerizing its operations. Issues and receipts of books/journals to members is a subsystem of this. This project addresses this problem.

The library has various classes of members such as staff, students/external members. The following table maybe used as a guideline regarding the details.

Category	*No. of Cards*	*Items which can be Borrowed*
Staff	4	B, J, C
M.E. student	3	Books against first two cards, journal/ book against third card
Research student	3	B, J, C
Any other student	3	Only books
External	2	B, J, C

B: books, J: journals, C: conference proceedings.

Assume that the library has roughly 1500 student members, 500 staff, and 2 lakh volumes in all. Further, staff tickets are always valid, while others have an expiry date marked on the card after which it has to be renewed in order to borrow material from the library. The items that can be loaned are given below. Items not mentioned in the table are not loaned.

Item	*Period*
Books	2 weeks
Journals	2 days
Conference proceedings	2 days
Reference textbooks	Overnight

The overdue charges are as follows: 25 paise for books for the first five days and 50 paise/day for the subsequent period. 50 paise is charged per day for journals/ conference proceedings. For overnight lending the overdue charge is ₹1 if returned within one hour after the due time and date; else ₹ 2 for every half-day.

(i) Classification of books. The library follows an international system for classification. Thus the coding scheme is fixed and cannot be changed arbitrarily. The classification is done subjectwise. For example, consider the number 000.05. the number 000 indicates Computer Science, and 05 a subdivision within Computer Science. Every book also has an accession number. This indicates a unique number of the book within the library, i.e., no two books can have the same accession number.

(ii) Reservation of books. Books/journals can be reserved in advance if they are already issued. The title, author, classification no. and accession number are required. When the book is returned, it will be impounded at the counter and will be available to the persons who reserved it in the order of their reservations.

Display items can also be reserved and will be issued after the display period is over. Textbooks are to be reserved between 2 to 4 p.m. only for the same day, and will be issued on a first reserved first served basis at 7.00 p.m. New arrivals shall be marked as display items and will be on display for one week.

(iii) Current mode of operation. The user picks up an item. This is issued to him against a card if that card permits it and the item may be loaned for the appropriate period. Details of the book, i.e. the classification no., accession no., and the due date are written on a slip and attached to the borrower's card and retained. The cards are all sorted on classification number, but they are divided periodically into two groups, overdue and not overdue cards. Within each group they are arranged according to classification no. Whenever a reservation is made, the corresponding card is located and a slip is inserted in the card, marking it reserved, so that it cannot be renewed.

(iv) Problems faced by the issue section

1. Given that a book/journal has the required details such as author, title, classification no., whether multiple copies are available, whether even one copy has not yet been issued and still available in the stacks? Who are all the holders of the desired book?
2. Reservation is done currently, based on accession number. If a book with multiple copies is to be reserved, all the copies have to be reserved, i.e. the person who reserves has to do so for all copies and the reservation slip has to be inserted into all the cards against which various copies have been borrowed. But the problem is that the person who reserved wants only one copy, the first one to be returned; the subsequent ones are not needed by him. But he has to reserve all because he does not know who would return first.
3. Which are the books frequently reserved? Can the library purchase more copies?
4. Who are those having books overdue by more than 2 months?
5. Who are habitually late in returning books? It is difficult to get this information currently because all fines are just serially entered in a register as and when a borrower pays fine.
6. Day-wise fines for various classes of books/journals/conference proceedings.
7. What books are held by a given user. Currently the system is based only on classification number.

18.9 Insurance

The Bangalore divisional office of the XYZ Insurance Corporation of India desires to have computerized information system and has approached you for developing the same. Assume that the Corporation offers two types of policies, namely, endowment and money-back. The endowment policy operates as follows: the policy holder pays premium until the maturity of the policy or his death, at which time the sum assured and bonus is paid to the holder or the nominee, respectively.

The money-back scheme differs in that 25% of the sum assured is paid back to the holder after expiry of every quarter of the policy term. For instance, consider a policy for 20 years and ₹ 50,000 sum assured. ₹ 12,500 would be paid back to the holder at the expiry of 5, 10 and 15 years, respectively. At the maturity of the policy the remaining ₹ 12,500 and the bonus would be payable. In case of death the remaining amount of the sum assured and bonus would be paid back to the nominee. Assume that the Bangalore division has 100,000 policy holders and 150,000 policies of which 20,000 are money-back.

(i) Endowment policy. The premium to be paid is worked out based on the age of the person proposed and the term of the policy. The sample table given below shows the annual premium payable (in ₹) for an assured sum of ₹ 1000.

The sum assured should be a multiple of ₹ 1000 and the minimum is ₹ 10,000. The minimum term is 15 years. The minimum age at entry is 18 and the maximum is 45. Premium can be paid monthly, quarterly, half-yearly, or annually. From tables for the above cases using these data as a guideline. While designing the tables keep in mind that the total premia paid should be slightly lower in case of annual mode of payments, and the highest premium should be for monthly payment.

Age at entry	18	25	30	35
Policy term				
15 years	70	72	74	76.25
20 years	52	53	54	56
25 years	41	42	42.75	44
30 years	34	34.5	35	–

(ii) Money-back policy. The premia are much higher than the endowment case. Typically they would be ₹ 10–20 more for every ₹ 1000 sum assured. The amount paid as bonus by the Corporation on maturity is currently ₹ 65 per ₹ 1000 sum assured. This amount may be revised every year by the Corporation depending on the profits. This holds for both types of policies. Policies may lapse if the holder does not pay the premia for more than two years. However, it can be revived by paying penalty charges if the period of lapse does not exceed five years.

The Corporation has divided Bangalore division into 15 zones for its convenience. Each zone has a Zonal Manager. The Corporation also has several agents to procure business. A proposer of a policy may do so only through an authorized agent. These agents report to the Zonal Manager. Every zone has a number of agents allotted, typically ranging from 10 to 20. Every agent must procure a minimum of 10 customers and a policy value (combined) of at least ₹ 1.5 lakhs in order to retain his agency. Agents get commission as follows: 25% of the premium paid by a policy holder whom he has introduced, in the first year, and 5% in every subsequent year. Your system should keep track of zones, agents, policy holders and their details. In addition to operational information, your system should handle queries such as the following:

- Which agents perform consistently well/badly?
- Which areas can be assigned to more agents and which have already too many?
- Which type of policies perform well/badly?

18.10 Hospital Automation

Consider the operation of a medium size hospital. Patients' data are stored and retrieved. Further, data about the various types of and number of wards/beds/operation theatres available and allotment of wards to patients are also maintained. The hospital authorities are interested in computerizing this information.

The hospital has three types of operation theatres: A, B and C, respectively used for major, minor and small operations, respectively. Wards are of two types: general and special. General wards have capacities of 4 and 8 beds while special wards are either single or double beds. The hospitals also has a maternity ward with 25 rooms, each of which can accommodate 2 beds. The hospital has 20 general wards of capacity 4 and 30 general wards of capacity 8. It has 20 special wards with 2 beds each and 10 special wards with single bed. Moreover, the hospital has an intensive care unit with a 5-bed capacity.

The charges are given in the table below (₹).

General		*Special*		*Maternity*
4-bed	8-bed	Single	Double	
200	150	400	300	250

Intensive care unit costs ₹ 1000/- per day.

The doctor notes down the clinical details on the patient's card, which contains the following details. The name of the patient, address, date of visit, doctor's name, etc.

If the patient has to be admitted, other details such as bed no., ward no., type, and date of admission are noted. If the patient has been recommended or desires a special ward but it is currently not available, he may be admitted in a general ward and later transferred when a bed is available.

Operation costs are not fixed and are determined on a case by case basis. When the patient is discharged, the bed charges should be computed. To this should be added costs of medicine, treatment, consultation fees, operation charges if any, and laboratory charges. Assume that another subsystem of the hospital information system gives you these details and is available.

Your system will have to compute equipment charges for those used by the patients. Whenever an equipment is used by the patient, the details are noted. Assume that the available equipment are X-ray machine, infra red radiation generation machine, CT scanner, ultrasound machine, etc. The charges are based on the number of exposures in case of X-ray at ₹ 50 per exposure. The others are based on the number of hours of usage, at ₹ 100, 150 and 200, respectively.

All the above costs are to be added and presented in the final bill. Your system should keep track of patients' data, history, ward/equipment usage to calculate other costs, ward/bed allotted or otherwise, and operation theatre used.

The following problems are encountered by the hospital:

- Currently the hospital is not able to keep track of the free/allotted wards/beds and their details due to frequent shifting of the patients, difficulty in tracking both wardwise and patientwise.
- Patient data, especially history, is also difficult to extract.

Your system should be able to provide the following information:

- Details of a patient such as when a patient was admitted, his or her charges, facilities used, date of discharge, etc.
- State of beds/wards, free/occupied, etc.

Appendix

NOTATIONS USED FOR ENTITY-RELATIONSHIP DIAGRAMS

In this book we used a modified Chen's notation for drawing E-R diagrams. There are two other notations which are in common use. One of them is called *Crow's feet notation* which is popular in U.K. and those who use Oracle database tools. The other is close to *Universal Modelling Language* (UML) used in object modelling. We will briefly explain them.

Crow's Feet Notation

The following is used in Crow's Feet modelling:

Notation	***Meaning***
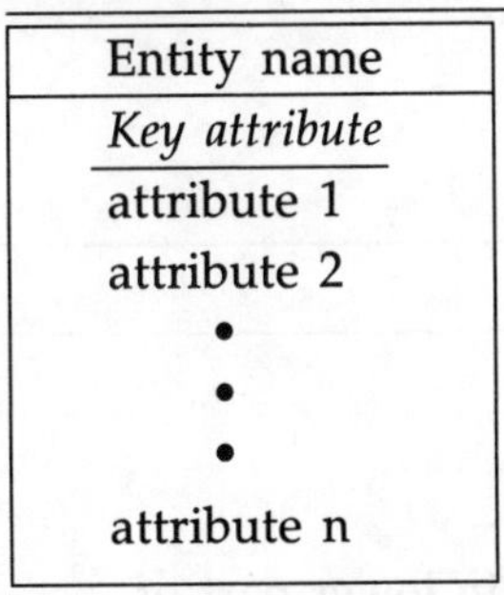	Entity Attributes listed in lower part of rectangle
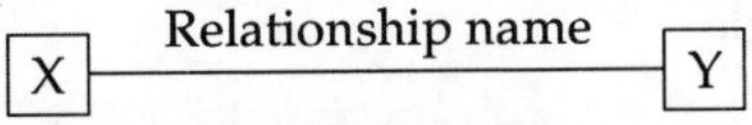	One-to-one relationship.
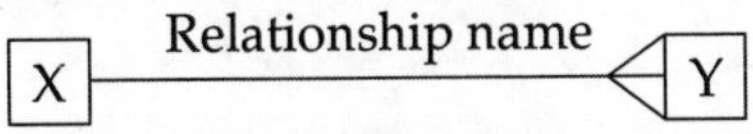	One-to-many relationship.
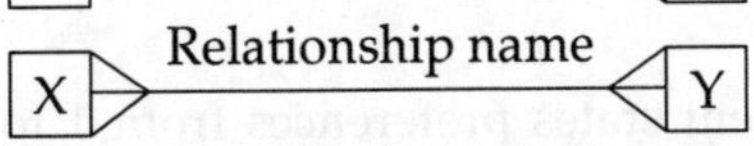	Many-to-many relationship.
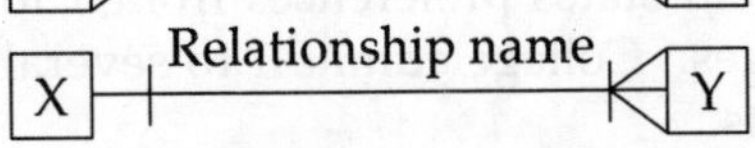	One-to-one relationship with mandatory participation for both entities X and Y.
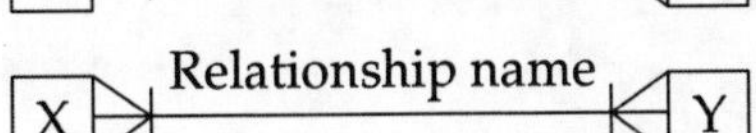	Many-to-many relationship with mandatory participation for both entities X and Y.

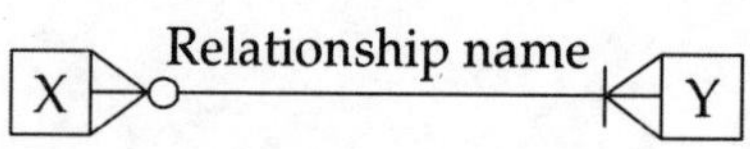

One-to-many relationship with optional participation of entity X and mandatory participation of entity Y.

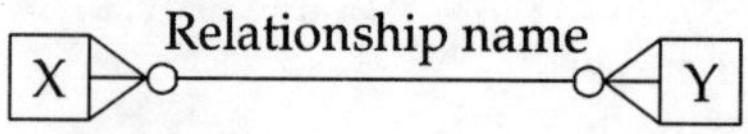

Many-to-many relationship with optional participation for both entities X and Y.

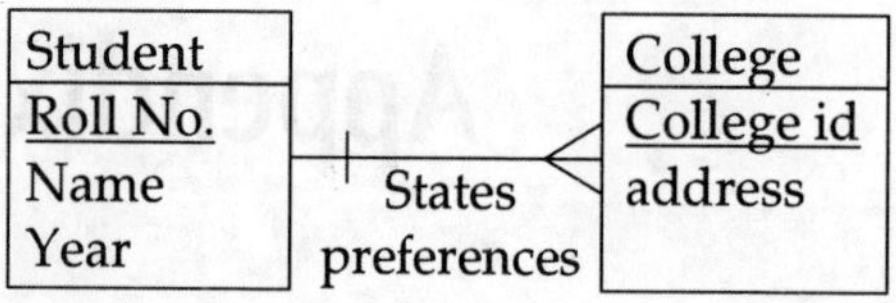

Each student must state preference for 1 to several colleges.

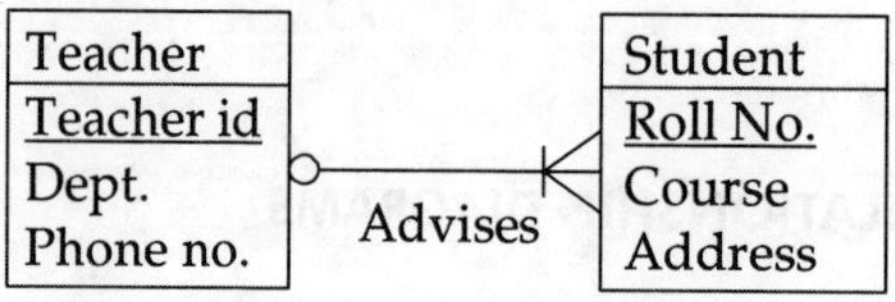

A teacher advises zero to several students. Each student must have one advisor.

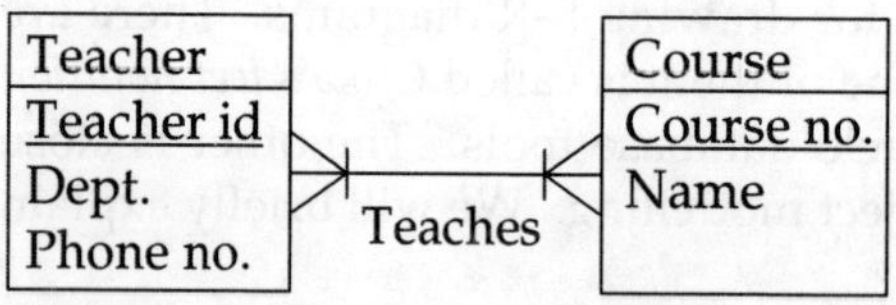

A teacher has to teach at least 1 to several courses. A course is taught by several teachers.

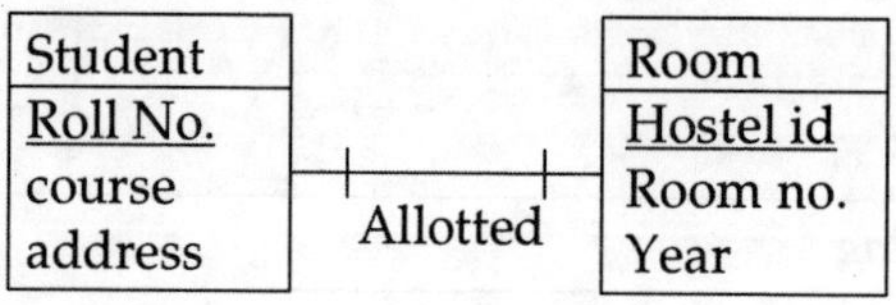

One student is allotted one room. Each room must be allotted to a student.

UML NOTATION

Notation	*Meaning*
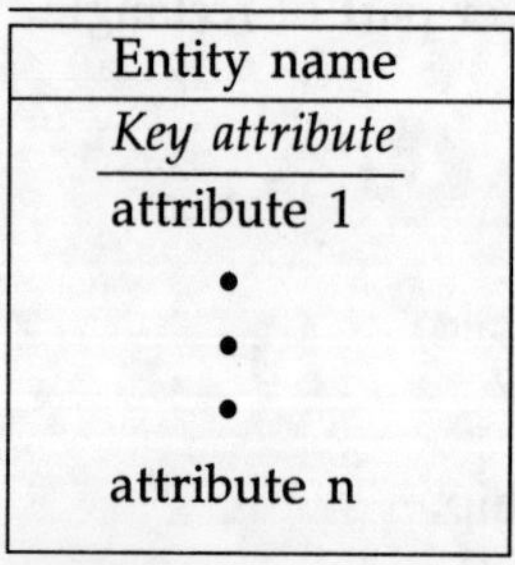	Entity Attributes listed in lower part of rectangle
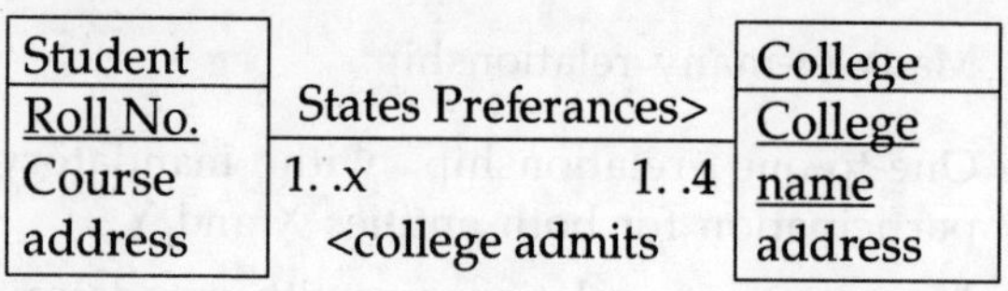	A student states preferences from 1 to 4 colleges. College admits 1 to several students.

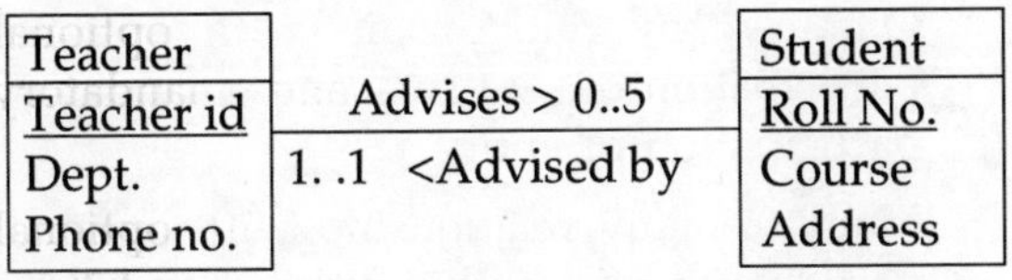

Teacher advises 0 to 5 students. A student advised by 1 teacher.

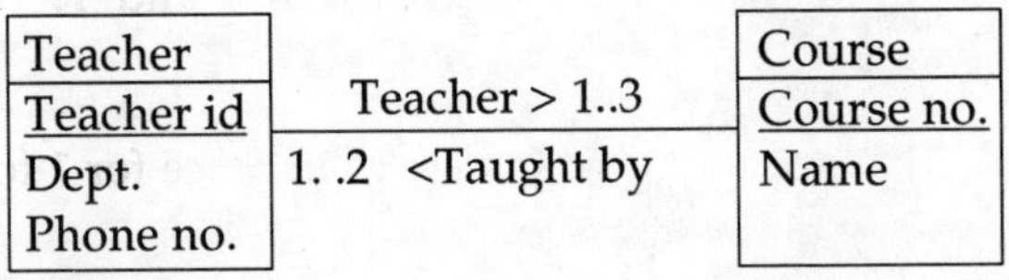

Teacher teaches 1 to 3 courses. A course is taught by 1 to 2 teachers.

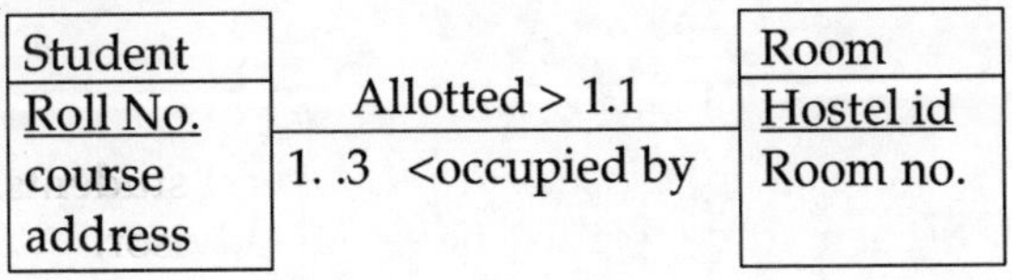

A student allotted 1 room. 1 room occupied by 1 to 3 students.

References

1. Hoffer, J.A., George, J.F. and Valacich, J.S., *Modern Systems Analysis and Design*, Third Edition, Pearson Education Asia, New Delhi, 2002.
2. Valacich, J.S., George, J.F. and Hoffer, J.A., *Essentials of Systems Analysis and Design*, Fourth Edition, PHI Learning, New Delhi, 2009
3. Jessup, L. and Valacich, J., *Information Systems Today*, Prentice Hall of India, New Delhi, 2003.
4. Laudon, K.C. and Laudon, J.P., *Management Information Systems*, Pearson Education Asia, New Delhi, 2004.
5. Senn, J.A., *Analysis and Design of Information Systems*, McGraw-Hill, Singapore, 1985.
6. Kendall, K.E. and Kendall, J.E., *Systems Analysis and Design*, Fifth Edition, Pearson Education Asia, New Delhi, 2003.
7. Atter, S., *Information Systems—A Management Perspective*, Addison Wesley, Reading, USA, 2000.
8. Pfleeger, S.L., *Software Engineering*, Second Edition, Pearson Education Asia, New Delhi 2001.
9. Galbraith, J., *Organization Design*, Addison-Wesley, Reading, Mass., 1977.
10. Yourdon, E. and Constantine, L.L., *Structured Design*, Prentice Hall Inc., Englewood Cliffs (N.J.), 1979.
11. Ziya Aktas, A., *Structured Analysis and Design of Information Systems*, Prentice Hall, Inc., Englewood Cliffs (N.J.), 1987.
12. Schultheis, R. and Summer, M., *Management Information Systems*, Tata McGraw Hill, New Delhi, 1999.
13. De Macro, T., *Structured Analysis and Systems Specifications*, Yourdon Press, New York, 1978.
14. Gane, C. and Sarson, T., *Structured Systems Analysis: Tools and Techniques*, Prentice Hall, Inc., Englewood Cliffs (N.J.), 1979.
15. Atter, S., *Information Systems*, Fourth Edition, Pearson Education Asia, New Delhi, 2004.

16. Hawryszkiewycz, I.T., *Introduction to Systems Analysis and Design*, Fourth Edition, Prentice-Hall of India, New Delhi, 2005.
17. Cockburn, A., *Writing Effective Use Cases*, Pearson Education Asia, New Delhi, 2003.
18. Ullman, J.D. and Widom, J., *A First Course in Database Systems*, Pearson Education Asia, New Delhi, 2001.
19. Date, C.J., *An Introduction to Database Systems*, Sixth Edition, Addison Wesley, Reading, U.S.A., 1995.
20. Kroenke, D.M., *Database Processing*, Eighth Edition, Pearson Education Asia, New Delhi, 2004.
21. Rajaraman, V., *Fundamentals of Computers*, Fifth Edition, PHI Learning, New Delhi, 2010.
22. Rajaraman, V. and Sahasrabuddhe, H.V., *Computer Programming in COBOL*, Prentice-Hall of India, New Delhi, 1987.
23. Watne, D.A. and Turney, P.B.B., *Auditing EDP Systems*, Prentice Hall Inc., Englewood Cliffs (N.J.), 1990.
24. Jacobson, I., Booch, G. and Rumbaugh, J., *The Unified Software Development Process*, Addison-Wesley Longman, 1999.
25. Elmasri, R. and Navatha, S.B., *Fundamentals of Database Systems*, Third Edition, Addison-Wesley, Reading MA, 2000.
26. Booch, G., Rumbaugh, J. and Jacobson, I., *The Unified Modelling Language—User Guide*, Addison-Wesley Longman Inc., Reading, MA, U.S.A., 1999.
27. Coad, P. and Yourdon, E., *Object-Oriented Analysis*, Second Edition, Yourdon Press, Englewood Cliffs, N.J., U.S.A., 1991.
28. Wirfs-Brock, R.J., Wilkerson, B. and Wiener, L., *Designing Object-Oriented Software*, Prentice Hall Inc., Englewood Cliffs, N.J., U.S.A., 1990.
29. Stallings, W., *Data and Computer Communications*, Seventh Edition, Prentice-Hall of India, New Delhi, 2003.
30. Denning, D.E., *Information Warfare and Security*, Addison-Wesley Longman, Reading, MA., U.S.A., 1999.
31. Rumbaugh, J., Blaha, M., Premerlani, W., Eddy, F. and Lorensen, W., *Object Oriented Modelling and Design*, Pearson Education, Asia, New Delhi, 2002.
32. Kalakota, R. and Whinston, A.P., *Frontiers of Electronic Commerce*, Addison Wesley Longman Inc., Reading, MA. , U.S.A., 1996.
33. Jacobson, I., Christerson, M., Jonsson, P. and Overgaard G., *Object-Oriented Software Engineering—A Use Case Approach*, Addison Wesley Longman Inc., Reading, MA., U.S.A., 1992.
34. Booch G., *Object-Oriented Analysis and Design*, Benjamin/Cummings Publishing Co., Redwood City, CA., U.S.A., 1994.
35. Rajaraman, V., *Essentials of E-commerce Technology*, PHI Learning, New Delhi, 2009.

Index